This book comes with access to more content online.

Watch videos, take practice tests,
and study with flashcards!

Register your book or ebook at
www.dummies.com/go/getaccess

Select your product, and then follow the prompts
to validate your purchase.

You'll receive an email with your PIN and instructions.

501

Practice Problems

Physics I

501

Practice Problems

Physics I

by The Experts
at Dummies

Physics I: 501 Practice Problems For Dummies®

Published by: **John Wiley & Sons, Inc.,** 111 River Street, Hoboken, NJ 07030-5774, www.wiley.com

Copyright © 2022 by John Wiley & Sons, Inc., Hoboken, New Jersey

Published simultaneously in Canada

For general information on our other products and services, please contact our Customer Care Department within the U.S. at 877-762-2974, outside the U.S. at 317-572-3993, or fax 317-572-4002. For technical support, please visit https://hub.wiley.com/community/support/dummies.

Wiley publishes in a variety of print and electronic formats and by print-on-demand. Some material included with standard print versions of this book may not be included in e-books or in print-on-demand. If this book refers to media such as a CD or DVD that is not included in the version you purchased, you may download this material at http://booksupport.wiley.com. For more information about Wiley products, visit www.wiley.com.

Library of Congress Control Number: 2022935115

ISBN 978-1-119-88371-5 (pbk); ISBN 978-1-119-88372-2 (ebk); ISBN 978-1-119-88373-9 (ebk)

SKY10034205_042222

Contents at a Glance

Table of Contents

CHAPTER 11: Rolling Around with Rotational Kinetics and Dynamics

CHAPTER 12: Bouncing with a Spring: Simple Harmonic Motion

CHAPTER 13: Heating Up with Thermodynamics and Heat Transfer

Introduction

Whether you want some extra practice for your college or high school physics class, you want to refresh your memory about a course you took long ago, or you're simply curious about the way the universe works, you've found the right book. After all, the best way to learn physics is to do physics, and the hundreds of problems in this book give you plenty of opportunity to do physics. You can practice as much as you like and become a pro at figuring out the right way to start out all sorts of problems that you'd expect to see in the first semester of a one-year physics course.

Why doesn't the moon crash into the earth? How is it possible to sleep on a bed of nails? Why does the water level in your glass stay the same when the ice melts? By working through the many problems in this book, you'll be better able to explain these and other mysteries of the universe to your friends.

What You'll Find

The Physics I practice problems in this book are divided into 15 chapters, beginning with foundational practice (such as calculating displacement and working with vectors); moving on to forces, energy, and momentum; and wrapping up with thermodynamics. Some of the questions require you to reference a diagram, but that instruction is always clear within the questions.

Chapter 16 contains the solutions to all of the practice problems, as well as detailed explanations that help you understand how to come up with the correct answer. If you get a particular question wrong, though, don't just read the answer explanation and move on. Instead, try solving the question again because you know that now you won't make the same mistake that got you to the original wrong answer in the first place. (After all, sometimes knowing what *not* to do is a great start in discovering what *to* do.)

Whatever you do, stay positive. The harder questions in this book aren't meant to discourage you. Rather, they're meant to prove to you just how well you can understand the many challenging concepts presented in a typical Physics I class.

How This Workbook Is Organized

This workbook is divided into two main parts: the questions and the answers.

Part 1: The Questions

The questions in this book cover the following topics:

>> **Math basics:** To learn physics, you need to know a little bit of math. (Just a little!) Chapter 1 checks your knowledge of basic algebra, trigonometry, units, and significant digits.

>> **Kinematics:** The basic quantities you use to describe motion are displacement, velocity, and acceleration. In Chapter 2 you practice one-dimensional motion problems. Chapter 3 deals with the two-dimensional case.

>> **Forces:** Newton's laws relate forces and motion. Chapter 4 has you applying Newton's laws; Chapter 5 questions you on friction and gravitational force.

>> **Angular motion:** The linear quantities you use to describe motion and forces have angular analogues. Chapter 6 checks your knowledge of angular velocity and angular acceleration. In Chapter 7 you practice solving circular motion problems. Chapter 11 deals with torque, angular momentum, and rotational kinetic energy.

>> **Energy and momentum:** You can discover a lot about the world around you by studying con-served quantities such as energy and momentum. Chapter 9 features work- and energy-related problems; Chapter 10 focuses on momentum and collisions.

>> **Simple harmonic motion:** Periodic motion occurs repeatedly in nature, which is why Chapter 12 has you practice working with springs and pendula.

>> **Liquids, gases, and thermodynamics:** Dealing with macroscopic properties is often easier than keeping track of the motion of each molecule. Chapter 8 focuses on density, pressure, and flow rates of liquids and gases. In Chapter 13 you examine temperature, heat, and heat transfer. Chapter 14 questions you on the ideal gas law, and Chapter 15 gets you applying the laws of thermodynamics to heat engines, heat pumps, and other situations.

Part 2: The Answers

Here's where you can find detailed answer explanations for every question in this book. Find out how to set up and work through all the problems so that you arrive at the correct solution.

Beyond the Book

In addition to what you're reading right now, this book comes with a free, access-anywhere Cheat Sheet that includes tips and other goodies you may want to have at your fingertips. To get this Cheat Sheet, simply go to www.dummies.com and type **Physics I Practice Problems For Dummies Cheat Sheet** into the Search box.

The online practice that comes free with this book offers you an addition 501 questions and answers, presented in a multiple-choice format. The beauty of the online problems is that you can customize your online practice to focus on the topic areas that give you trouble. If you're short on time and want to maximize your study, you can specify the quantity of problems you want to practice, pick your topics, and go. You can practice a few hundred problems in one sitting or just a couple dozen, and whether you can focus on a few types of problems or a mix of several types. Regardless of the combination you create, the online program keeps track of the questions you get right and wrong so you can monitor your progress and spend time studying exactly what you need.

To gain access to the online practice, you simply have to register. Just follow these steps:

1. Register your book or ebook at Dummies.com to get your PIN. Go to www.dummies.com/go/getaccess.

2. Select your product from the dropdown list on that page.

3. Follow the prompts to validate your product, and then check your email for a confirmation message that includes your PIN and instructions for logging in.

If you don't receive this email within two hours, please check your spam folder before contacting us through our Technical Support website at http://support.wiley.com or by phone at 877-762-2974.

Now you're ready to go! You can come back to the practice material as often as you want — simply log in with the username and password you created during your initial login. No need to enter the access code a second time.

Your registration is good for one year from the day you activate your PIN.

Where to Go for Additional Help

The solutions to the practice problems in this book are meant to walk you through how to get the right answers; they're not meant to teach the material. If certain physics concepts are unfamiliar to you, you can find help at www.dummies.com. Just type **physics** I into the Search box to turn up a wealth of physics-related articles.

If you need more detailed instruction, check out *Physics I For Dummies*, 3rd Edition; *Physics I Workbook For Dummies*, 3rd Edition; and *Physics Essentials For Dummies*, all published by John Wiley & Sons.

1 The Questions

Chapter **1**

Reviewing Math Fundamentals and Physics Measurements

Physics explains how the world works. You can use physics to predict how objects move and interact. This process often involves some basic algebra and trigonometry. To check these predictions, you can make a measurement. Sometimes you need to convert units to compare different measurements.

The Problems You'll Work On

Here are some of the things you'll do in this chapter:

>> Solving for an unknown variable with basic algebra

>> Using basic trigonometry to determine side lengths and angles

>> Converting between different types of units

>> Writing numbers in scientific notation

>> Understanding unit prefixes in the metric system

>> Rounding to the correct number of significant digits

What to Watch Out For

Be sure to remember the following:

>> Making sure your answer has the right units

>> Using conversion factors correctly

>> Checking that your answers make sense physically

Equipping Yourself with Basic Algebra

1–3

1. Solve the equation $y = 2m + 3$ for m.

2. You are given that $I = \frac{1}{2}mr^2$ and $m = m_0 + m_1$. Solve this expression for $m_0 + m_1$.

3. Solve the equation $\gamma = \dfrac{1}{\sqrt{1 - \dfrac{v^2}{c^2}}}$ for v.

Tackling a Little Trigonometry

4–5

4. If $\cos\theta = 0.8$ and the hypotenuse of a right triangle is 8 meters long, how long is the adjacent side of the right triangle?

5. A pool is 2.0 meters deep. You dive in at an angle of 35 degrees to the surface of the water. If you continue in this direction, how far from the edge of the pool will you hit the bottom?

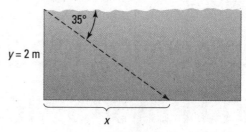

Illustration by Thomson Digital

Converting between Units

6–10

6. A jet plane flies at about 10,000 meters. If 1 meter is 3.3 feet, how high does the plane fly in feet?

7. When you drive into Canada, the speed limit sign says 100 kilometers per hour. What is the speed limit in miles per hour? One mile consists of 1.6 kilometers.

8. You hire 5 painters to paint your house, which has a surface of 200 square meters. If each painter can paint 10 square meters per hour, how long does it take for them to paint the entire house?

9. Your dog eats a quarter pound of dog food each day. How often (in days) do you have to buy a 10-pound bag of dog food?

10. Your grandmother can knit a sweater in 3 days if she works 8 hours per day. She uses one bobbin of yarn every 2 hours, and each bobbin has 10 yards of yarn. How long is all the yarn in the sweater?

Practicing Scientific Notation

11–14

11. The speed of light is about 300,000,000 meters per second. Write this speed in scientific notation.

12. A femtosecond is 1 millionth of 1 billionth of a second. What is a femtosecond in scientific notation?

13. The radius of the sun is $r = 6.955 \times 10^8$. Using the formula $V = \frac{4}{3}\pi r^3$ for the volume of a sphere of radius r, what is the volume of the sun in scientific notation?

14. Assume there are 7 billion people on Earth. If each person has a mass of 70 kilograms and the Earth has a mass of 6×10^{24} kilograms, what fraction of the mass of Earth is due to humans?

Understanding Unit Prefixes

15–16

15. A recipe calls for 500 milligrams of salt. You're serving a large group, so you want to triple the portions of the recipe. How many grams of salt are required?

16. How many milliwatts are in 1 megawatt? Give the answer in scientific notation.

Spotting the Number of Significant Digits

17–20

17. How many significant digits are in the number 303.4?

18. Calculate the sum $21.21 + 4.8 + 2.33$ to the correct number of significant digits.

19. How many significant digits are in the number 5,003?

20. How many significant digits can you retain for $\sqrt{45.365 + 29.821}$?

Rounding to the Correct Number of Digits

21–25

21. Calculate the sum $98.374 + 28.56$ to the correct number of significant digits.

22. A physicist adds 0.25 gallons of paint to a container that already holds 10 gallons of paint. Given the significant digits of the quantities added, how much paint can the physicist say is in the container?

23. Evaluate the equation $t = (5.01 \times 4.4) + (3.2 \times 18)$ to the correct number of significant digits, and express the answer using the appropriate notation.

24. You measure the height of your apartment to be 2.6 meters high. If the apartment building has 8 floors, what is the height of the apartment building?

25. What is the result of $63.005 \times \left(18.54 + \dfrac{65}{4}\right)$ to the correct number of significant digits?

Chapter 2

Moving along with Kinematics

To describe motion, you can use terms like displacement, speed, velocity, and acceleration. *Displacement* is a distance in a particular direction. *Speed* is the distance traveled in a certain amount of time. If you combine speed with a direction, you get *velocity*. *Acceleration* measures how quickly velocity changes.

The Problems You'll Work On

In this chapter you'll move through the following topics:

- ❯❯ Finding the displacement in one and two dimensions
- ❯❯ Using velocity to determine the displacement
- ❯❯ Taking the average of the instantaneous speed
- ❯❯ Determining the change in velocity using acceleration
- ❯❯ Relating displacement, velocity, acceleration, and time

What to Watch Out For

You'll speed through these questions if you keep the following in mind:

- ❯❯ Using the distance traveled, not the displacement, to determine average speed
- ❯❯ Remembering that velocity is the change in displacement in a certain amount of time
- ❯❯ Remembering that acceleration is the change in velocity in a certain amount of time

Determining Displacement Using Positions in One Dimension

26–29

26. You leave your apartment and walk 2 blocks north, only to realize that you forgot your keys. You turn around and walk back 2 blocks south to get them. What is your total displacement?

27. A car is driven north for 5 miles, then south for 3 miles, and then north again for 2 miles. What is its displacement?

28. A pair of figure skaters skates together for 10 meters. Then skater A launches their partner, skater B, through the air. Skater B lands 15 meters ahead of where the pair began skating. What is skater B's displacement with respect to the launching point after landing?

29. An elevator is at the ground floor. It goes once to the first floor, twice to the second floor, three times to the third floor, and then four times to the fourth floor. What is the displacement of the elevator when Ms. Smith gets on at the third floor?

Getting Displacement in Two Dimensions with Axes

30–35

30. You move a marker on a board from one point (2 centimeters, 4 centimeters) to another point (5 centimeters, 8 centimeters). What is the magnitude of the displacement of the marker?

31. To get to your friend's house, you walk 4 blocks north and 1 block east. What is the direction of your displacement with respect to the direction east?

32. A basketball player shoots the ball, releasing it at 8 feet above the floor and 5 feet from the basket. The ball goes straight through the basket, which is 10 feet above the floor. What is the magnitude of the displacement of the ball from the point at which it is released and the point at which it passes through the hoop?

33. A chess board is 8 squares by 8 squares. You move your bishop from one square (3,1) to another square (7,5). What is the magnitude (in squares) and angle of the displacement?

34. A child is scooting around on a toy truck. The child scoots 5 meters down the hall, then turns 90 degrees to the right and scoots 3 meters, then turns again 90 degrees to the right and scoots 2 more meters. What is the child's displacement over this trip?

35. You want to shoot a laser beam from the edge of a stage to a disco ball hanging from the ceiling. The stage is 1 meter above the floor, the disco ball is 1 meter below the ceiling, and the height of the ceiling is 4 meters. The horizontal distance from the edge of the stage where the laser is mounted to the point directly under the disco ball is 5 meters. At what angle above the horizontal should you aim the laser?

Traveling with Average Speed and at Instantaneous Speed

36–40

36. You run from your house to the grocery store in 1.0 minute, and the store is 300 meters from your house. What is your average speed in meters per second for the trip?

37. In a traffic jam, you drive at 10 miles per hour for 10 minutes, at 20 miles per hour for 1 minute, at 15 miles per hour for 5 minutes, at 30 miles per hour for 2 minutes, and at 5 miles per hour for 15 minutes. What is your maximum instantaneous speed?

38. A ball is dropped from the top floor of your five-floor apartment building. You're on the bottom floor and see the ball go past your small window. If you measure the ball's speed at this point, is it the average speed, the instantaneous speed, both the average and the instantaneous speed, or neither the average nor the instantaneous speed?

39. The average speed of a car being driven in London is 11 miles per hour. If you have to drive 15 miles from your home to work in London, how long do you expect it to take?

40. The average speed of runner A is 10 percent greater than that of runner B. If runner B is given a 10-meter head start in a 100-meter dash, which runner will finish first?

Distinguishing between Average Speed and Average Velocity

41–45

41. You travel north for 80 miles and then east for 30 miles. What is the magnitude of your average velocity if the entire trip takes 4 hours?

42. You travel north for 80 miles and then east for 30 miles. What is your average speed if the entire trip takes 4 hours?

43. You travel 35 miles north and 20 miles east. If the trip takes 30 minutes, what is the magnitude (in miles per hour) and direction of your average velocity?

44. A mail carrier walks 10 blocks north, then 3 blocks east, and then south for an unknown number of blocks. The time for the trip is 1.0 hour, and each block is 100 meters long. If the mail carrier's average speed is 1.0 meter per second, what is the magnitude of the average velocity in meters per second?

45. You travel 40 miles north, then 30 miles east, then 20 miles north, and then 10 miles south. If your trip takes 2 hours, what is your average speed?

Speeding Up and Down with Acceleration

46–50

46. It takes you 2.0 seconds to accelerate from a standstill to a running speed of 7.0 meters per second. What is the magnitude of your acceleration?

47. The acceleration due to gravity at the surface of Earth is about 9.8 meters per second per second. If you drop a small heavy ball from the fourth floor of a building, how fast is the ball moving after 0.5 seconds?

48. Your infant daughter has a maximum crawling velocity of 0.3 meters per second. If she accelerates at 2 meters per second per second, how long does it take her to reach her maximum velocity when she starts from rest?

49. A plane's takeoff speed is 300 kilometers per hour. If it accelerates at 2.9 meters per second per second, how long is it on the runway after starting its takeoff roll?

50. You ride your bicycle at 10 meters per second and accelerate at −2.3 meters per second per second for 10 seconds. What is your final velocity?

Finding Displacement with Acceleration and Time

51–54

51. Starting from rest, you accelerate at 2 meters per second per second for 2 seconds to get up to full speed on your bicycle. How far do you travel during this time?

52. A car accelerates northward at 4.0 meters per second per second over a distance of 30 meters. If it starts at rest, for how long does it accelerate?

53. Starting from rest, a motorcycle rider covers 200 meters in 10 seconds. What was the rider's acceleration?

54. A tennis player serves a ball at 100 miles per hour. If the ball accelerates over a period of 0.05 seconds from essentially a standstill, how far (in meters) does the ball travel during its acceleration? One mile consists of 1,609 meters.

Finding Displacement with Acceleration and Velocities

55–58

55. You're driving at 20 meters per second northbound and brake to slow to 10 meters per second. During that time, you cover 50 meters. What was your acceleration?

56. You ski along at 3.0 meters per second, and your friend whizzes by at a greater speed. You have to accelerate at 2.0 meters per second per second for 20 meters to attain the same speed. At what speed was your friend skiing?

57. A baseball pitcher throws a fastball at 90 miles per hour. The pitcher accelerates the ball over a distance of 2.0 meters. What is the acceleration of the ball?

58. In a spaceship, you accelerate from 200 meters per second to 500 meters per second at 10 meters per second per second. How many kilometers do you travel during this acceleration?

Finding Acceleration with Displacement and Time

59–62

59. A speed skater accelerates from a standstill to full speed over a distance of 12 meters. If the speed skater takes 2.6 seconds to do this, what is the magnitude of skater's acceleration?

60. Your car can accelerate at 3.4 meters per second per second. You are stopped at a red light and have 20 meters to accelerate onto the freeway when the light turns green. How long will it take you to accelerate over this distance?

61. You're driving at 18 meters per second when you apply the brake for 4 seconds. If the magnitude of your acceleration is 2.8 meters per second per second, how far did you travel in this time?

62. A ferry boat is traveling east at 1.3 meters per second when the captain notices a boat in its path. The captain engages the reverse motors so that the ferry accelerates to the west at 0.2 meters per second per second. After 20 seconds of this acceleration, what is the boat's position with respect to its initial position?

Finding Acceleration with Velocities and Displacement

63–66

63. A cheetah can accelerate from 0 miles per hour to 60 miles per hour in 20 meters. What is the magnitude of its acceleration?

64. A speedboat can accelerate from an initial velocity of 3.0 meters per second to a final velocity that is 3 times greater over a distance of 42 meters. What is the magnitude of its acceleration?

65. You drop a feather from your balcony, which is 4.5 meters above the ground. After falling 0.20 meters, it moves at the speed of 0.30 meters per second. What is the magnitude of its acceleration?

66. A boat moving north at 2.3 meters per second undergoes constant acceleration until its speed is 1.2 meters per second northward. With respect to its initial position, its final position is 200 meters northward. What is its acceleration?

Finding Velocities with Acceleration and Displacement

67–70

67. A boat accelerates at 0.34 meters per second per second northward over a distance of 100 meters. If its initial velocity is 2.0 meters per second northward, what is its final velocity?

68. A train brakes to a stop over a distance of 3,000 meters. If its acceleration is 0.1 meter per second per second, what is its initial speed?

69. To pass another race car, a driver doubles the car's speed by accelerating at 4.5 meters per second per second for 50 meters. What are the initial and final speeds?

70. From a position 120 meters above a pigeon, a falcon dives at 9.1 meters per second per second, starting from rest. After diving 25 meters, the falcon stops accelerating. What is the falcon's speed when it strikes the pigeon?

Chapter **3**

Moving in a Two-Dimensional World

The basic quantities you use to describe motion in two dimensions — displacement, velocity, and acceleration — are vectors. A *vector* is an object that has both a magnitude and a direction. When you have an equation that relates two vectors, you can break each vector into parts, called *components*. You end up with two equations, which are usually much easier to solve.

The Problems You'll Work On

In this chapter on two-dimensional vectors and two-dimensional motion, you work with the following situations:

» Adding and subtracting vectors

» Multiplying a vector by a scalar

» Taking apart a vector to find its components

» Determining the magnitude and direction of a vector from its components

» Finding displacement, velocity, and acceleration in two dimensions

» Calculating the range and time of flight of projectiles

What to Watch Out For

While you zig and zag your way through the problems in this chapter, avoid running into obstacles by:

» Identifying the correct quadrant when finding the direction of a vector

» Finding the components before trying to add or subtract two vectors

» Breaking the displacement, velocity, and acceleration vectors into components to turn one difficult problem into two simple problems

» Remembering that the vertical component of velocity is zero at the apex

» Recognizing that the horizontal component of acceleration is zero for freely falling objects

Getting to Know Vectors

71–72

71. How many numbers are required to specify a two-dimensional vector?

72. Marcus <u>drives</u> <u>45 kilometers</u> at a <u>bearing</u> of <u>11 degrees</u> <u>north of west</u>. Which of the underlined words or phrases represents the magnitude of a vector?

Adding and Subtracting Vectors

73–75

73. Vector **U** points west, and vector **V** points north. In which direction does the resultant vector point?

74. If vectors **A**, **B**, and **C** all point to the right, and their lengths are 3 centimeters, 5 centimeters, and 2 centimeters, how many centimeters long is the resultant vector formed by adding the three vectors together?

75. Initially facing a flagpole, Jake turns to his left and walks 12 meters forward. Jake then turns completely around and walks 14 meters in the opposite direction. How many meters farther away from the flagpole would Jake have ended had his journey started by turning to the right and walking 14 meters and then turning completely around and walking the final 12 meters?

Adding Vectors and Subtracting Vectors on the Grid

76–79

76. If $A = (2, 4)$ and $B = (3, 8)$, what is the value of $A + B$?

77. If $V = (6, -4)$, what is the value of $\frac{1}{2} V$?

78. Given that $A = (-2, 2)$ and $B = (3, 1)$, calculate $3A + 5B$.

79. Given the three vectors $A(7, -3)$, $B(0, 4)$, and $C(-3, -3)$, solve for **D** if $2A - 3B = D - 3C$.

Breaking Vectors into Components

80–83

80. Vector **A** has a magnitude of 28 centimeters and points at an angle 80 degrees relative to the x-axis. What is the value of A_y? Round your answer to the nearest tenth of a centimeter.

81. Vector **C** has a length of 8 meters and points 40 degrees below the x-axis. What is the vertical component of **C**, rounded to the nearest tenth of a meter?

82. Jeffrey drags a box 15 meters across the floor by pulling it with a rope. He exerts a force of 150 newtons at an angle of 35 degrees above the horizontal. If work is the product of the distance traveled times the component of the force in the direction of motion, how much work does Jeffrey do on the box? Round to the nearest ten newton-meters.

83. Three forces pull on a chair with magnitudes of 100, 60, and 140, at angles of 20 degrees, 80 degrees, and 150 degrees to the positive x-axis, respectively. What is the component form of the resultant force on the chair? Round your answer to the nearest whole number.

Reassembling a Vector from Its Components

84–87

84. What are the magnitude and direction of the vector **W** $= (6, 3)$? Round your answers to the nearest tenth place and give your angle (direction) in units of degrees.

85. Given vectors **A** $= (3, -3)$ and **W** $= (-2, 4)$, what angle would vector **C** make with the x-axis if **C** $=$ **A** $+$ **W**? Round your answer to the nearest tenth of a degree.

86. If you walk 12 paces north, 11 paces east, 6 paces south, and 20 paces west, what is the magnitude (in paces) and direction (in degrees relative to the positive x-axis) of the resultant vector formed from the four individual vectors? Round your results to the nearest integer.

87. After a lengthy car ride from a deserted airfield to Seneca Airport, Candace finds herself 250 kilometers north and 100 kilometers west of the airfield. At Seneca, Candace boards a small aircraft that flies an unknown distance in a south-westerly direction and lands at Westsmith Airport. The next day, Candace flies directly from Westsmith to the airfield from which she started her journey. If the flight from Westsmith was 300 kilometers in distance and flew in a direction 15 degrees south of east, how many kilometers was Candace's flight from Seneca to Westsmith? Round your answer to the nearest integer.

Describing Displacement, Velocity, and Acceleration in Two Dimensions

88–96

88. Hans drives 70 degrees north of east at a speed of 50 meters per second. How fast is Hans traveling northward? Round your answer to the nearest integer.

89. If you walk 25 meters in a direction 30 degrees north of west and then 15 meters in a direction 30 degrees north of east, how many meters did you walk in the north-south direction?

90. Jake wants to reach a postal bin at the opposite corner of a rectangular parking lot. It's located 34 meters away in a direction 70 degrees north of east. Unfortunately, the lot's concrete was recently resurfaced and is still wet, meaning that Jake has to walk around the lot's edges to reach the bin. How many meters does he have to walk? Round your answer to the nearest whole meter.

91. If $D = A + B + C$, use the following information to determine the components of D. Use ordered-pair notation rounded to the nearest tenth of a meter for your answer. (All angles are measured relative to the x-axis.)

A: 45 meters at 20 degrees

B: 18 meters at 65 degrees

C: 32 meters at −20 degrees

92. To walk from the corner of Broadway and Park Place to the corner of Church and Barkley in Central City, a person must walk 150 meters west and then 50 meters south. How many meters shorter would a direct route be? Round your answer to the nearest meter.

93. If Jimmy walks 5 meters east and then 5 meters south, what angle does the resultant displacement vector make with the positive x-axis (assuming the positive x-axis points east)?

94. A swimmer can move at a speed of 2 meters per second in still water. If the athlete attempts to swim straight across a 500-meter-wide river with a current of 8 meters per second parallel to the riverbank, how many meters will the swimmer traverse by the time she reaches the other side? Round your answer to the nearest tenth of a kilometer.

95. A basketball rolling at a rate of 10 meters per second encounters a gravel patch that is 5 meters wide. If the basketball is moving in a direction 15 degrees north of east, and if the sides of the gravel patch are aligned parallel to the north-south axis, will the basketball still be rolling by the time it reaches the far side of the patch if the gravel gives it an acceleration of −3 meters per second squared in the same direction it started when it entered the patch? If so, what will be its speed in the easterly direction upon exiting the patch? If not, how many meters (measured perpendicularly to the western edge of the patch) will the basketball roll on the gravel before stopping? Round your numerical response to the nearest integer.

96. Partway through a car trip, a dashboard compass stops working. At the time it broke, Bill had driven 180 kilometers in a direction 70 degrees north of west. Bill then proceeds to drive 45 kilometers due south on the highway, before turning right and driving 18 kilometers west on Sunset St. When he stops the car, how far is Bill from the location where he began the trip? Round your answer to the nearest kilometer.

Moving under the Influence of Gravity: Projectile Launched Horizontally

97–99

97. A marble rolls off a 2-meter-high, flat tabletop. In how many seconds will it hit the floor? Round your answer to the nearest tenth of a second.

98. Mark rolls a boulder off a cliff located 22 meters above the beach. If Mark is able to impart a velocity of 0.65 meters per second to the boulder, how many meters from the base of the cliff will the boulder land? Round your answer to the nearest tenth of a meter.

99. A car flies off a flat embankment with a velocity of 132 kilometers per hour parallel to the ground 45 meters below. With what velocity does the car ultimately crash into the ground? Round your answer to the nearest meters per second.

Moving under the Influence of Gravity: Projectile Launched at an Angle

100–105

100. Alicia kicks a soccer ball with a velocity of 10 meters per second at a 60-degree angle relative to the ground. What is the horizontal component of the velocity? Round your answer to the nearest tenth of a meter per second.

101. The punter for the San Diego Chargers kicks a football with an initial velocity of 18 meters per second at a 75-degree angle to the horizontal. What is the vertical component of the ball's velocity at the zenith (highest point) of its path? Round your answer to the nearest tenth of a meter per second.

102. A cannonball fired at a 20-degree angle to the horizontal travels with a speed of 25 meters per second. How many meters away does the cannonball land if it falls to the ground at the same height from which it launched? Round your answer to the nearest meter.

103. Launching from a 100-meter-high ski jump of unknown inclination at 40 meters per second, an Olympic athlete grabs 8.2 seconds of hang time before landing on the ground. How far away from the jump does she land?

104. A cannon tilted at an unknown angle fires a projectile 300 meters, landing 11 seconds after launch at a final height equal to its starting one. At what angle was the cannon fired?

105. Will a baseball struck by a bat, giving the ball an initial velocity of 35 meters per second at 40 degrees to the horizontal, result in a home run if it must clear a 1.8-meter-high fence 120 meters away? If it will, by how many centimeters will the ball clear? If not, how many centimeters short will it be? Assume that the ball is struck at a height of 0.8 meters, and round your answer to the nearest 10 centimeters.

Chapter **4**

Pushing and Pulling: The Forces around You

Newton's laws of motion describe how objects move when forces are applied to them. The first law states that an object's velocity won't change unless you apply a force to it. One of the most famous and important equations in physics is Newton's second law: Force equals mass times acceleration. The third law says that whenever you exert a force on an object, that object will exert an equal force on you.

The Problems You'll Work On

In this chapter you'll apply Newton's laws to the following types of problems:

>> Using Newton's second law to relate force and acceleration

>> Drawing free-body diagrams

>> Determining equal and opposite forces with Newton's third law

>> Redirecting forces with pulleys

What to Watch Out For

You'll be forced to try the problem again unless you keep the following in mind:

>> Drawing a free-body diagram to make sure you include all the forces

>> Determining the components of the forces on your free-body diagrams with the correct signs

>> Remembering that the equal and opposite forces in Newton's third law always act on different objects

>> Recalling that the magnitude of the tension of a massless rope is the same all along the rope

Resisting Motion with Newton's First Law

106–107

106. What are the SI units of mass?

107. What property of an object does mass measure?

Forcing a Massive Object to Accelerate

108–109

108. What is the acceleration of a 0.25-kilogram particle subject to a single force of 10 newtons eastward?

109. A 300-gram block slides across a ceramic floor at a speed of 13.5 meters per second. If no forces act on the block along the axis of its motion, what is the block's speed 1 second later? Round your answer to the nearest tenth of a meter per second.

Drawing Free-Body Diagrams

110–113 Use the following force diagram of a mass hanging from a pulley by a massless rope to answer Questions 110–113. Letters on the diagram signify vectors.

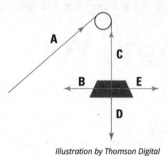

Illustration by Thomson Digital

110. Which vector represents the gravitational force Earth exerts on the mass?

111. Which vector(s) represent the force of tension?

112. Which two vectors always have equal magnitudes?

113. In terms of B, C, D, and E — the magnitudes of their respective vectors — what inequality must be true if the mass accelerates downward?

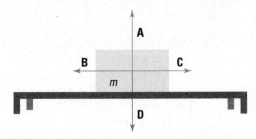

Illustration by Thomson Digital

114. Which vector represents the normal force exerted on the box?

115. Assuming the table completely supports the box's weight, what equality must be true in terms of A, B, C, and D — the magnitudes of the vectors displayed?

116. A cat sees the box on the table and hops up to start pushing it to the right. After it starts moving, the cat exerts a force **K** on the box that keeps it moving at a constant velocity. If **K** is parallel to the table's surface, write an expression for the magnitude of **K** in terms of A, B, C, and D — the magnitudes of the four vectors shown.

Adding Forces Together to Obtain the Net Force

117–119

117. One force pulls a brick with a force of 12 newtons due west, while another force pulls the brick with a force of 8 newtons due east. What is the magnitude of the net force exerted on the brick?

118. What is the net force in the east-west direction on a crate being pushed with 58 newtons of force 12 degrees north of east by one worker and with 30 newtons of force 64 degrees south of east by a second worker? Round your answer to the nearest integer.

119. Two forces act on a cardboard box: a 340-newton force directed 25 degrees south of east and a 300-newton force directed 85 degrees north of west. What net force does the cardboard box experience? Round your answer to the nearest integers, in units of newtons and degrees.

Moving a Distance with the Net Force

120–122

120. A force of 50 newtons provides a constant acceleration to a 25-kilogram crate originally at rest. How many meters does the crate move in 3 seconds? Round your answer to the nearest tenth.

121. Given a box of mass M initially at rest, write an expression for the displacement the box experiences when a force F constantly accelerates it to a speed of v meters per second. Your answer should contain only variables given here, but it does not have to use all of them.

122. Starting from rest, a 5,850-kilogram sports car goes from 0 to 200 kilometers per hour. It proceeds at that same speed for 10 seconds. The driver then slams on the brakes, producing a constant acceleration until the car is once again stationary. If the engine and brakes provided forces of 40,600 newtons and 31,800 newtons, respectively, what is the sports car's displacement during the trip? Round your answer to the nearest tenth of a kilometer.

Finding the Needed Force to Speed Up

123–125

123. Johnny guns the engine on a 180-kilogram vehicle currently rolling backward at 10.8 kilometers per hour. If the engine provides 450 newtons of force (10 percent of which is lost to frictional forces), what is the vehicle's velocity 4 seconds later? Round your answer to the nearest meter per second.

124. A 7,200-kilogram car is accelerated from rest at a constant rate by an engine producing 96 kilonewtons of force for 7.5 seconds. How fast is the car traveling — in kilometers per hour — after that time? Round your answer to two significant digits.

125. An 800-kilogram elevator cab is attached to a cable rising high into a skyscraper. The sides of the cab contain brake pads that provide the force for slowing the cab down when it approaches its destination. Taking the elevator down from the tenth floor to the fifth floor, Floyd obtains a maximum vertical velocity of 5 meters per second downward before the cab starts braking halfway between the sixth and seventh floors. Assume that the tension in the cable when the brakes are off is twice the tension in the cable when the brakes are on. If each floor is 20 meters apart, how many newtons of force do the elevator's brakes exert? Round your answer to three significant digits.

Pairing Up Equal and Opposite Forces

126–128

126. Joe pushes against a brick wall with a force of 12 newtons. If Joe's mass is 120 kilograms and the brick wall's mass is 12,000 kilograms, with how much force does the brick wall push against Joe? Give your answer in newtons using two significant digits.

127. An angry, 120-kilogram astronaut punches a 3,040-kilogram space shuttle with 45 newtons of force. What is the magnitude of the astronaut's acceleration — in meters per second squared — as a result of the punch? Round your answer to two significant digits.

128. Four boxes of varying masses are dragged along a frictionless surface by a force of 85 newtons, as shown here:

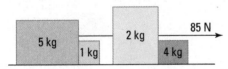

Illustration by Thomson Digital

How many more times powerful is the contact force between the 2- and 4-kilogram boxes than the contact force between the 5- and 1-kilogram boxes? Round your answer to the nearest integer.

Overcoming Friction by Pulling Hard

129–131

129. Two cartons of milk are dragged across a table. They are identical except for their masses: One has a mass of 1.2 kilograms and the other has a mass of 0.6 kilograms. If the magnitude of the force of friction acting on the more massive carton is denoted z, what is the magnitude of the force of friction on the other carton, in terms of z?

130. Playing street hockey, Neil strikes a puck weighing 5.2 newtons with 44 newtons of force. If the frictional force on the puck from the concrete street is 3 newtons, what is the puck's acceleration along the concrete after Neil hits it? Round your answer to the nearest meter per second squared.

131. Hans pushes a 312-kilogram tackling dummy across the football field with a force F. The dummy accelerates at a rate of 0.15 meters per second squared. If Hans doubles his force, the dummy accelerates at 0.75 meters per second squared. How many newtons of friction force is the field exerting on the tackling dummy? Round your answer to the nearest integer.

Pulling in a Different Direction with Pulleys

132–136

132. A massless rope connecting two masses — one of 5 kilograms and one of 10 kilograms — is draped over a pulley hanging from the ceiling. If the force of tension pulling upward on the 5-kilogram mass is T, what is the amount of tension pulling upward on the 10-kilogram mass, in terms of T?

133. Farmer Dell uses a pulley system to lift bales of hay up to the top story of the barn. The system consists of a massless rope draped over a frictionless pulley; the bale of hay is attached to one end of the rope, and Dell pulls on the other end as shown in the following figure. With how many newtons of force does Farmer Dell have to pull on the rope to move a 15-kilogram hay bale at constant velocity? Round your answer to the nearest tenth.

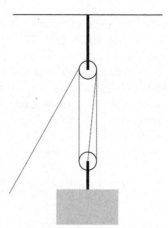

Illustration by Thomson Digital

134. Calculate the acceleration of m_2 in the following diagram if $m_1 = 300$ grams and $m_2 = 750$ grams. Assume that the pulley is frictionless and that the rope is massless. Round your answer to the nearest tenth of a meter per second squared.

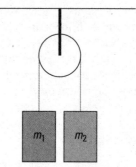

Illustration by Thomson Digital

Use the following diagram, which shows three masses and two (massless) strings hanging from a frictionless pulley attached to the ceiling, to answer Question 135.

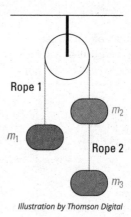

Illustration by Thomson Digital

135. If m_1 has a mass of 3 kilograms, m_2 has a mass of 8 kilograms, and m_3 has a mass of 5 kilograms, what is the tension in rope 2? Round your answer to the nearest tenth of a newton.

136. If m_2 is twice as massive as m_1 in the following diagram, write an expression for the tension in the massless rope connecting the two blocks in terms of m_1 and g — the gravitational acceleration on Earth's surface.

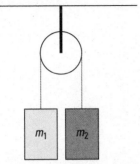

Illustration by Thomson Digital

Balancing the Forces to Find Equilibrium

137–140

137. Fill in the blank: An object's _____ always equals 0 when the object is in equilibrium.

138. When an object is in equilibrium, which of its physical properties *must* remain constant and *can* equal a value other than 0?

139. Three cords hold up an 85-kilogram banner, one at each end and one in the middle. The two at the ends, which are angled at 45 degrees to the horizontal, can each provide 100 newtons of tensile force. How much tension does the cord in the middle, which hangs perfectly vertical, need to provide for the banner to remain hanging? Round your answer to the nearest ten newtons.

140. While looting an abandoned cave, two bandits — Benny and Buddy — discover a crate of gold with handlebars on opposite sides of the crate. The 8.4-kilogram crate contains 106 bars of gold, each with a mass of 1.05 kilograms. When it is lifted to a certain height, Benny — the taller bandit — holds the crate with an angle of 40 degrees to the horizontal. Buddy holds the crate with an angle of 25 degrees to the horizontal. With how much more force does Benny have to hold the crate than does Buddy if the crate is to stay in equilibrium? Round your answer to the nearest ten newtons.

Chapter **5**

Slipping and Sliding: Motion and Forces

wo of the most common forces you encounter every day are gravity and friction. *Weight* is the gravitational force that Earth exerts on an object. Near the surface of Earth, weight is equal to mass times 9.8 meters per second squared. The *magnitude* of the force of friction is equal to the coefficient of friction times the magnitude of the normal force. Friction always opposes the relative motion between two surfaces.

The Problems You'll Work On

In this chapter, you'll deal with the following types of problems involving gravity and friction:

>> Finding the component of the weight parallel and perpendicular to a slope

>> Using the coefficient of friction to determine the force of friction

>> Determining the force needed to push or pull an object up a slope

>> Calculating the trajectory of projectiles with additional forces

What to Watch Out For

You won't encounter any resistance when solving these problems if you keep the following in mind:

>> Recalling that weight always points down and the normal force is always perpendicular to the surface of contact

>> Using the coefficient of static friction if the surfaces of contact are stuck together

>> Using the coefficient of kinetic friction if the surfaces of contact are sliding

>> Finding the force of friction using the normal force, not the weight

Finding the Gravity along a Slope

141–144

141. A 100-kilogram crate sits on a ramp elevated 10 degrees with respect to the horizontal. Rounded to the nearest newton, what is the magnitude of the force of Earth's gravitational pull on the crate?

142. What is the normal force on a motorcycle with a weight of 400 newtons if it sits on the side of a frictionless hill banked at an angle of 40 degrees with the horizontal? Round your answer to the nearest 10 newtons.

143. A box sits on a frictionless ramp with a 30-degree inclination. How large is the acceleration of the box along the axis of the plane's surface? Round your answer to the nearest tenth of a meter per second squared.

144. Audrey weighs 400 newtons. As she starts walking up a hill of constant slope, the amount of force she feels from the ground on her feet drops to 350 newtons. By how many degrees is the hill inclined with respect to the horizontal? Round your answer to the nearest integer.

Moving Fast with Gravity down a Slippery Slope

145–147

145. A 25-kilogram sled slides down a hill of ice-covered snow. If the sled accelerates down the hill at a rate of 4.8 meters per second squared, what is the hill's angle of inclination with respect to the horizontal? Assume the snow is a frictionless surface and round your answer to the nearest whole degree.

146. Samantha pushes her daughter, who is sitting on a sled, down a snowy hill of constant slope with a force of 50 newtons. The sled accelerates at 1.9 meters per second squared. If the combined mass of her daughter and the sled is 40 kilograms, what is the hill's angle of inclination? Round your answer to the nearest degree.

147. A 12-meter-long wheelchair ramp rises 1.8 meters off the ground from beginning to end. A healthcare official pushes a patient in a chair up the ramp with an acceleration of 0.7 meters per second squared by exerting a force of 75 newtons. How many newtons of force does the *ramp* exert on the wheelchair, including the patient riding in it, if you ignore the friction between the chair's wheels and the ramp? Round your answer to two significant digits.

Getting Unstuck with the Coefficient of Friction

148–150

148. What inequality is always true regarding the magnitudes of the normal force, F_N, and the force of friction, F_F, on an object, assuming the coefficient of friction is smaller than 1?

149. Marshall pulls a chair weighing 140 kilograms across a wooden floor. If the coefficient of friction is 0.2, what amount of frictional force is resisting Marshall's efforts? Round your answer to the nearest newton.

150. A hotel porter pulls a 795-kilogram trunk by its strap with 5,000 newtons of force, moving it across the lobby rug toward the elevator bay. The trunk's owner has offered a good tip if the trunk is at the elevators — 40 meters away — in less than 10 seconds. If the strap makes a 20-degree angle with the horizontal, and if the coefficient of friction is 0.65, does the porter reach the elevator bay in time to earn the tip? If so, how many seconds does the porter have to spare? If not, by how many meters does the porter miss? In either case, round your numerical solution to the nearest tenth.

Starting Motion with Static Friction

151–154

151. A car is parked on a banked road. What type of friction prevents it from moving?

152. Using a spring scale, Paul measures the amount of force required to start sliding a cabinet across his dorm-room floor as 3,400 newtons. If the cabinet has a mass of 800 kilograms, what is the coefficient of static friction between the cabinet and the floor? Round your answer to the nearest hundredth.

153. Franz pushes a 60-kilogram table with 500 newtons of force. If the coefficient of static friction between the table's feet and the floor below is 0.3, what is the table's initial acceleration in meters per second squared? Round your answer to the nearest tenth.

154. A self-storage employee uses a rope to pull a crate along the floor, but a force of 890 newtons is needed at an angle of 28 degrees with respect to the horizontal to get it moving initially.

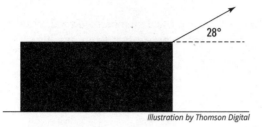

If $\mu_s = 0.18$ and $\mu_k = 0.14$, what is the crate's mass? Round your answer to the nearest kilogram.

Keeping on Moving with Kinetic Friction

155–157

155. A car's engine provides 12,000 newtons of force east, keeping the car moving at a constant speed. If friction is the only other force acting on the car along the east-west axis, what is the magnitude and direction of the force of kinetic friction? Give your answer in newtons, rounded to two significant digits.

156. A skater glides with a velocity of 12 meters per second in a southeasterly direction on an ice-covered pond. What is the coefficient of kinetic friction between the skate blades and the ice if the skater's acceleration is −0.8 meters per second squared? Round your answer to the nearest hundredth.

157. A girl weighing 95 kilograms is roller-skating 8 meters per second along a level street. If the coefficient of kinetic friction between her skates' wheels and the pavement is 0.7, how many more meters will she coast before coming to a complete stop? Round your answer to the nearest tenth of a meter.

Pushing and Pulling on a Non-Slippery Slope

158–161

158. Jillian pushes a crate up a slope with a constant velocity. Taking "up the slope" as the positive direction, what is the force of friction on the crate if Jillian pushes with 50 newtons of force and gravity exerts 40 newtons of force on the crate along the axis of the slope?

159. Exercising for a boxing match, Rocky pushes a crate up a mountain with a 21-degree angle of inclination and a coefficient of kinetic friction of 0.36. His opponent, Bob, does a similar routine using a crate of the same mass on a mountain of identical slope, but a layer of snow on the mountain drops the coefficient of friction to 0.18. How many times larger is the force with which Rocky pushes than the force with which Bob pushes if each boxer accelerates his crate at a rate of 2.6 meters per second squared up his respective mountain? Round your answer to the nearest tenth.

160. An 18-kilogram toy car rolls down a slanted driveway at a constant speed of 3.3 meters per second. If the coefficient of friction between the car and the driveway is 0.29, at what angle is the driveway inclined with respect to the horizontal? Round your answer to the nearest degree.

161. A 98-kilogram snowboarder accelerates down a mountainside of constant slope at a rate of 2.1 meters per second squared. If the coefficient of kinetic friction between the snowboard and the snow is 0.1, at what angle is the mountainside inclined relative to the horizontal? Round your answer to the nearest degree.

Covering the Distance on a Non-Slippery Slope

162–165

162. A 5-kilogram crate is given an initial push down a 42-degree ramp to overcome the force of static friction and then is immediately released. After it's moving, how far does it slide in 4 seconds if the force of friction is 30 newtons up the ramp? Round your answer to the nearest tenth of a meter.

163. A block with an initial speed of 1 meter per second slides 5.8 meters down a ramp inclined at 72 degrees in 1.1 seconds. What is the coefficient of friction between the block and the ramp, rounded to the nearest hundredth?

164. An Olympic skier slides down from the top of a snow-covered hill angled 70 degrees with respect to the horizontal. The hill is 18 meters high and bottoms out on a flat plain, still covered with snow. How far from the base of the hill does the skier's momentum carry her if the coefficient of friction between her skis and the snow is 0.08? Round your answer to the nearest 10 meters.

165. At a skate park, a 2.5-kilogram skateboard is released from the top of a 6-meter-high hill that has a 55-degree angle of inclination. It travels down the hill, rolls along a 3.5-meter-long flat section, and continues up a second hill until it stops at a height of h meters above the ground, as pictured in the following diagram. What is the value of h if the coefficient of kinetic friction between the board's wheels and the floor is 0.18? Round your answer to the nearest meter.

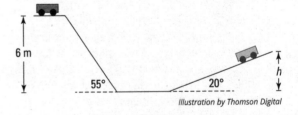
Illustration by Thomson Digital

Shooting Objects Straight Up to the Maximum Height

166–169

166. A baseball is thrown into the air, landing several meters away. What is the ball's vertical velocity at the highest point of its arc? Give your answer in meters per second.

167. A 50-gram Frisbee is launched at an initial velocity of 20 meters per second straight up with its flat side parallel to the ground. The Frisbee experiences a constant force of air resistance of 10 newtons. What is the maximum height in meters that the Frisbee can reach above its original launch point? Round your answer to two significant digits.

Use the following diagram showing each section's length and mass data, as well as the subsequent information, to solve Questions 168 and 169: A 13,500-kilogram rocket has two stages, measuring 42 meters in total length. The engines on the lower stage provide 150 kilonewtons of force. Sixty seconds after launch from rest, the 34-meter section falls off (its engine is now dormant), and the engines on the upper stage kick in, providing 25 kilonewtons of force. Thirty seconds after that, the upper stage's engines stop operating. During the entire flight, air resistance equals $\dfrac{m}{\left(0.06\,\text{s}^2/\text{m}^2\right)l}$, where m is the mass of the object and l is its length.

8 m (1,850 kg)

34 m (11,650 kg)

Illustration by Thomson Digital

168. What is the maximum height reached by the lower, 34-meter-long section? Round your answer to the nearest tenth of a kilometer.

169. What is the maximum height reached by the upper, 8-meter-long section? Round your answer to the nearest tenth of a kilometer.

Taking the Time to Go Up and Down

170–172

170. A bottle rocket fires straight up off the ground, landing 2.8 seconds later at the same spot from which it launched. How long after launch does the rocket reach its highest point, assuming there is no air resistance? Round your answer to the nearest tenth of a second.

171. Sam tosses a 1.2-kilogram billiard ball straight up into the air, where it is subjected to an air resistance equivalent to 10 percent of its weight. If the ball reaches the apex of its flight after 2.7 seconds, with what speed does Sam toss the billiard ball into the air? Round your answer to the nearest meter per second.

172. The engine on a 1,300-kilogram rocket provides a force of 15 kilonewtons. When it is launched, the rocket travels 150 meters straight up, at which point the engine is deactivated. How long (from launch) does it take for the rocket to reach its apex? Round your answer to the nearest second.

Shooting at an Angle: Separating the Motion in Components

173–177

173. A basketball shot at an angle of 70 degrees with respect to the horizontal experiences 3 newtons of air resistance while in the air. What is the magnitude of the air resistance in the vertical direction immediately after the ball is shot? Round your answer to the nearest tenth of a newton.

174. A cannonball is fired with a horizontal component of velocity equal to 20 meters per second. Neglecting air resistance, what is the cannonball's velocity at the top of the arc it makes through the air? Give your answer in meters per second.

Use the following information and diagram to answer Question 175 (ignore air resistance): Mariska pushes a 50-kilogram buoyant shipping pallet from rest up a ramp that is inclined at an angle of 30 degrees with respect to the horizontal. Mariska exerts a constant force of 600 newtons, increasing the pallet's height by 3.8 meters by the time it "launches" off the ramp. It then proceeds to splash into the water below.

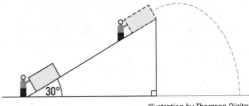

Illustration by Thomson Digital

175. From the instant Mariska starts pushing the pallet to the time it strikes the water, how much time elapses? Ignore friction, and round your answer to the nearest tenth of a second.

Use the following diagram and accompanying information to solve Questions 176–177: A woman who is 1.6 meters tall tosses a 285-gram disc across a field. The disc's center of mass has a velocity of 5 meters per second, and the disc travels with an angle of θ_v degrees with respect to the horizontal. The disc is tilted with an angle of θ_{disc} degrees with respect to the horizontal and experiences a wind-resistance force of 1.2 newtons perpendicular to its flat surface. Assume the disc starts its flight at the top of the woman's body.

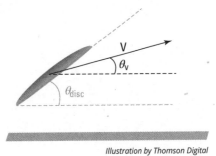

Illustration by Thomson Digital

176. If $\theta_{\text{disc}} = 35°$ and $\theta_v = 20°$, what is the maximum height the disc reaches above the ground? Round your answer to the nearest centimeter.

177. If $\theta_{\text{disc}} = 12°$ and $\theta_v = 33°$, how far across the field does the disc land if the wind stops blowing 1.3 seconds into the disc's flight? Round your answer to the nearest tenth of a meter.

Reaching Far with a Projectile

178–180

Use the following information to answer Question 178: A 0.42-kilogram baseball thrown at an angle of 23 degrees to the horizontal with a velocity of 12 meters per second encounters a stiff wind blowing in the opposite direction with a resistance of 1.4 newtons parallel to the ground. The baseball starts its flight 1.25 meters above the ground.

178. How long is the baseball in the air? Round your answer to the nearest tenth of a second.

Use the following information and diagram to answer Questions 179–180: A block slides up a ramp with an initial velocity v meters per second, eventually launching off the ramp and landing elsewhere. The ramp is h meters high and angled θ degrees with respect to the horizontal. Assume that the ramp's surface is frictionless unless otherwise stated.

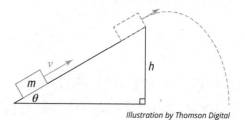

Illustration by Thomson Digital

179. Given $m = 5$ kilograms, $v = 6.3$ meters per second, $h = 1.3$ meters, and $\theta = 18$ degrees, calculate the block's horizontal component of velocity upon leaving the ramp's surface. Round your answer to the nearest tenth of a meter per second.

180. If the block has an initial velocity of 10 meters per second, traveling 3.5 meters along the ramp that is angled 25 degrees above the horizontal, how far from the base of the ramp does the block land after launch? Round your answer to the nearest tenth of a meter.

Chapter **6**

Describing Rotational Motion

ircular motion is all around you, particularly if you're at the center of a merry-go-round. Circular motion is best described using angular displacement, angular velocity, and angular acceleration. The equations that these angular quantities satisfy are the same as the linear motion equations if you replace displacement with angular displacement, velocity with angular velocity, and acceleration with angular acceleration.

The Problems You'll Work On

When you get around to working on this chapter, you'll see the following types of problems:

>> Measuring angles in radians

>> Using angular velocity and angular acceleration to understand circular motion

>> Finding the centripetal acceleration and centripetal force of objects moving in a circle

What to Watch Out For

These problems won't make your head spin if you keep the following in mind:

>> Converting all angles to radians

>> Remembering that for an object in uniform circular motion the acceleration points toward the center of the circle, and the velocity is perpendicular to the acceleration

>> Recalling that the speed is constant but the velocity is changing for an object in uniform circular motion

Keeping a Constant Speed around a Circle

181–183

181. The second hand on a clock is 0.10 meters long. At what speed in meters per second does the tip of the second hand travel?

182. One parent stands on one side of a merry-go-round and the other parent stands on the opposite side. Their child rides the merry-go-round and takes 15 seconds to travel from one parent to the other parent. What is the child's angular speed in radians per second?

183. The blades of an electric blender turn at 15 radians per second. How many revolutions do the blades make in 1 minute?

Measuring Angles in Radians

184–187

184. You cut a cake into 8 equal pieces. What is the angle, in radians, of the pointed end of each piece of cake?

185. A drawbridge pivots upward from the horizontal position and moves through two-thirds of the angle between the horizontal and vertical positions. Through what angle (in radians) has the drawbridge moved?

186. Cleaning up after a party, you find leftover pizza slices. The pizzas were round, and the slices were all cut through the center. You find two slices that together make a quarter of a pizza and three slices that are each an eighth of a pizza. If you put all the slices side by side as in a pizza, what is the total angle made (in radians)?

187. A roller coaster enters a curve heading down. At this point the track makes an angle of 15 degrees with the ground. The track then curves up again, and the roller coaster leaves the curve on a track heading up at 35 degrees above the horizontal. Through what angle (in radians) does the roller coaster travel?

Traveling in a Circle with Angular Velocity

188–191

188. The moon rotates once on its axis in about 27 days. If the moon's radius is 1,734 kilometers, how far in kilometers does an astronaut standing on the moon's surface move in 1 day?

189. You drive along the freeway at 60 miles per hour. You go through a curve with a radius of one quarter mile. What is your angular speed in radians per hour in this curve?

190. The outer edge of a roulette wheel has 34 evenly spaced separators that divide the wheel into little compartments. When the wheel is spun, each separator makes a click as it passes a bump on the table. If the wheel spins at 8.3 radians per second, what is the time in seconds between clicks?

191. An airplane propeller rotates at 3,000 revolutions per minute. If one blade of the propeller is 1.3 meters from the axis of rotation to the tip, how far in meters does that tip travel in 10 seconds?

Speeding Up and Down around a Circle with Angular Acceleration

192–196

192. When you switch your room fan from medium to high speed, the blades accelerate at 1.2 radians per second squared for 1.5 seconds. If the initial angular speed of the fan blades is 3.0 radians per second, what is the final angular speed of the fan blades in radians per second?

193. The radius of a car tire is about 0.35 meters. If the car accelerates in a straight line from rest at 2.8 meters per second squared, what is the angular acceleration, both magnitude and direction, of the front passenger-side tire?

194. A merry-go-round has an angular acceleration of 0.30 radians per second squared. After accelerating from rest for 2.8 seconds, through what angle in radians does the merry-go-round rotate?

195. Driving down a straight road at 22 meters per second, you accelerate at 2.7 meters per second squared in the direction opposite to your velocity. After 2.6 seconds, you continue forward at a constant velocity. If the radius of your tires is 0.37 meters, through what angle in radians do your tires rotate during this acceleration?

196. You're part of a relay team in the 4-x-100-meter relay race. You take the baton at the beginning of the curve in the track when you're not moving and accelerate at 3.4 meters per second squared. If the curve radius is 25 meters, what is your angular speed in radians per second after 1.4 seconds?

Accelerating toward the Center with Centripetal Acceleration

197–200

197. A ferry boat makes a 180-degree turn in 12 minutes. The radius of the turn is 0.50 miles. If the boat's speed is 1.2 meters per second, what is its centripetal acceleration in radians per second squared during the turn? There are 1,609 meters in 1 mile.

198. You twirl a lasso over your head at a constant angular speed of 3.8 radians per second. What is the centripetal acceleration in meters per second squared of the lasso's tip, 1.4 meters from your hand?

199. The documentation for your slot-car set says that the maximum centripetal acceleration the cars can withstand without being ejected from the track is 3.8 meters per second squared. You notice that the slot cars fly off the track if they exceed 1.1 meters per second. What is the radius in meters of the curve in the track?

200. You're driving your car around a 70-meter-radius curve, and you want to keep your centripetal acceleration below 1g (where $g = 9.8$ meters per second squared is the acceleration due to gravity on Earth). At what maximum speed in meters per second can you travel through the curve?

Providing the Centripetal Force Based on Mass, Velocity, and Radius

201–205

201. You sit on a stool, stick your legs straight out in front of you, and spin around. If you complete 3 revolutions in 9.0 seconds and your legs are 0.85 meters long, what is the centripetal force in newtons on your big toe? Assume that the mass of your big toe is 0.035 kilograms.

202. The mass of Earth is 6.0×10^{24} kilograms and its distance from the sun is about 1.5×10^{11} meters. What is the magnitude of the force in newtons that causes Earth to complete an orbital rotation?

203. In a prototype helicopter, a jet is expelled out of the helicopter's side to make it turn in the opposite direction. If the jet supplies 10,000 newtons of force and the helicopter's forward speed remains at 20 meters per second, what is the radius in meters of the helicopter's turn? Assume the mass of the helicopter is 2,000 kilograms.

204. A centrifuge spins a 0.050-kilogram test tube at 1,000 radians per second. If the radius of the centrifuge is 0.15 meters, what force in newtons does the centrifuge apply on the test tube?

205. To turn, a jet ski ejects a spray of water perpendicular to its velocity. If this water spray applies a force of 2,000 newtons on the jet ski, and the mass of the jet ski is 400 kilograms, what is the difference in the jet ski's speed in meters per second between when it enters a turn and when it exits the turn?

Chapter **7**

Rotating Around in Different Loops

W hether you're riding in a bus around a roundabout or orbiting Earth in a space capsule, you require a force to maintain your circular motion. For uniform circular motion, this force points toward the center of the circle.

The Problems You'll Work On

In this chapter you work through circular motion problems of the following types:

>> Figuring out how fast you can go and how sharply you can turn on a flat road without skidding

>> Solving banked turn problems

>> Using Newton's law of universal gravitation to examine circular orbits

>> Applying Kepler's laws to non-circular orbits

>> Computing the speed needed to go around a vertical loop

What to Watch Out For

To keep from going in circles, be sure you watch out for the following:

>> Remembering that for uniform circular motion, the velocity is tangential and the acceleration is radial

>> Never including the fictitious centrifugal force in your equations and diagrams

>> Using the distance between a satellite and the center of the planet it orbits as the radius in orbital motion problems

Turning on a Flat Road with a Little Help from Friction

206–209

206. If the coefficient of friction on a circular curve of road is doubled, by what factor will the maximum speed that a car can drive on the curve without slipping be multiplied?

207. What is the maximum speed that a 1,500-kilogram car can maintain without slipping when driving along a section of frictionless road with a radius of curvature of 25 meters? Round your answer to the nearest meter per second.

208. Driving around a level bend, you notice that your car starts to skid just as you hit a speed of 50 kilometers per hour. If the coefficient of friction between the road and your car's tires is 0.33, what is the curve's radius of curvature — in meters — rounded to three significant digits?

209. Driving along the perfectly flat Mons Super-highway on Mars, NASA Secret Agent Mully encounters a curved stretch of road when the rover begins to skid when the speedometer hits 120 kilometers per hour. Because secret agent helped create the material from which the road is constructed, Mully knows that the coefficient of friction between the road and the rover's wheels is 0.72. What is the road's radius of curvature — rounded to the nearest 10 meters — at Mully's current location? Use the following data to help you solve this problem.

$$m_{\text{Mars}} = 6.42 \times 10^{23} \text{ kg}$$
$$r_{\text{Mars}} = 3.37 \times 10^{3} \text{ km}$$
$$G = 6.67 \times 10^{-11} \frac{\text{N} \cdot \text{m}^{2}}{\text{kg}^{2}}$$

Making a Banked Turn in Debt to Normal Force

210–214

210. At what angle should a banked road be tilted if the magnitude of the force exerted by the ground on the car is the same as that exerted by gravity on the car? Give your answer in degrees.

211. What is the critical speed that a 1,500-kilogram car must maintain to avoid slipping while driving along a section of frictionless road with a radius of curvature of 25 meters and an inclination of 10 degrees? Round your answer to the nearest meter per second.

212. A 180-kilogram snowmobile runs on a snow-covered, curved bank of radius 6 meters. What is the curve's angle of inclination if the snowmobile can reach a speed of 15 meters per second before skidding? Assume that the surface is frictionless. Round your answer to the nearest degree.

213. Trial runs along a racetrack located on a mysterious planet have shown that a maximum speed of 130 kilometers per hour can be maintained around the 80-meter-radius turns if those turns are banked at 42 degrees. What is the magnitude of the gravitational acceleration on this planet's surface if the coefficient of friction between a vehicle's tires and the track is 0.18? Round your answer to the nearest tenth of a meter per second squared.

214. What is the minimum velocity a 450-kilogram car must travel on a banked road ($\mu = 0.08$) with a radius of curvature of 18 meters and an inclination of 25 degrees to prevent it from slipping down the bank? Round your answer to the nearest kilometer per hour.

Applying the Law of Universal Gravitation to the Stars

215–218

215. The gravitational force between objects A and B is 4 newtons. If the mass of B were one-half as large as it currently is while A's mass remains the same, how large is the gravitational force?

216. Calculate the force of gravity between two 3-kilogram ball bearings separated by a distance of 10 centimeters. Round your answer to two significant digits.

217. A 9,000-kilogram starship is pulled toward Planet X, a 6.2×10^{33} kilogram behemoth with a radius of 65,000 kilometers. When the starship is 2,500 kilometers from the planet's surface, what is the starship's acceleration (providing that its engines are turned off)? Round your answer to two significant digits.

218. The starship *Orion* is stuck between a rock (a large asteroid named Cerberus) and a hard place (an even larger asteroid named Hades). How many kilometers from the surface of Cerberus is *Orion* located if the starship has no net acceleration? Use the following data and round your answer to two significant digits.

$$m_{\text{Orion}} = 850 \text{ kg}$$
$$m_{\text{Cerberus}} = 1.88 \times 10^6 \text{ kg}$$
$$r_{\text{Cerberus}} = 1.8 \times 10^4 \text{ km}$$
$$m_{\text{Hades}} = 8.2 \times 10^7 \text{ kg}$$
$$r_{\text{Hades}} = 2.5 \times 10^5 \text{ km}$$

The distance between the surface of Cerberus and the surface of Hades is 5.78×10^6 kilometers.

Accelerating with Gravity Near a Planet's Surface

219–223

219. In terms of g, the acceleration of gravity near Earth's surface, what is the acceleration of gravity near the surface of a planet with the same mass as Earth but twice the radius?

220. Researchers at NASA load a 100-kilogram package onto a rocket on Earth. When the rocket lands on the surface of Neptune, where the acceleration of gravity is approximately 1.2 times as great as that on Earth, what will be the package's mass, rounded to the nearest integer?

221. How many times greater is the acceleration due to gravity at Jupiter's "surface" than at Earth's? Use the following data and round your answer to the nearest integer.

$$m_{\text{Earth}} = 5.98 \times 10^{24} \text{ kg}$$
$$r_{\text{Earth}} = 6.4 \times 10^{6} \text{ m}$$
$$m_{\text{Jupiter}} = 1.9 \times 10^{27} \text{ kg}$$
$$r_{\text{Jupiter}} = 6.99 \times 10^{7} \text{ m}$$

222. A circular space station in the shape of a bicycle wheel has a mass of 1,800 kilograms. It rotates about a central axis 450 meters away from the station's outer ring and takes 30 minutes to make one revolution. What is the acceleration of an object located on the outer ring? Round your answer to two significant digits.

223. How much longer does it take a rock to fall from a 100-meter-high cliff on Mars than to fall from the same height on Earth? Mars's mass is 6.42×10^{23} kilograms, and its volume is 1.63×10^{11} kilograms. Assume that neither planet has air resistance, and round your answer to the nearest tenth of a second.

Finding the Speed of the Satellites in Circular Orbits

224–227

224. Let v represent Mars's orbital velocity about Sol, the star at the center of our solar system. If Jupiter is located 3 times farther away from Sol than is Mars, what is Jupiter's orbital velocity in terms of v?

225. A satellite orbits Earth at an altitude of 400 kilometers above the planet's surface. What is its speed in meters per second? Round your answer to three significant digits.

226. The GMVX satellite is "pulled" along by the force of Earth's gravity at a speed of 1,200 meters per second. How many kilometers from Earth's center is the GMVX located? Round your answer to two significant digits.

227. Find the speed of a satellite in geosynchronous orbit above Venus if a Venusian day is 243 times longer than a day on Earth. Give your answer in units of kilometers per hour, rounded to three significant digits. Venus has a mass of 4.88×10^{24} kilograms.

Taking the Time to Travel around Celestial Bodies

228–231

228. Earth is located 1 a.u. (astronomical unit — a measure of distance) from its sun. In units of "Earth years," how long does Mars take for its own solar revolution if it's located 1.5 a.u. from the sun? Round your answer to the nearest tenth.

229. Earth's moon completes an orbit of Earth in approximately 28 days. How many kilometers separate the centers of these two celestial bodies? Use the following data and round your answer to three significant digits.

$$m_{Earth} = 5.98 \times 10^{24} \text{ kg}$$
$$m_{Moon} = 7.36 \times 10^{22} \text{ kg}$$

230. If the International Space Station is located 420 kilometers above Earth's surface, how many hours does it take to make a complete orbit? Use the following data and round your answer to the nearest hundredth of an hour.

$$r_{Earth} = 6.38 \times 10^{3} \text{ km}$$

231. Use the following information (where m is the mass of a particular body and r is its average distance to the sun) to determine the period (in Earth days) of Neptune's orbit about the sun. Round your answer to three significant digits.

$$r_{Neptune} = 4.5 \times 10^{12} \text{ m}$$
$$m_{Neptune} = 1.03 \times 10^{26} \text{ kg}$$

$$m_{Earth} = 5.98 \times 10^{24} \text{ kg}$$
$$m_{Neptune} = 1.03 \times 10^{26} \text{ kg}$$
$$m_{Sun} = 1.991 \times 10^{30} \text{ kg}$$
$$r_{Earth} = 1.496 \times 10^{8} \text{ km}$$
$$r_{Neptune} = 4.5 \times 10^{9} \text{ km}$$

Moving Fast to Avoid Falling Off in a Vertical Loop

232–235

232. Traveling upside down, a cart on a roller coaster reaches the top of a circular loop with a speed of V kilometers per hour, barely maintaining contact with the track. With the same cart and track on Earth's moon, where the acceleration due to gravity is approximately one-sixth that of Earth, what minimum speed — in terms of V — must the cart maintain at the top of the loop to just barely maintain contact with the track?

233. A 20-gram mouse is running in a stationary, vertical wheel of diameter 15 centimeters. How fast must the mouse run to make it all the way around the loop without falling upside down to the bottom of the wheel? Round your answer to the nearest hundredth of a meter per second.

234. The Circle o' Death roller coaster has a loop-the-loop with a minimum speed requirement at the top of 126 kilometers per hour to prevent the cart from losing contact with the coaster's track. What is the diameter of the loop? Round your answer to the nearest hundredth of a kilometer.

235. Fred swings a bucket of water in a vertical loop at a constant speed of 3 meters per second — the minimum speed required to keep the water from spilling. What is the positive difference between the tension in Fred's arm at the top of the loop and the tension in his arm at the bottom of the loop if the combined mass of bucket and water is 3.5 kilograms and Fred's arm is 80 centimeters long? Round your answer to the nearest tenth of a newton.

Chapter **8**

Going with the Flow: Fluids

Fluids exert a buoyant force on submerged objects. The magnitude of this buoyant force is equal to the weight of the fluid displaced by the object. The pressure exerted by a fluid tends to decrease where the fluid is moving faster or has a greater height.

The Problems You'll Work On

In this chapter you'll get your feet wet with the following topics:

>> Going from density to specific gravity and back

>> Finding pressure at different depths

>> Understanding hydraulic systems with Pascal's principle

>> Applying Archimedes's principle to floating and submerged objects

>> Determining the flow rate from the pipe size

>> Using Bernoulli's equation to relate speed, pressure, and height

What to Watch Out For

If you're under pressure to get these problems right, remember the following:

>> Recalling that specific gravity is the ratio of the density of the object to the density of water

>> Remembering that the buoyant force depends on the weight of the fluid displaced, not the weight of the object

>> Decreasing the radius of a pipe increases the fluid speed

>> Increasing the speed of a fluid decreases its pressure

Getting Denser with More Mass Packed Together

236–237

236. You make a cake that has a mass of 300 grams and fits in a cake pan that is 30 by 10 by 6.0 centimeters cubed. What is the density of the cake?

237. A box of cough drops has a mass of 1.0 gram, and its dimensions are 1.0 by 5.0 by 8.5 centimeters cubed. It contains 30 cough drops, each of which has a mass of 2.2 grams. What is the density of the box when it is full of cough drops?

Comparing Densities Using Specific Gravity

238–239

238. The density of gasoline is 721 kilograms per cubic meter. What is its specific gravity?

239. A pile of pillows has a density of 55 kilograms per cubic meter. You cram them into a box and close the lid, and the specific gravity of the pillows triples. What is the density of the pillows inside the box?

Applying Pressure with a Force

240–241

240. What is the magnitude of the force exerted on a piston of cross-sectional area 0.3 square meters by a fluid at a pressure of 1.2 pascals?

241. A table weighs 300 newtons. If each of its four legs has a cross-sectional area of 10 square centimeters, what pressure does each leg exert on the floor?

Working under Pressure: Calculating the Pressure at a Depth

242–244

242. The pressure at the top of a pipe full of water is 101 pascals. What is the change in pressure between the top and the bottom of the pipe, 3.4 meters lower?

243. One end of a 50-meter-long hose is attached to the bottom of a large basin full of water. How many meters below the top of the basin must the hose outlet be positioned for the water pressure at the outlet to be 18,000 pascals?

244. While diving with a friend, you note that the water pressure at your current depth is 130,000 pascals. If you swim up another 2.5 meters, what is the water pressure?

Passing on Pressure with Pascal's Principle

245–247

245. In a hydraulic system, a piston with a cross-sectional area of 21 square centimeters pushes on an incompressible liquid with a force of 38 newtons. The far end of the hydraulic pipe connects to a second piston with a cross-sectional surface area of 100 square centimeters. What is the force on the second piston?

246. Consider a hydraulic system with two pistons. Piston 1 applies a force F_1 over area A_1, and piston 2 applies a force F_2 over area A_2. If you double the force applied by piston 1 and reduce the area of piston 2 by a factor of 3, what is the new force F_2' in terms of the original force F_2?

247. A piston that is part of a hydraulic system has a surface area of 0.025 square meters. The hydraulic fluid pushes on the piston with a pressure of 20,000 pascals. What pressure pushes on another piston in the same system?

Floating with Archimedes' Principle

248–250

248. A block of wood with the dimensions 0.12 by 0.34 by 0.43 cubic meters floats along a river with one broad face facing down. The wood is submerged to a height of 0.053 meters. What is the mass of the piece of wood?

249. You plunge a basketball beneath the surface of a swimming pool until half the volume of the basketball is submerged. If the basketball has a radius of 12 centimeters, what is the buoyancy force on the ball due to the water?

250. A 4,000-kilogram boat floats with one-third of its volume submerged. If two more people get into the boat, each of whom weighs 690 newtons, what additional volume of water is displaced?

Distinguishing Different Types of Flow

251–252

251. You have two identical cups filled to the same height with different liquids (call them liquids A and B). You tip both cups in the same way, and liquid B spills out faster than liquid A. Which of the following statements do you know is true?

    (A) Liquid A is more compressible than liquid B.

    (B) Liquid A undergoes irrotational flow.

    (C) Liquid A is less viscous than liquid B.

    (D) Liquid A is less compressible than liquid B.

    (E) Liquid A is more viscous than liquid B.

252. In an experiment, a marker floats in a liquid that flows down a curved channel. In the curve of the channel, the same point on the marker always points toward the center of the curve. Which of the following statements do you know is true?

    (A) The flow is irrotational.

    (B) The liquid is incompressible.

    (C) The liquid is more viscous in the curve.

    (D) The liquid is less viscous in the curve.

    (E) The flow is rotational.

Flowing Faster with a Smaller Pipe

253–256

253. Water travels through a hose at 0.8 meters per second. If the cross-sectional area of the exit nozzle is one-fifth that of the hose, at what speed does water exit the hose?

254. You are watching leaves float past you in a stream that is 0.76 meters wide. At one point, you estimate that the speed of the leaves triples. If the stream has a constant depth, what is the width of the stream at this point?

255. The volume flow rate through pipe 1 is 2.5 times that of pipe 2. If the cross-sectional area of pipe 1 is one-half that of pipe 2, what is the ratio of the flow speed in pipe 1 to that in pipe 2?

256. If the water that exits a pipe fills a pool that is 3 meters deep, 20 meters long, and 5 meters wide in 3 days, what is the flow rate?

Relating Pressure and Speed with Bernoulli's Equation

257–260

257. A dam holds back the water in a lake. If the dam has a small hole 1.4 meters below the surface of the lake, at what speed does water exit the hole?

258. A hose lying on the ground has water coming out of it at a speed of 5.4 meters per second. You lift the nozzle of the hose to a height of 1.3 meters above the ground. At what speed does the water now come out of the hose?

259. An unknown liquid flows through a pipe with a constant radius. The pipe runs horizontally, then rises 2.8 meters, and then continues on horizontally. If the speed of the fluid in the pipe changes from 3.9 meters per second to 1.2 meters per second, and the pressure in the pipe changes from 110,000 to 101,000 pascals, what is the density of the liquid?

260. To push water up a pipe, a pump creates a pressure of 115,000 pascals and a water speed of 2.5 meters per second. The water rises to a certain height, at which point the pressure is 110,000 pascals and the water speed is 1.0 meter per second. How high does the water rise?

Putting It All Together with Pipes

261–265

261. Blood in the aorta exerts a pressure of 12,000 pascals on the walls of the aorta. The blood moves at 0.30 meters per second. What fraction of the aorta must be blocked to reduce the blood pressure to 0? Assume that the blood flows horizontally. The density of blood is about 1,060 kilograms per cubic meter.

262. An irrigation channel narrows from 1.0 meter wide to 0.33 meters wide. If its depth increases from 0.20 meters to 0.80 meters, by what percent does the water speed change in going from the wider section to the narrower section?

263. Pipe 1 and pipe 2 join in parallel (side by side) and then connect to pipe 3. Water flows through pipes 1 and 2 at the same speed toward pipe 3. The radius of pipe 1 is 3.0 centimeters, and the radius of pipe 2 is 4.0 centimeters. What does the radius of pipe 3 need to be to keep the water speed the same as in pipes 1 and 2?

264. You want to collect water from a mountain lake by inserting a pipe 1.2 meters deep into the lake and collecting the water 25 meters lower. What radius pipe must you use to get a volume flow rate of 1 liter per second?

265. Water flows through a 3.0-centimeter-radius pipe at 0.20 meters per second. What is the volume flow rate?

Chapter 9

Getting Some Work Done

When a force is applied to an object and that object is displaced, *work* is done on that object. If a net force does work on an object, the kinetic energy of that object will change. Kinetic energy can be converted into other types of energy, such as potential energy. In the absence of friction, the total energy is conserved.

The Problems You'll Work On

In this chapter, you'll work on work (along with kinetic and potential energy) problems:

>> Determining the work done from force and displacement

>> Using the work-energy theorem to relate the work done and the change in kinetic energy

>> Figuring out the gravitational potential energy

>> Converting kinetic energy into potential energy and back again using conservation of energy

>> Finding the rate of doing work with power

What to Watch Out For

If you keep the following in mind you'll only have to work through the problems once:

>> Remembering to multiply by the cosine of the angle between the force and the displacement when calculating the work

>> Recalling the minus sign if the force has a component opposite to the direction of motion

>> Using the work done by the net force when finding the change in kinetic energy

>> Applying conservation of energy only in the absence of friction

Applying Force in the Direction of Movement

266–268

266. What is the net amount of work done when a librarian lifts a 925-gram book from the floor to a point 1.5 meters above the floor at a constant speed? Round your answer to the nearest joule.

267. A bookshelf has five shelves, each 40 centimeters apart. If Roger lifts a 2.8-kilogram dictionary at a constant velocity from the second shelf to the fifth shelf, how much work does Roger do? Round your answer to three significant digits.

268. How much work is done by an automotive engine that accelerates an 800-kilogram car from 25 kilometers per hour to 40 kilometers per hour in 5 seconds? The coefficient of friction between the car's tires and the ground is 0.72. Round your answer to the nearest kilojoule.

Applying Force at an Angle

269–272

269. What range of angles between a force vector and a displacement vector results in a negative amount of work being done? Give your answer in interval notation [a, b], where a is the lower boundary and b is the upper boundary.

Use the following information to answer Questions 270 and 271: A father pushes a sled carrying his child down a 200-meter-high hill with a 32-degree slope. Using a constant force of 500 newtons directed parallel to the hill's surface, he increases the sled's velocity from 0 to 4.2 meters per second in 5 seconds, at which point he lets go. The combined mass of the child and the sled is 65 kilograms.

270. How much work does the father do pushing his child? Round your answer to the nearest ten joules.

271. How many kilojoules of work does friction do between the time that the father lets go of the sled and the time that the sled stops at the bottom of the hill? The coefficient of kinetic friction between the sled's runners and the snow is 0.13. Round your answer to the nearest integer.

272. A hockey goal sitting on the ice is suddenly subjected to two impacts that combine to do 23 joules of work on the goal. One impact pushes it with 25 newtons of force due west, and the other pushes it with 2 newtons of force toward the southern half of the ice, but the exact direction is unknown. If the impacts move the goal 89 centimeters, in what direction did the 2-newton force point? Round your answer to the nearest degree.

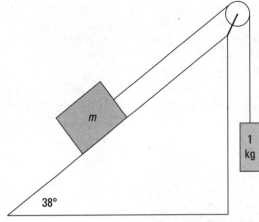

Applying Force in the Opposite Direction of Motion

273–275

273. What is the work done by a westward-directed force of 12.8 newtons that moves an object 3.1 meters to the east? Round your answer to the nearest joule.

274. What amount of work does gravity do on a 5-kilogram crate filled with 32 0.8-kilogram lemons if the crate is lifted 1.25 meters off of the ground? Round your answer to the nearest ten joules.

Use the following diagram of two blocks — connected by a rope — and — pulley system — that start from rest and move along an inclined plane to answer Question 275. Note that the coefficient of friction between the block of mass m and the inclined plane is 0.2.

275. If the block of unknown mass accelerates down the inclined plane at a rate of 1.3 meters per second squared for 3 seconds, how much work does gravity perform on the two-block system? Round your answer to the nearest joule.

Finding the Kinetic Energy of a Moving Object

276–278

276. Two birds, a sparrow and a cardinal, fly east with the same speed. If the sparrow's kinetic energy is denoted K, what is the cardinal's kinetic energy in terms of K if it is four times as massive as the sparrow?

277. How fast is a 28-gram bullet moving if it has a kinetic energy of 90 joules? Give your answer in kilometers per hour, rounded to two significant digits.

Use the following information and diagram to solve Question 278: Two blocks are connected by a string that is pulled taut over a frictionless pulley at the edge of a table. The coefficient of friction between the tabletop and the sitting block, m_1, is 0.35. The 1-kilogram hanging block does not touch the side of the table. Ignore the masses of the pulley and the string.

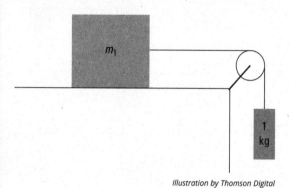

Illustration by Thomson Digital

278. If the sitting block is twice the mass of the hanging one, what is the kinetic energy of the hanging block 1.2 seconds after the system is released from rest? Round your answer to the nearest tenth of a joule.

Turning Work into Kinetic Energy

279–283

279. Sam and Eric race each other while each is pushing a 12-kilogram shopping cart from rest, but Sam performs three times the amount of work as Eric. If Eric's speed after 5 seconds is v_E, what is Sam's speed after 5 seconds? Ignore friction and give your answer as an exact multiple of v_E (no decimals).

280. An engine provides 500 joules of work to change a car's kinetic energy from 220 joules to 670 joules. How much energy is lost to friction?

281. Sammy catches a 480-gram baseball traveling 130 kilometers per hour, bringing it to a complete stop in 0.8 seconds. How many joules of work does Sammy do on the ball? Round your answer to three significant digits.

282. A 105-kilogram skier's speed down a frictionless hill doubles in the span of 8 seconds. How many kilojoules of work is done on the skier moving 48 meters in that time span? Round your answer to two significant digits.

283. A 100-kilogram soldier jumps out of a plane 1.1 kilometers above the ground. At some point before hitting the ground, the soldier opens the parachute, eventually hitting the ground with a speed of 18 kilometers per hour. If the force of air resistance against soldier's descent is 70 newtons before the chute opens and 2,500 newtons after it opens, how many meters above the ground is the soldier when the parachute opens? Round your answer to two significant digits.

Banking on Potential Energy by Working against Gravity

284-287

284. A basketball and baseball are both shot straight up in the air so that each ends up with the same amount of gravitational potential energy. The baseball has a mass of 200 grams and reaches a maximum height of four times that of the basketball. What is the basketball's mass, rounded to the nearest gram? Ignore air resistance.

285. Stan carries a tablet computer with a mass of 200 grams in a backpack while walking up a ramp with a slope of 25 degrees to the horizontal. How much does the computer's potential energy increase in the amount of time that it takes Stan to walk 6 meters along the ramp's surface? Round your answer to the nearest tenth of a joule.

Use the following information and diagram to answer Questions 286–287: Two masses made of identical material — one atop a flat plateau and one on a ramp — are connected by a massless string stretched taut over a massless, frictionless pulley.

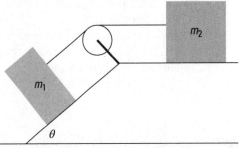

Illustration by Thomson Digital

286. If m_2 moves 1.8 meters to the left, what is the change in m_1's gravitational potential energy if $m_1 = 18$ kilograms, $m_2 = 4$ kilograms, and $\theta = 30$ degrees? Round your answer to the nearest ten joules.

287. Imagine that m_1 slides down the ramp, losing 30 joules of potential energy in 6 seconds. If m_1 has a mass of 60 kilograms, m_2 has a mass of 23 kilograms, and the ramp is angled 16 degrees with respect to the horizontal, what is the coefficient of friction between the masses and the surface covering the ramp and plateau? Round your answer to the nearest hundredth.

Cashing Out Potential Energy to Kinetic Energy

288–292

288. A bowling ball is lifted to a height such that its gravitational potential energy is 20 joules relative to the ground. If released from rest, how much kinetic energy does the ball have just before striking the ground? Ignore air resistance.

289. A gust of wind shakes loose a football that was stuck in a tree. Ignoring air resistance, if the football falls from a height of 10.8 meters, what is its speed just before hitting the ground? Round your answer to the nearest tenth of a meter per second.

290. A ball rolls down a frictionless slide that has an odd attachment at the bottom, which keeps the ball moving and elevates it 80 centimeters before the ball leaves the slide:

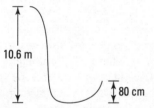

10.6 m

80 cm

Illustration by Thomson Digital

If the distance from the top of the slide where the ball is released from rest to the lowest point of the dip at the bottom of the slide is 10.6 meters, what is the ball's speed when it loses contact with the slide? Round your answer to the nearest tenth of a meter per second.

291. A cannonball is fired at an angle of 47 degrees relative to the horizontal, landing 208 meters away from the cannon after spending 6.8 seconds in the air. Ignoring air resistance, how high is the cannonball at the top of its arc? Round your answer to the nearest tenth of a meter.

292. A wheeled desk rolls down a long ramp with an inclination of 3 degrees to the horizontal and reaches the bottom with a speed of 2.3 meters per second. If the desk starts from rest, and the coefficient of friction between the desk's wheels and the ramp is 0.05, what fraction of the desk's initial mechanical energy is used to counteract the frictional force? Give your answer as a percentage between 0 and 100, rounded to the nearest 5 percent.

Maintaining the Total Mechanical Energy

293–296

293. A particle has 37.5 joules of kinetic energy and 12.5 joules of gravitational potential energy at one point during its fall from a tree to the ground. An instant before striking the ground, how much mechanical energy — rounded to the nearest joule — does the particle have? Ignore air resistance.

Use the following picture of a cart of mass m moving on a roller coaster, as well as the accompanying information, to answer Questions 294-296: The section of track displayed is frictionless and composed entirely of semicircular pieces of radius r. The letters A through E refer to specific locations along the track.

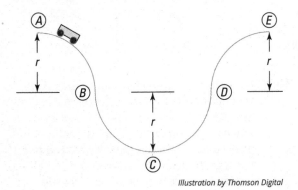

Illustration by Thomson Digital

294. If $r = 12$ meters, and if the cart's velocity at point A is 3 meters per second, what is the cart's velocity at point D? Round your answer to the nearest tenth of a meter per second.

295. If the cart has a mass of m kilograms and a velocity at point A of 0 meters per second, write an expression for the normal force on the cart at point C using only m, g, and r. Combine your answer into a single fraction if necessary.

296. A speedometer on the cart records the speed at point C as 20 meters per second and the speed at point E as 8 meters per second. What is the cart's speed at point B? Round your answer to the nearest tenth of a meter per second.

Gaining Power by Doing More Work Faster

297–300

297. The rate at which energy is changed is called what?

298. What is the average power in watts delivered to a light bulb that uses 120 kilowatt hours of energy in 1 minute? Round your answer to two significant digits.

299. How many watts of power does a car engine perform in accelerating a 1,300-kilogram car from 25 kilometers per hour to 50 kilometers per hour in 8 seconds? Ignore friction and round your answer to two significant digits.

300. Matt drags an 80-kilogram stone up a ramp using an average of 870 watts of power. The ramp's angle of elevation is 25 degrees, and the coefficient of kinetic friction between the stone and the ramp is 0.4. If the stone's acceleration is 0.1 meters per second squared, how long does it take Matt to drag the stone 11.2 meters up the ramp? Round your answer to the nearest second.

Powering Up with Speed

301. The product of force and what quantity is equivalent to power?

302. At what velocity is a car moving at the instant its engine is using 2,000 watts to exert 100 newtons of force on the car's wheels?

303. Thirteen kilowatts of power are used to accelerate an 850-kilogram speedboat at a rate of 1.6 meters per second squared. If the boat's speed is 8 meters per second before the power kicks in, what is its velocity afterward? Ignore friction and round your answer to the nearest tenth of a meter per second.

304. A 950-gram, radio-controlled toy car accelerates at a rate of 0.8 meters per second squared for 5 seconds. If its velocity doubles in that time, how many watts of power do the car's batteries provide during the process? Round your answer to the nearest hundredth. Ignore friction.

305. Paul pushes a 400-kilogram cabinet from rest up from the bottom of a ramp using an average of 600 watts of power. Just as Paul reaches the top of the ramp, the cabinet has a speed of 0.9 meters per second. The ramp has a coefficient of friction of 0.2 and an angle of elevation of 7 degrees. How long is the ramp? Round your answer to the nearest tenth of a meter.

Chapter **10**

Picking Up Some Momentum with Impulse

Momentum is the product of mass and velocity. Momentum is a conserved quantity, which means that the total momentum of an isolated system does not change. If the objects in the system interact — in a collision, for example — the momentum of each object can change even though the total momentum does not change. The momentum of an object changes when you apply an *impulse* to it — that is, when you apply a force to it for a certain amount of time.

The Problems You'll Work On

In this chapter you'll run into the following problems:

>> Relating impulse to force and time

>> Determining the change in momentum using the impulse

>> Using conservation of momentum to deal with inelastic collisions

>> Using conservation of momentum and conservation of kinetic energy to deal with elastic collisions

What to Watch Out For

To avoid colliding with an incorrect answer, pay close attention to the following:

>> Remembering that impulse and momentum are vector quantities

>> Recalling that the total momentum is conserved in a collision

>> Recognizing that the total kinetic energy can change in a collision

Applying Force for a Period of Time: Impulse

306–308

306. Impulse is the product of force and what quantity?

307. If the interaction force between two objects in a collision remains unchanged, doubling the length of time over which the collision takes place does what to the original impulse J? Your answer should be the value of the new impulse in terms of J.

308. A baseball scout watching a pitcher and catcher warming up records the following data with a stopwatch:

Ball leaves pitcher's hand: 0.00 seconds

Ball initially hits catcher's glove: 1.03 seconds

Ball completely stops in glove: 1.08 seconds

If the catcher's glove exerts an average force of 400 newtons on the baseball, calculate the impulse on the ball, rounded to the nearest ten newton-seconds.

Gathering Momentum in All Directions

309–312

309. Tripling an object's momentum would require either tripling its mass or multiplying its velocity by what?

310. Dividing an object's momentum by its mass and then multiplying by an amount of time results in what type of measurement?

311. A 650-kilogram automobile cruising along at 80 kilometers per hour suddenly accelerates to a speed of 120 kilometers per hour. What amount of change in momentum is required to achieve this acceleration? Use units of kilogram-meters per second for your answer, which should have two significant digits.

312. A 2,350-kilogram airplane maintains a momentum of 6.3×10^5 kilogram-meters per second 20 degrees south of east for 1 hour before turning in a direction 80 degrees north of east with a momentum of 1.01×10^6 kilogram-meters per second, which it maintains for another 2 hours. What is the airplane's net distance from its starting point — in thousands of kilometers — after these 3 hours? Round your answer to three significant digits.

Supplying Impulse to Change Momentum

313-314

313. Dividing the units of impulse by those of velocity results in units describing what quantity?

314. A baseball weighs W newtons on the surface of Earth. The same ball is then brought into interstellar space, far from any significant gravitational attraction, where a pipe strikes it from rest. It then travels D kilometers at a constant velocity for T minutes. Write an expression for the amount of force with which the pipe struck the ball if the two items were in contact for 1 second. Your answer should be in the form of a single, reduced fraction and may contain any variables listed here as well as any universal constants.

Finding Force from Impulse and Momentum

315-317

315. An oxygen molecule whose mass is approximately 2.7×10^{-26} kilograms travels at a speed of 50 meters per second. If no force acts on the molecule, what is its new speed 10 seconds later? Round your answer to the nearest meter per second.

316. Two bowling balls roll toward each other and collide. The larger one — labeled A — has a mass that is twice that of the smaller one — labeled B. Write an expression for the magnitude of the change in velocity of ball A (Δv_A) in terms of the change in velocity of ball B (Δv_B).

317. A 200-gram billiard ball traveling 16 meters per second in the positive x-direction strikes a stationary, 18-kilogram bowling ball. The impulse from the 0.15-second collision causes the billiard ball to travel back the way it came at 6 meters per second. What average force — in newtons — does the bowling ball exert on the billiard ball during the collision? Round your answer to three significant digits.

Conserving Momentum during Collisions

318-321

318. What quantity is always conserved during elastic collisions but never during inelastic ones?

319. What quantity is always conserved during a collision?

320. An astronaut does an experiment on the International Space Station in which a glob of polymer with velocity v runs into a second, stationary, equal-massed glob of polymer. If the two globs stick together after the collision, what is the velocity of the combined glob in terms of v?

321. An astronaut hovers motionless far above Earth's atmosphere, while holding a 20-kilogram wrench. If the astronaut throws the wrench, in which direction will the astronaut travel? For purposes of the answer let the direction that the wrench is thrown be the positive x-axis, and give your answer as a degree measurement relative to the positive x-axis.

Sticking Together: Finding the Velocity in Inelastic Collisions

322–326

322. Object A rams into object B at a speed of 3 meters per second in a perfectly inelastic collision. Object B was originally motionless, and both A and B have the same mass. How fast are the objects moving after the collision? Round your answer to the nearest tenth of a meter per second.

323. A stationary, 32-kilogram ice skater catches a 0.5-kilogram snowball thrown at a speed of 45 meters per second. How fast — in meters per second — does the skater move afterward? Round your answer to two significant digits.

324. A 0.156-kilogram bullet enters a 4.25-kilogram block of rubber at a speed of 400 kilometers per hour. Hanging from a taut string, the rubber block is initially at rest on the frictionless surface of a table. What is the maximum height above the table that the block of rubber reaches after the collision? Round your answer to the nearest 10 centimeters.

325. Tarzan steps off his treehouse balcony while grasping a 6.5-meter-long rope that starts parallel to the ground below and remains taut as he swings down off the balcony. At the lowest point of his swing, Tarzan is struck by a watermelon traveling toward him at 6 meters per second. Holding onto the fruit but letting go of the rope, how far away (horizontally) from his starting point does Tarzan land if his treehouse is 20 meters off the ground? Tarzan has a mass of 108 kilograms, and the watermelon's mass is 22 kilograms. Round your answer to the nearest tenth of a meter.

326. A 1.5-kilogram brick with a small magnet attached to its side slides 125 centimeters along a 20-degree ramp where the coefficient of friction is 0.08. At the end of the ramp, the brick immediately collides with a stationary, 2.5-kilogram box, which also has a connected magnet. The two items magnetically attach to each other and proceed along a completely horizontal surface where the coefficient of friction is 0.2. How many meters do the connected items slide before stopping? Round your answer to the nearest tenth.

Finding the Initial Velocity of Collisions

327–330

327. A novice ice skater, Molly, realizes she's moving too fast to stop safely and heads toward her stationary friend, Jésus. Molly's mass is 42 kilograms, and Jésus's is 62 kilograms. Molly's speed drops from 15 meters per second before the collision to 3 meters per second afterward (her direction stays the same). What is Jésus's speed after Molly skates into him? Round your answer to the nearest meter per second.

328. A 91-kilogram baseball player is caught between third base and home plate as the opposing catcher jogs toward him at 2 meters per second, attempting to make the tag. The runner decides to speed toward the 100-kilogram catcher at a rate of 5 meters per second and collide with the catcher. How fast — and toward which base — will the runner be moving after the collision if the catcher's velocity is 1 meter per second toward home plate? Round your answer to the nearest tenth of a meter per second.

329. Two astronauts — Kate and Axel — are about to collide head-on during a space walk. Kate's mass is 45 kilograms, Axel's is 80 kilograms, and their initial velocities are 4 meters per second and 3.2 meters per second toward each other, respectively. After the collision, Axel continues moving in his original direction, but at only 0.2 meters per second. If Kate's original direction of motion is considered the positive direction of travel, what is her final velocity in meters per second? Round your answer to the nearest tenth.

330. In an inelastic collision, a rock thrown at a block of wood (initially at rest, hanging by a 5-meter-long rope) causes the block to swing through 13.6 degrees of arc. If the block is 5 times as massive as the rock, what is the rock's initial velocity if it bounces backward at a speed of 1.4 meters per second upon colliding with the block? Round your answer to the nearest tenth of a meter per second.

Colliding Elastically along a Line

331–335

331. Two equal masses collide elastically, one moving with velocity v and the other stationary. What is the speed of the stationary mass following the collision, in terms of v?

332. Two round pieces of hard bubble gum (one grape-flavored, the other orange-flavored) elastically collide. The grape gum is originally stationary, and the orange gum runs into it with a speed of 1.2 meters per second. If the mass of the orange gum is 1 kilogram and the velocities of the grape gum and orange gum after the collision are 0.1 meters per second and 0.8 meters per second (in the same direction as the orange gum's initial motion), respectively, what is the mass in kilograms of the grape gum? Round your answer to two significant digits.

333. Given an elastic, head-on collision between masses A (initial speed v_{ai}) and B (initial speed v_{bi}), what is the sum of the initial and final velocities of mass A ($v_{ai} + v_{af}$) in terms of the initial and final velocities of mass B (v_{bi} and v_{bf}, respectively)?

334. An astronaut located far above Earth's surface swings a 650-gram rubber ball attached to a thin rope in a circular path. The rope is 80 centimeters long, and the astronaut's arm exerts 100 newtons of force on the rope. The ball suddenly detaches and travels a short distance before colliding elastically with a piece of space debris heading in the same direction at 4 meters per second. What is the momentum of the debris after the collision if the ball is moving along its initial path at 3 meters per second after the collision? Round your answer to the nearest tenth of a kilogram-meter per second.

335. A 2-kilogram, duckpin bowling ball is rolled down an eastward-facing frictionless ramp 7 meters long at a 25-degree inclination. Immediately after clearing the ramp, it collides elastically with a ten-pin bowling ball traveling west at 10 meters per second. As a result of the collision, the ten-pin ball heads east at 5 meters per second. How far back along the ramp does the duckpin bowling ball travel? Round your answer to the nearest tenth of a meter.

Colliding Pool Balls: Elastic Collisions in Two Dimensions

336–340

336. A 1-kilogram particle moving along the positive x-axis at 2 meters per second collides elastically with a stationary, 2-kilogram particle. The 2-kilogram particle's velocity post-collision is 0.5 meters per second at a −68-degree angle relative to the positive x-axis. At what speed is the 1-kilogram particle moving after the collision if its motion is 30 degrees to the positive x-axis? Round your answer to the nearest tenth of a meter per second.

337. Two 8-kilogram bowling balls collide elastically when a white ball traveling due east at 40 meters per second runs into a stationary black ball. The white ball's final velocity is 20 meters per second 60 degrees north of east. What is the direction of the black ball after the collision? Round your answer to the nearest degree.

338. Block 1 has a mass of 0.6 kilograms and is traveling at 3 meters per second along the positive x-axis. It collides elastically with block 2, which was stationary before the collision. If block 1's path after the collision takes a direction of 45 degrees with respect to the positive x-axis, and if block 2's velocity is 0.735 meters per second with a direction of −61 degrees to the same axis, what is the mass of block 2, rounded to the nearest tenth of a kilogram?

339. Rolling east, a 0.475-kilogram ball bearing (BB1) collides with a stationary, 0.182-kilogram ball bearing (BB2). After the collision, BB1 moves off with a speed of 2.22 meters per second in a direction 20 degrees north of east, and BB2 heads in a direction 21.6 degrees south of east. How fast was BB1 rolling before the collision? Round your answer to the nearest tenth of a meter per second.

340. A 620-gram block slides due east down a ramp where the coefficient of friction is 0.1. The ramp is tilted 50 degrees to the horizontal. The ramp ends on a flat, frictionless surface, and the block eventually collides (elastically) with another 620-gram block. If the first block initially starts 65 centimeters higher than the flat surface, and if, following the collision, it proceeds in a direction 12 degrees north of east, what is the velocity of the second block after the collision? Round your speed to the nearest tenth of a meter per second and your direction to the nearest degree, relative to due east.

Chapter **11**

Rolling Around with Rotational Kinetics and Dynamics

Rotational motion shouldn't make your head spin. There are close analogies between the quantities you deal with in linear motion — such as velocity, acceleration, mass, force, and momentum — and those you deal with in rotational motion, including angular velocity, angular acceleration, moment of inertia, torque, and angular momentum.

The Problems You'll Work On

In this chapter you'll wrap your head around the following problems:

>> Relating linear and tangential motion

>> Determining angular velocity and angular acceleration

>> Computing torques and moments of inertia

>> Setting the torque equal to zero to solve rotational equilibrium problems

>> Using torques and moments of inertia to solve rotational motion problems

>> Finding rotational kinetic energy

>> Conserving angular momentum

What to Watch Out For

To avoid going in circles, keep the following in mind:

>> Remembering the relations and analogies between linear and rotational motion

>> Including both the linear and rotational kinetic energy

>> Using the correct formula for moment of inertia for each shape

>> Finding the lever arm when calculating the torque

>> Recalling that friction does no work on an object that rolls without slipping

Linking Linear and Tangential Motion with Radius

341–343

341. A bug sitting on the outer rim of a rotating gear has an angular velocity of 3 radians per second. If the gear's radius is 0.2 meters, what is the bug's linear velocity? Round your answer to the nearest tenth of a meter per second.

342. A speck of dust at the edge of a spinning disc has an angular acceleration of 7 radians per second squared and a linear acceleration of 0.18 meters per second squared. What is the disc's radius? Round your answer to the nearest centimeter.

343. A 12-gram ball hangs from the end of a rope that can withstand a maximum tension of 31 newtons. The ball is swung in a vertical circle. What is the length of the rope if the ball's maximum angular velocity is 20 radians per second? Answer in meters with your solution rounded to two significant digits.

Finding Centripetal Acceleration

344–345

344. Riding at the edge of a carousel (radius = 12 meters), Sally has an angular velocity of 0.3 radians per second. What is the magnitude of Sally's centripetal acceleration in meters per second squared? Round your answer to two significant digits.

345. It takes Franz 1 minute and 12 seconds to travel two complete rotations on a Ferris wheel. If the wheel rotates at a constant rate, and if Franz experiences a centripetal acceleration of 0.38 meters per second squared, what is the diameter of the wheel? Round your answer to the nearest meter.

Figuring Out Angular Velocity and Acceleration

346–348

346. What is the magnitude of the angular velocity — in radians per second — of a carousel rider who makes 12 revolutions in 140 seconds? Round your answer to two significant digits.

347. What is the linear velocity of a fly sitting on the very end of a clock's 20-centimeter-long minute hand? Round your answer to the nearest tenth of a meter per second.

348. A gear with a radius of 2.3 centimeters starts from rest and accelerates at a rate of 4.8 radians per second squared. How much farther does a point on one of the gear's teeth travel in the second minute of acceleration than it did during the first? Round your answer to the nearest 10 meters.

Twisting Around with Torque

349–352

349. Torque always points in the direction of what other quantity?

350. How much torque is produced by opening a jar of pickles if the lid on the jar has a radius of 3.8 centimeters and the force exerted tangentially to the lid is 150 newtons? Assume that the force is concentrated at one point on the lid. Round your answer to the nearest tenth of a newton-meter.

351. A 78-gram pendulum swings in a clock, never swinging past an angle of 8 degrees with the vertical.

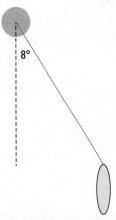

Illustration by Thomson Digital

If the pendulum swings on a 2.8-meter-long string attached to a pivot, what is the maximum amount of torque about the pivot? Round your answer to the nearest tenth of a newton-meter.

352. The bar in the following diagram is 180 centimeters long, and the white pivot point is exactly in the middle. $F_1 = 50$ newtons, $F_2 = 40$ newtons, and $F_3 = 30$ newtons. F_1 and F_3 are exerted at the bar's edges, and F_2 is exerted 40 centimeters from the bar's center. Calculate the net torque — rounded to the nearest newton-meter — about the pivot, and determine whether the bar's rotation is clockwise or counterclockwise.

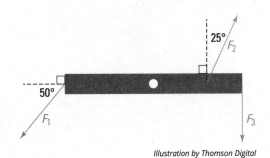

Illustration by Thomson Digital

Lifting Some Weights: An Exercise in Rotational Equilibrium

353–356

353. Besides angular acceleration, what net quantity must equal 0 if a system is in rotational equilibrium?

354. What value must remain constant but does not have to equal 0 for an object to be in rotational equilibrium?

355. A 95-kilogram skier positions himself facing up a 22-degree slope. Locking his arms straight, perpendicular to his ski poles, the skier pushes the poles into the snow at a 45-degree angle to the hill. Doing so, the skier is just able to maintain perfect equilibrium. If his arms are 0.58 meters long, how much torque does his shoulder joint experience? Ignore friction, and round your answer to the nearest newton-meter. The coefficient of static friction between the ski blades and the snow is 0.08.

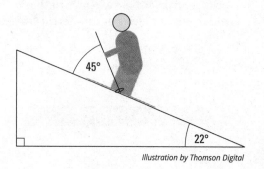

Illustration by Thomson Digital

356. A bodybuilder holds a 30-kilogram barbell at arm's length from her body with her arm extended parallel to the ground. The bodybuilder's biceps muscle exerts a tension force directed 8 degrees above the horizontal from a position one-quarter of the way down her arm, and her shoulder joint exerts a force on the bones in her arm directed in an unknown direction *below* the horizontal, as shown in the following figure. Assume that the bodybuilder's arm has a mass of 3.5 kilograms, evenly distributed across the length of the arm, and calculate the magnitude of the force exerted by her shoulder joint. Round your answer to the nearest 10 newtons.

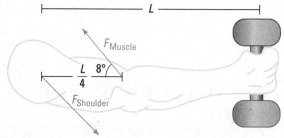

Illustration by Thomson Digital

Hanging Signs up and Keeping the Torques Balanced

357–360

Use the following information and diagram to answer Question 357: The plank shown here is subjected to two forces —F_1 and F_2— located at distances r_1 and r_2 from the axis of rotation, respectively.

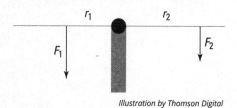

Illustration by Thomson Digital

357. The plank is in rotational equilibrium. If $F_1 = 25$ newtons and is 0.8 meters away from the axis of rotation, what is the magnitude of F_2 if it's located 0.5 meters from the axis? Round your answer to the nearest newton.

Use the following information and diagram to answer Question 358: Two window washers stand on a platform supported by two ropes, as shown, such that the entire system is in equilibrium. Person A has a mass of 100 kilograms, and person B has a mass of 60 kilograms. Ignore the mass of the platform, which is 2.5 meters in length.

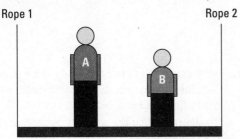

Rope 1   Rope 2

Illustration by Thomson Digital

358. Rope 1 exerts a tension of 960 newtons, and rope 2 exerts a tension of 608 newtons. If person A stands 50 centimeters from rope 1, how far from rope 2 is person B standing? Round your answer to the nearest hundredth of a meter.

359. In the diagram here, how much force must be exerted by a nail hammered down into the pole at A to keep the system in rotational equilibrium? The crate hanging at the end of the horizontal bar is 120 kilograms. The bar is 154 centimeters long and has a mass of 18 kilograms, which is evenly distributed over the entire length. Assume that the entire weight of the system is supported by the pole at B, and round your answer to the nearest 10 newtons.

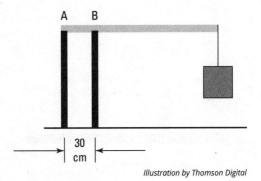

A   B

30 cm

Illustration by Thomson Digital

Use the following information and diagram to answer Question 360: The following diagram shows a 90-kilogram worker standing on a 4-meter-long beam attached to a brick wall by a wall joint on its left end, and by a 6-meter rope on its right end. A 100-kilogram crate hangs a quarter-meter from the right end of the beam. The mass of the beam is 30 kilograms, uniformly distributed.

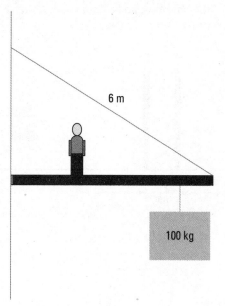

6 m

100 kg

Illustration by Thomson Digital

360. The maximum amount of tension this rope can handle is the value required to keep the system in equilibrium when the wall joint exerts 4 times as much force in the horizontal direction as it does in the vertical direction. How far from the wall can the worker walk before the rope snaps? Round your answer to the nearest tenth of a meter.

Leaning against the Wall with Help from Friction

361–364

361. A 2.8-meter-long ladder leaning against a smooth wall at an angle of 75 degrees to the horizontal is subjected to 120 newtons of force from contact with the wall. If the coefficient of static friction between the ladder and the floor is 0.134 and the ladder is just on the verge of losing equilibrium, what is its mass? Round your answer to the nearest kilogram.

362. Joey stands near the top of a ladder that's leaning against the side of a smooth-sided building. The ladder forms an angle of 70 degrees with the ground. Joey's mass is 95 kilograms, and the ladder's mass is 35 kilograms. If Joey is located 1 meter from the top of the ladder and the coefficient of static friction between the ladder and the ground is 0.23 and the ladder is just on the verge of losing equilibrium, how long is the ladder? Round your answer to the nearest tenth of a meter.

363. A rope attached to the top of a 6-kilogram, 14-meter-tall sign is pulled taut when a gust of wind blows the sign so that the sign makes an angle of 80 degrees with the ground.

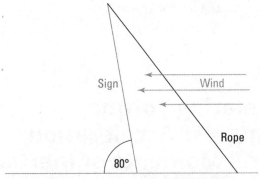

Illustration by Thomson Digital

If the force of the wind is 180 newtons and the coefficient of static friction between the sign and the ground is 0.24, what is the force of tension in the rope that keeps the sign in equilibrium? Round your answer to the nearest 10 newtons. Assume that the wind and the force of gravity both act on the exact center of the sign.

364. Lenny stands on a 26-kilogram, 3.4-meter-long ladder that leans up against a wall, touching it 2 meters above the ground. The ladder has 15 steps, and Lenny — whose mass is 110 kilograms — stands on the 13th step. If the system is in equilibrium and the ground provides a normal force of 820 newtons, which coefficient of static friction has a greater value: the one between the wall and the ladder, or the one between the ground and the ladder? By how much? Assume that the steps are equally spaced, with the 1st step at the very bottom of the ladder and the 15th step at the very top. Round the numerical portion of your answer to the nearest tenth.

Converting Tangential Acceleration to Angular Acceleration in Newton's Second Law

365–367

365. What is the tangential acceleration of a bug sitting on the edge of a rotating disc of radius 0.12 meters if its angular acceleration is 2 radians per second squared? Answer in units of meters per second squared.

366. An ant standing at the edge of a rotating disc experiences a torque of 0.02 newton-meters. If the disc has a diameter of 16 centimeters and an angular acceleration of 49 radians per second squared, what is the ant's mass? Round your answer to the nearest gram.

367. Jennifer rides a Ferris wheel ride at the carnival, traveling a distance of 260 meters while making 4 complete revolutions around the central axis. During the time it takes to make those revolutions, Jennifer's angular velocity increases from 3 radians per second to 12 radians per second. If Jennifer weighs 480 newtons, how many newton-meters of torque does she experience during those 4 revolutions? Round your answer to two significant digits.

Looking into Mass Distribution to Find Moments of Inertia

368–370

368. In addition to a particle's distance from an axis of rotation, what other physical quantity of the particle determines its moment of inertia?

Use the following scenario to answer Questions 369–370: Two ball bearings are swung on massless strings around a tetherball pole. The black bearing has a mass of 2.8 kilograms, and the silver bearing has a mass of 3.3 kilograms.

369. If the black ball bearing swings counterclockwise in a circular path of radius 1.8 meters, and if the silver ball bearing swings clockwise with a radius of 1.2 meters, what is the total moment of inertia of the two-bearing system? Assume that each bearing's string is stretched out parallel to the ground below, and round your answer to the nearest kilogram-square meter.

370. The black ball bearing swings around the pole on a string of unknown length and unknown angle to the horizontal:

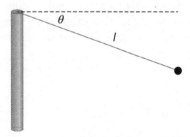

If the tension in the string is 260 newtons and the black ball bearing's moment of inertia is 25 kilogram-square meters, how much torque does the force of gravity apply to the black ball bearing? Round your answer to the nearest newton-meter.

Relating Torque, Angular Acceleration, and Moments of Inertia

371–374

371. Tripling the force used to loosen a screw changes the screw's angular acceleration by what factor? Phrase your answer as a multiple of the original angular acceleration, α_i.

372. Friction operates tangentially on the rim of a wheel, producing an angular acceleration of magnitude 1.9 radians per second squared. If the wheel has a radius of 0.43 meters and a moment of inertia equal to 12 kilogram-square meters, what is the magnitude of the force of friction, rounded to the nearest newton?

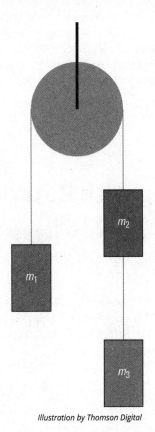

Illustration by Thomson Digital

373. What is the pulley's moment of inertia if it experiences a torque of 12 newton-meters from a net force of 20 newtons rotating the pulley counterclockwise? Round your answer to the nearest kilogram-square meter.

374. If $m_1 = 50$ kilograms, $m_2 = 25$ kilograms, $m_3 = 40$ kilograms, and the pulley's radius is 12 centimeters, what is the pulley's angular acceleration? Round your answer to the nearest tenth of a radian per second squared.

Putting a Spin on Work

375–378

375. Two wheels of different radii experience the same tangential force. Wheel A has a radius twice as large as that of wheel B. If the same amount of work is done rotating each wheel, which one — if either — rotates more, and by how many more times?

376. A lawnmower with a "pull start" has a cord wrapped around a motor. The user pulls sharply on the cord to start the engine. If the motor is approximated as a solid cylinder with a mass of 1.8 kilograms and a radius of 9 centimeters, how much work must a person do to accelerate the motor from rest to an angular velocity of 12 radians per second before the 130-centimeter-long cord is fully unraveled? Answer in units of joules, rounded to two significant digits.

377. A pebble caught in a tire's tread travels 400 meters during the time that a car's engine does 6.8 kilojoules of work rotating the tire. If the tire's mass is 11 kilograms, what is the pebble's tangential acceleration? You may assume that the tire is a solid disc and that the pebble's moment of inertia is negligible. Round your answer to the nearest tenth of a meter per second squared.

Use the following scenario and diagram to answer Question 378: A special merry-go-round has two rings on which children can ride. The inner ring is 80 percent as far from the central axis as is the outer ring. Assume that the rings have no width, for the purpose of solving the problems. The mass of the inner ring is 144 kilograms, and the mass of the outer ring is 180 kilograms.

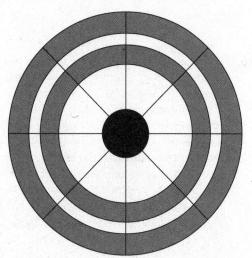

Illustration by Thomson Digital

378. An adult performs 2,790 kilojoules of work pushing tangentially on the outer ring of the merry-go-round while two children ride: An 80-kilogram child sits on the inner ring, and a 90-kilogram child sits on the outer ring. If the merry-go-round has a radius of 4.8 meters, how much farther does the 90-kilogram child travel than the 80-kilogram child if the merry-go-round's angular acceleration is 0.4 radians per second squared? Round your answer to the nearest 10 meters.

Rolling with Rotational Kinetic Energy

379–381

379. A vinyl album spins on a turntable. If its rotational velocity is divided in half, by what factor is its total kinetic energy multiplied?

380. A hollow, 5.2-kilogram globe spins with an angular velocity of 2.7 radians per second. If the globe has 12 joules of kinetic energy, what is the longest piece of string that can encircle the globe without overlap? Round your answer to the nearest tenth of a meter.

Use the following information and diagram to answer Question 381: *A 1.8-kilogram block is dropped 1.2 meters onto a small platform attached to a solid rotating disk with a mass of 10 kilograms and a radius of 15 centimeters. The platform extends 3 centimeters beyond the disk's rim.*

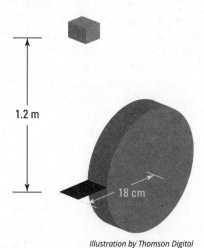

1.2 m

18 cm

Illustration by Thomson Digital

381. If the block strikes the outermost edge of the platform in an elastic collision, bouncing off as the wheel starts spinning, what is the magnitude of the wheel's resulting angular velocity? Round your answer to the nearest radian per second. Neglect the mass of the platform.

Finding Rotational Kinetic Energy on a Ramp

382–386

382. A solid sphere (SS), a hollow sphere (HS), a solid cylinder (SC), and a hollow cylinder (HC) each have the same mass and the same radius. Initially at rest in the same spot, all four roll down a ramp one at a time without slipping. Rank their final rotational kinetic energies at the bottom of the ramp from least to greatest, using the two-letter abbreviations listed.

383. A 12-kilogram bowling ball rolls down a ramp with a translational speed of 5 meters per second without slipping. What is the ball's amount of rotational kinetic energy? Round your answer to the nearest joule.

384. A 14-kilogram wheel rolls 20 meters down a plank angled with respect to the horizontal. If the wheel's translational velocity is 7 meters per second, what angle does the plank make with the horizontal? Assume that the wheel is a solid disc that starts at rest and rolls without slipping. Round your answer to the nearest degree.

Use the following diagram of a solid disk traveling along two ramps and a flat middle section to answer Questions 385–386:

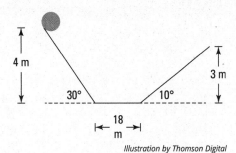

Illustration by Thomson Digital

385. What is the minimum coefficient of static friction required on the first ramp to prevent the disk from slipping? Round your answer to the nearest hundredth.

386. If the middle section of the ramp is frictionless, how long does it take the disk to travel from the top of the left ramp to the top of the right ramp? Assume that the disk rolls without slipping on both of the ramps, and round your answer to the nearest tenth of a second.

Conserving Angular Momentum

387–390

387. A figure skater does a tight spin with his arms pulled in close to his body. If his current angular velocity is z, what will be his angular velocity in terms of z if the skater suddenly shoots his arms outward, doubling the average distance of his body's mass from the axis of rotation?

388. A porcelain tree standing at the edge of a spinning plate on top of a music box has a tangential velocity of 12 centimeters per second. The plate has a radius of 7 centimeters and a mass of 800 grams. What is the plate's angular momentum in kilogram-square meters per second? Ignore the mass of the tree, and round your answer to three significant digits.

389. Calculate Earth's angular momentum around the sun in kilogram–square meters per second. Use an orbital period of exactly 365 days, and approximate Earth's orbit as a circle. The mass of both Earth and the sun, as well as the distance between them, is:

$$m_{\text{Earth}} = 5.98 \times 10^{24} \text{ kg}$$
$$m_{\text{Sun}} = 1.991 \times 10^{30} \text{ kg}$$
$$r = 1.496 \times 10^{8} \text{ km}$$

390. An 80-kilogram man spins himself starting from rest in a chair with his arms and legs near the axis of rotation. After 15 revolutions at a constant angular acceleration of 3 radians per second squared, he sticks out both his arms and his legs, cutting his angular velocity in half. If his arms, which contain 10 percent of his body's mass, are each 80 percent as long as each of his legs, which contain 20 percent of his body's mass, what is the tangential velocity of a point on the man's fingertips after he extends his arms and legs? Approximate the man's body as a solid cylinder with a 20-centimeter radius, and approximate his arms and legs as narrow rods. Round your final answer to the nearest tenth of a meter per second.

Chapter **12**

Bouncing with a Spring: Simple Harmonic Motion

I
f you look around, you'll see simple harmonic motion again and again. If you displace a spring or pendulum from its equilibrium point, it will oscillate. Springs exert a restoring force that is proportional to the displacement from equilibrium. The period of oscillation of a mass on a spring depends on the mass and spring constant. The period of oscillation of a pendulum is independent of the mass of the pendulum but depends on the length of the pendulum.

The Problems You'll Work On

Solving the problems in this chapter will put a spring in your step:

» Determining the force exerted by a spring

» Finding the equilibrium position of a spring

» Understanding sine waves

» Finding the period and frequency of oscillation of springs and pendulums

» Using conservation of energy to compute the motion of springs and pendulums

What to Watch Out For

Be sure to keep the following in mind as you work your way through this chapter:

» Making sure you get the factors of 2π right

» Remembering the minus sign in Hooke's law

» Checking your units so you don't mix up the frequency and period formulas

» Including the elastic potential energy when using conservation of energy

Compressing and Stretching Springs: The Spring's Restoring Force

391–393

391. A spring compressed a distance of 3 centimeters exerts a force of 100 newtons on an object. What is the spring constant of the spring?

392. A spring has a spring constant of 4,000 newtons per meter. You stretch the spring 1.2 centimeters from equilibrium in the positive *x* direction. What force does the spring apply on you?

393. Two springs are attached horizontally in a line, with one at the end of the other. The spring constant of the first spring is 2,500 newtons per meter, and that of the second spring is 1,500 newtons per meter. The first spring is attached on the left to a wall. What distance does the combination of these two springs stretch when you pull the end of the second spring to the right with a force of 500 newtons?

Getting around the Equilibrium

394–395

394. A spring with a spring constant of 1,200 newtons per meter hangs from the ceiling. What mass should you hang on it to make it stretch 4.5 centimeters?

395. A spring hanging from the ceiling stretches 4.8 centimeters when you hang a 2.2-kilogram mass on it. What is the spring constant?

Riding the Sine Wave of Simple Harmonic Motion

396–400

396. The tip of an oscillating spring follows the sine wave $y = (1.3 \text{ cm}) \sin\left(2.2 \text{ s}^{-1} \times t\right)$. How long does it take for the tip of the spring to first reach its maximum position?

397. The distance between two atoms in a vibrating molecule is $x = x_0 + A \sin(\omega t)$. If the angular frequency of the oscillation is $1.8 \times 10^{13} \text{ s}^{-1}$, how many times do the atoms reach their maximum separation during a time interval of $t_0 = 4.6 \times 10^{-12} \text{ s}$?

398. An electrical circuit oscillates with an angular frequency of $3.5 \times 10^5 \, \text{s}^{-1}$. If you describe this oscillator using the concept of a reference circle, through what angle in the reference circle has the oscillator moved after 3.2 seconds?

399. The tip of an oscillating spring follows the cosine wave $y = (1.3 \, \text{cm}) \cos\left(2.7 \, \text{s}^{-1} \times t\right)$. What is the distance between the maximum and minimum positions of the tip of the spring?

400. The tip of a spring oscillates according to $y = (4.9 \, \text{cm}) \sin\left(98 \, \text{s}^{-1} \times t\right)$. How much time elapses before the second time the tip of the spring is at 3.9 centimeters?

Finding Period and Frequency

401–403

401. Your child makes 13 turns on a merry-go-round in 10 minutes. What is the period of rotation for the merry-go-round?

402. A cube has a mirror on each side. The cube rotates around a vertical axis so that the base of the cube is perpendicular to the vertical axis. A horizontal laser beam hits the cube and is reflected horizontally as the cube spins. At what angular frequency should the cube spin so that it sends a reflected laser beam signal back toward the laser source with a frequency of 3.5 hertz?

403. A branch oscillates up and down in the wind with a period of 3.5 seconds. How many times does it move through its equilibrium position in 1 minute?

Linking Position, Velocity, and Acceleration

404–407

404. The speed of the tip of an oscillating spring is given by $v = 2.8 \, \text{m} / \text{s} \times \cos\left(3.5 \, \text{s}^{-1} \times t\right)$. What is its speed at $t = 45$ seconds?

405. If the position of an oscillator is given by $y = 3.7 \, \text{cm} \times \sin\left(14 \, \text{s}^{-1} \times t\right)$, what is the oscillator's maximum acceleration?

406. The angular frequency of an oscillator is 54 radians per second. The maximum acceleration of the same oscillator is 13 meters per second squared. What is the oscillator's maximum displacement?

407. An oscillator's displacement is $y = 4.5 \text{ m} \times \sin\left(23 \text{ s}^{-1} \times t\right)$. What is its acceleration at 4.8 seconds?

Finding the Period of a Mass on a Spring

408–411

408. A mass of 3.6 kilograms oscillates on a spring with a spring constant of 1,300 newtons per meter. What is the period of oscillation?

409. A mass oscillates on a spring with a frequency of 4.5 hertz. If the spring constant is 650 newtons per meter, what is the mass?

410. With a 10-kilogram mass attached, a spring takes 0.50 seconds to complete half of an oscillation. What is the spring constant of the spring?

411. You have two springs to which you attach identical masses. If the frequency of oscillation of spring one is 5.5 times greater than that of spring two, what is the ratio of the spring constant of spring one to that of spring two?

Taking Energy into Account in Simple Harmonic Motion

412–416

412. An oscillating spring is moving through its equilibrium position at a speed of 3.5 meters per second. If its mass is 2.1 kilograms, what is its total energy, in joules?

413. What is the potential energy of a spring with a spring constant of 1,300 newtons per meter if it is compressed a distance of 0.55 meters from equilibrium?

414. At maximum compression, the potential energy of a spring-mass system is 12 joules. What is the potential energy when the spring is at three-quarters of its maximum compression?

415. A spring-mass system oscillates with a frequency of 35 hertz. The mass of the system is 2.3 kilograms. What is the potential energy of the system when the mass is 2.2 meters from equilibrium?

416. Two spring-mass systems have the same mass but different spring constants. The speed of mass one is 13 meters per second at the equilibrium position, and the speed of mass two is 8.5 meters per second at equilibrium. What is the ratio of the spring constant of spring one to that of spring two if the maximum extension of spring one is 3.3 meters and that of spring two is 1.3 meters?

Swinging with Pendulums

417–420

417. The frequency of a pendulum is 2.1 hertz. If you double the mass and reduce the length of the pendulum arm by a factor of 3, what is the new frequency?

418. The acceleration of gravity on the moon is 1.6 meters per second squared. If a pendulum has a period of 4.8 seconds on the moon, what is its period on Earth?

419. The maximum angle of oscillation of a pendulum is 0.13 radians. If the angular frequency of oscillation is 33 radians per second, what is the pendulum's maximum angular acceleration?

420. To reduce the period of a pendulum by one-third, by what factor do you have to multiply the acceleration due to gravity on Earth?

Chapter 13

Heating Up with Thermodynamics and Heat Transfer

Temperature is a measure of how fast the molecules of an object are moving. Most objects expand when their temperature increases. When two objects of different temperatures touch, *heat* flows from the hot object to the cold object. When heat flows into or out of an object, the *phase* (solid, liquid, or gas) of that object can change. Heat can be transferred by convection, conduction, or radiation.

The Problems You'll Work On

You can keep yourself warm on a winter day by working on the following problems in front of a fire:

- » Calculating the change in size of objects as they are heated or cooled
- » Using heat transfer to determine change in temperature
- » Finding how much heat energy is transferred in a phase change
- » Transferring heat through convection and conduction
- » Applying the law of black body radiation

What to Watch Out For

You may be burned if you don't keep the following in mind:

- » Using different coefficients of linear expansion, heat capacities, latent heats, and thermal conductivities for different objects
- » Recalling that objects expand as they warm up and contract as they cool down
- » Remembering that heat flows from the warm object to the cool object, and the temperatures of both objects may change
- » Including the heat radiated by the environment when using the Stefan-Boltzmann law of radiation

Measuring Temperature in Different Ways

421–423

421. If it's 72 degrees Fahrenheit in your house, what is the temperature in degrees Celsius?

422. Most refrigerators maintain a temperature of about 40 degrees Fahrenheit. What temperature is that in Celsius?

423. At what temperature does pure water freeze in degrees Celsius, at sea level?

Getting to the Coldest Zone

424–426

424. At what temperature in kelvins does pure water freeze at sea level?

425. Iron melts at approximately 2,800 degrees Fahrenheit. What is that temperature in kelvins?

426. Why is 0 on the Kelvin scale known as *absolute zero?*

Expanding with the Heat: Getting Longer

427–429

427. The formula for calculating thermal expansion is $\Delta L = \alpha L_0 \Delta T$. The coefficient of linear expansion of aluminum is $2.22 \times 10^{-5} \, \mathrm{K}^{-1}$. If an aluminum rod is 10 meters long at 273 kelvins, how long will it be if its temperature increases to 423 kelvins? (Round as necessary to two decimal places.)

428. New homeowners are installing a concrete sidewalk in front of their house. They know they should leave space between the blocks for thermal expansion. The coefficient of linear expansion of concrete is $1.2 \times 10^{-5} \, {}^{\circ}\mathrm{C}$. If they leave a 5-millimeter gap between blocks that are 2 meters long, how great a temperature range can they expect the blocks to handle before they run out of room?

429. What was the original length of a steel rod if it increases in length by 1.72×10^{-2} when heated by 115 °C? (Recall that the coefficient of linear expansion of steel is $1.20 \times 10^{-5}\,°C^{-1}$.)

Expanding with the Heat: Taking Up More Space

430–432

430. When factories fill soda bottles, they leave space at the top, and consumers may feel cheated if they don't consider the thermal volume expansion of the liquid in the bottle. How much extra volume should a factory leave available for expansion if the bottle contains 2 liters of soda and if shipping and storage is expected to result in a possible temperature increase of 40 degrees Celsius?

Assume a coefficient of volume expansion of $5.26 \times 10^{-4}\,°C^{-1}$ for the soda, and round to two decimal places as necessary. Don't worry about accounting for the expansion of the plastic.

431. A plastic bottle has a volume of 12 ounces of liquid at the factory, where the room is kept at a constant temperature of 22 degrees Celsius. The same bottle contains 11.92 ounces of liquid in the cooler at the retail store. What temperature is the cooler?

Don't worry about accounting for the expansion of the plastic. Use $5.26 \times 10^{-4}\,°C^{-1}$ as the coefficient of thermal volume expansion for the soda. Round to two decimal places as necessary.

432. A steel paint can has a volume of 1 gallon at the factory, where it is filled with 0.93 gallons of paint and where the temperature is a constant 23 degrees Celsius. How much unused volume is in the can when the consumer gets ready to open it outside, where the temperature is 33 degrees Celsius? Use a value of $3.6 \times 10^{-5}\,°C^{-1}$ for the coefficient of thermal volume expansion of steel and $4.05 \times 10^{-4}\,°C^{-1}$ for the paint.

Changing the Temperature with Energy Flow

433–436

433. How much thermal energy is needed to increase the temperature of 2 kilograms of aluminum by 40 degrees Celsius? Round your answer to the nearest joule. (Remember that aluminum has a specific heat capacity, c, of $\frac{903\,J}{kg \bullet °C}$.)

434. If 12,800 joules of heat energy are added to a 1-kilogram block of solid material, and its temperature increases by 15 degrees Celsius, what is the specific heat capacity of the material?

435. How much pure water, by mass, requires 8.79×10^7 J of heat energy to increase in temperature by 72 degrees Celsius? Round your answer to three significant figures. (Note that pure water has a specific heat capacity, c, of $\frac{4,180 \text{ J}}{\text{kg} \cdot {}^\circ\text{C}}$.)

436. A barista adds 0.5 kilograms of coffee at 88 degrees Celsius to 1.2 kilograms of coffee at 46 degrees Celsius. Assuming no heat energy is lost, what is the temperature of the mixture when it reaches equilibrium? Round to the nearest whole degree. Use the specific heat capacity of water, $\frac{4,180 \text{ J}}{\text{kg} \cdot {}^\circ\text{C}}$, for the coffee.

Gaining or Losing Energy to Change into a New Phase

437–441

437. How much heat energy is necessary to turn 1.77 kilograms of water into steam? (Note that the latent heat of vaporization of water is 2.26×10^6 J/kg.)

438. It takes 1.55×10^6 J of heat energy to melt 7.40 kilograms of copper. What is the latent heat of fusion of copper?

439. How many kilograms of boiling temperature water vaporize with 5.126×10^7 J of heat energy? (Note that the latent heat of vaporization of water at boiling temperature is $\frac{2.258 \times 10^6 \text{ J}}{\text{kg}}$.)

440. You place a 3-kilogram block of ice at 0 degrees Celsius into a pot on a stove and turn on the heat. As the pot warms up, it feeds heat energy into the ice. If the stove adds 1 million joules of heat energy to the ice, what is the final temperature of the ice? (Note that the latent heat of fusion is 3.35×10^5 joules per kilogram.)

441. Assuming no heat is lost elsewhere, how many kilograms of 0-degree-Celsius ice can you melt with 2.5 gallons of boiling temperature (100 degrees Celsius) water?

Note that water has a specific heat capacity of $\frac{4,180 \text{ J}}{\text{kg} \cdot {}^\circ\text{C}}$; the latent heat of fusion of water, c, is 3.35×10^5 J/kg; and pure water weighs about 3.8 kilograms per gallon.

Rising with the Hot Fluid in Convection

442–445

442. Convection is the transfer of _____ through _____.

443. When fluids heat, they (generally) become less dense, resulting in the upward motion of the warmer fluid because of a specific force. What is that force called?

444. Why do fluids generally become less dense as they are heated?

445. What is the primary functional difference between a standard oven and a convection oven?

Getting in Touch with Thermal Conduction

446–448

446. What is actually "conducted" when something is heated via conduction?

447. You cut a piece from a pie that is right out of the oven. Why is the crust much less likely to burn your mouth than the filling?

448. A pan conducts 600,000 joules of heat energy to the water inside in a 42-second time period. If nothing else changes, how much heat energy does the pan conduct in 70 seconds?

Working with Thermal Conductivity

449–452

449. A steel rod 16 centimeters long with a 2-centimeter-squared cross-section area conducts 1.7×10^4 J of heat energy across its length in a given time. If the rod is replaced with a smaller steel bar that is only 8 centimeters long and 1-centimeter-squared in the cross-section area, how much heat is transferred in the same time period?

450. One end of an oak dowel that is 16 centimeters long with a 2-centimeter-squared cross-section area is 150 degrees Celsius, and the other end is only 50 degrees Celsius. How much heat is conducted through the dowel in 90 seconds? Use a value of $\frac{0.17\text{J}}{\text{s} \cdot \text{m} \cdot {}^\circ\text{C}}$ for k, the thermal conductivity of oak.

451. The thermal conductivity of glass is about 0.8 joules per second per meter per degree Celsius. You keep your house at 22 degrees Celsius, and it is 0 degrees Celsius outside. If the window in your bedroom has an area of 1 meter squared and is 0.5 centimeters (0.005 meters) thick, how much thermal energy is lost through it every minute? Round your answer to two significant digits.

452. One end of a solid steel (thermal conductivity of 14 joules per second per meter per degree Celsius) fireplace poker is frozen into a 2-kilogram block of ice at 0 degrees Celsius, and the other end is placed into a fire burning at 1,100 degrees Celsius. The poker has a length of 0.5 meters and a cross-sectional area of 2 square centimeters $\left(2.0 \times 10^{-4}\,\text{m}^2\right)$. How long does it take the heat conducted through the poker to melt the ice? (Note that the latent heat of water is 3.35×10^5 J/kg.)

Radiating with Black Bodies

453–455

453. What portion of the heat energy that Earth receives from the sun is due to radiation?

454. A certain star has a surface temperature of 2,600 kelvins and radiates energy at a rate of 3.2×10^{30} watts. How much energy does an otherwise identical star radiate if it has a surface temperature of 3,400 kelvins?

455. A 6-foot-tall, 200-pound athlete has a body surface area of approximately 2.15 square meters. Using the Stefan–Boltzmann law of radiation, $\frac{Q}{t} = e\sigma A \left(T_{hot}^4 - T_{cold}^4\right)$, calculate the estimated rate at which such a person radiates heat in a room that is 22 degrees Celsius. Round your answer to two significant digits.

Hint: The Stefan–Boltzmann constant is $\dfrac{5.67 \times 10^{-8}\ \text{J}}{\text{s} \cdot \text{m}^2 \cdot \text{K}^4}$. The emissivity e of human skin is approximately 0.97, and a healthy person has an average skin temperature of 33 degrees Celsius. Don't forget to convert the temperature to kelvins!

Chapter **14**

Living in an Ideal World with the Ideal Gas Law

deal gases have two properties: The average distance among molecules is much larger than the size of the molecules, and the only interactions among molecules are elastic collisions. Many real gases behave like ideal gases when the temperature is high and the pressure is low. For an ideal gas, the product of the pressure and volume is proportional to the number of molecules times the temperature.

The Problems You'll Work On

In this chapter, you expand your mind (at constant temperature) using the following types of problems:

>> Using Avogadro's number to find the number of molecules

>> Relating the pressure and volume at constant temperature with Boyle's law

>> Relating the volume and temperature at constant pressure with Charles's law

>> Using the ideal gas law to find the temperature, pressure, or volume

>> Finding the kinetic energy in an ideal gas

What to Watch Out For

If you're under pressure to get these problems right, a large number of them will be easier if you keep the following in mind:

>> Remembering that one mole of any ideal gas has the same number of molecules

>> Distinguishing the number of moles from the number of molecules in the two versions of the ideal gas law

>> Converting all temperatures to kelvins

Finding Number of Molecules in Moles with Avogadro's Number

456–457

456. What is Avogadro's number, N_A, used for?

457. What is the mass of 1 mole of water molecules? (Hint: Water molecules each consist of two hydrogen atoms and one oxygen atom. Hydrogen has an average atomic mass of 1 AMU, and oxygen has an average atomic mass of 16 AMU.)

Working with Boyle's Law When Temperature Is Constant

458–461

458. Boyle's law is commonly expressed as $P_1 \bullet V_1 = P_2 \bullet V_2$. What does this mean, in English?

459. A sample of an ideal gas is compressed from 3.0 liters to 1.25 liters. If the final pressure of the gas is 3.0×10^5 pascals, what was the initial pressure? (Hint: Boyle's law states that $P_1 \bullet V_1 = P_2 \bullet V_2$.)

460. A sample of an ideal gas at 327 kelvins occupies 0.3 cubic meters at 4.2 atmospheres of pressure. What would be the pressure of the same sample in a 2.6 cubic-meter container at 327 kelvins?

461. A sample of helium occupies 22.40 liters at 273.0 kelvins and 1.000 atmosphere of pressure. What volume, in cubic meters, would it occupy at 273.0 kelvins and a pressure of 1.500×10^6 pascals? (Hint: 1 atm $= 1.013 \times 10^5$ Pa, 1 L $= 1.000 \times 10^{-3}$ m^3.)

Working with Charles's Law When Pressure Is Constant

462–464

462. A sample of an ideal gas occupies a volume of 0.02 cubic meters when it's at a temperature of 225 kelvins. Assuming constant pressure, what would be the volume at a temperature of 395 kelvins? Round to two decimal places as necessary. (Hint: Charles's law states that $\frac{V_f}{T_f} = \frac{V_i}{T_i}$.)

463. A 6.04-liter sample of helium at −56 degrees Celsius is heated until it has a volume of 8.23 liters. Assuming constant pressure, what is the final temperature of the sample, to the nearest 1 degree Celsius?

464. A sample of oxygen occupies 300.0 milliliters at −32.00 degrees Celsius. Assuming constant pressure, what is the volume, in cubic meters, of the sample at −64.00 degrees Celsius? Round to the nearest ten-thousandth of a cubic meter. (Hint: 1 L = 0.001 m³.)

Relating Pressure, Volume, and Temperature with the Ideal Gas Law

465–468

465. A given volume of an ideal gas has a temperature of 290 kelvins. What would be the temperature of the same gas if it were compressed into one-half the volume?

466. A given volume of an ideal gas contains 56.0 moles of gas molecules at 272 kelvins. How many moles of molecules of the same gas would there be at a temperature of 449 kelvins, if the volume is held constant?

467. 5.4 moles of an ideal gas are contained in a 2.3-liter vessel and held at a temperature of 150 kelvins. What would you need to do to increase the amount of the gas to 10.8 moles without changing the temperature?

468. 11.6 moles of an ideal gas are at 1.0 atmosphere of pressure in a 4.6-liter container. If the amount of gas increases to 17.4 moles and the container compresses to 3.3 liters, what would the final pressure be if the temperature is held at constant?

Calculating the Kinetic Energy of the Ideal Gas Molecules

469–470

469. A sample of nitrogen gas has an average molecular kinetic energy of 4.55×10^{-21} joules. Nitrogen molecules (two atoms each) have an average mass of 4.65×10^{-26} kilograms. What is the average speed, in meters per second, of the molecules in the sample? Hint: Use $v = \sqrt{\dfrac{2KE}{m}}$, where v is velocity (speed, in this case), KE is kinetic energy, and m is mass.

470. How much thermal energy do 8.5 moles of helium contain at 125 degrees Celsius? (Hint: $KE = \left(\dfrac{3}{2}\right)nRT$, and the universal gas constant, R, is 8.31 joules per mole-kelvin.)

Chapter 15

Experiencing the Laws of Thermodynamics

The first law of thermodynamics states that the change in internal energy is equal to the heat absorbed minus the work done. The second law implies that heat naturally flows from hotter objects to colder ones. A heat engine does work by absorbing heat from a high-temperature heat source and dumping heat into a low-temperature heat sink.

The Problems You'll Work On

Writing the questions in this chapter on a balloon and inflating the balloon may help you understand the following topics:

>> Using the first law of thermodynamics to relate heat and work

>> Understanding isobaric, isochoric, isothermal, and adiabatic processes

>> Determining the amount of work done by heat engines

>> Finding the efficiency of heat engines

>> Using heat engines in reverse as heat pumps

What to Watch Out For

You'll work through this chapter without breaking a sweat if you don't overlook the following:

>> Using the correct signs for heat and work in the first law of thermodynamics

>> Remembering that a gas does work only when it expands

>> Keeping track of which quantities are held constant

Conserving Energy Using the First Law: Heat and Work

471–474

471. What does the first law of thermodynamics, $\Delta U = Q - W$, state in English?

472. Consider a motor that does 2,500 joules of work while releasing 1,200 joules of heat. By how much does the internal energy of the system change?

473. Consider a system that emits 2,150 joules of heat during a process that decreases the internal energy of the system by 2,850 joules. How much work was done on or by the system?

474. Consider two different systems, system x and system y. System x emits 1,235 joules of heat as 828 joules of work are done on it. System y emits 2,120 joules of heat as 1,548 joules of work are done on it. What is the change in internal energy of each system?

Staying at Constant Pressure in Isobaric Processes

475–477

475. When the temperature of a given amount of a gas increases, the volume, the pressure, or some combination also increases. If you increase the temperature of a given amount of gas in an isobaric process, what other change must occur?

476. A quantity of gas is heated, and its volume increases by 4.35×10^{-5}. If pressure is kept constant at 1.35×10^{6}, how much work is done?

477. A sample of steam is maintained at 2 atmospheres (2.02×10^{5} pascals) pressure in an isobaric chamber as it is cooled. The volume is reduced by 2.30×10^{-3} cubic meters. If the chamber absorbs 220 joules of heat, what is the change in internal energy of the sample? (Hint: $\Delta U = Q - W$, where $Q = -220$ J.)

Staying at Constant Temperature in Isothermal Processes

478–481

478. You apply heat energy totaling 2,335 joules to a sample of gas with an initial volume of 3.2 liters. The sample volume increases to 7.4 liters as the temperature is held constant. What type of system is this?

479. An isothermal cylinder holds a sample of oxygen gas. If 5.94×10^4 joules of heat energy are added to the gas, how much work is done on the cylinder by the gas as it expands, assuming an ideal system? (Hint: $\Delta U = Q - W$, where ΔU is the change in internal energy, which is zero in an isothermal process.)

480. A 7.00-mole sample of helium does 1.12×10^4 joules of work as it expands. If the volume of the sample expands from 1.00 cubic meter to 2.00 cubic meters, what is the temperature of the gas?

(Hints: $W = nRT \cdot \ln\left(\dfrac{V_f}{V_i}\right)$;

R = the gas constant: $\dfrac{8.31\text{ J}}{(\text{mol} \cdot \text{K})}$).

481. Assuming an ideal system, how much heat energy do 3.5 moles of helium gas (maintained at 55 degrees Celsius) emit when compressed from 4,200 cubic centimeters to 1,700 cubic centimeters?

(Hints: $W = nRT \cdot \ln\left(\dfrac{V_f}{V_i}\right)$ and $\Delta U = Q - W$;

R = the gas constant: $\dfrac{8.31\text{ J}}{(\text{mol} \cdot \text{K})}$;

$0\ ^\circ\text{C} = 273.15\text{ K}$; in an isothermal system, $\Delta U = 0$.)

Staying at Constant Volume in Isochoric Processes

482–485

482. How much work is done by a 3-mole sample of ideal gas as 1,400 joules of heat energy are added to it during an isochoric process?

483. A 5.0×10^{-5} cubic meters sample of helium is held at a constant 1.52×10^5 pascals as it is heated from 345 kelvins to 395 kelvins. Is this process isochoric? Why or why not?

484. A can of compressed air is placed in an ice bath. If the gas emits 237 joules of heat energy as it cools by 32 kelvins, what is the resultant change in internal energy of the air?

485. A sealed rigid tank contains 3.26×10^{-3} kilograms of H_2O steam held under 1 atmosphere of pressure. The gas is heated from 392 kelvins to 470 kelvins. What is the change in internal energy of the helium? (Hints: $Q = cm\Delta T$; c, the specific heat capacity of steam, is $\dfrac{2{,}020 \text{ J}}{\text{kg} \bullet \text{K}}$.)

Staying at Constant Heat in Adiabatic Processes

486–489

486. A cylinder is surrounded by a perfect insulator, preventing any heat from entering or leaving the system. It is $Q = 0$ in the $\Delta U = Q - W$ formula, so $\Delta U = -W$. Is this system an isothermal system? If not, what kind of system is it?

487. A total of 2.2 liters of helium at 1.0 atmosphere pressure undergoes an adiabatic process, resulting in an increase in volume to 3.5 liters. What is the pressure of the gas after the expansion?

488. A sample of hydrogen at 2.8 atmospheres pressure undergoes an adiabatic expansion resulting in a volume of 7.8 liters at 0.65 atmospheres pressure. What was the initial volume of the sample?

489. A sample of helium with volume 1.18×10^{-6} cubic meters starting at 3.04×10^{5} pascals pressure is tripled in pressure via an adiabatic process. What is the final volume, in liters? (Hints: 1.013×10^{5} Pa $= 1$ atm; $1 \text{ m}^3 = 1{,}000$ liters.)

Putting Heat to Work with Heat Engines

490–493

490. How do you calculate the efficiency of a heat engine?

491. A certain heat engine is 25 percent efficient and uses 2.8×10^{4} joules of heat. How much work does it do? (Hint: efficiency $= \dfrac{W}{Q_h}$, where $W =$ work and $Q_h =$ heat input.)

492. High-tech marine diesel engines can achieve 60 percent thermal efficiency. If such an engine can do 8.22×10^{7} joules of work per gallon of diesel, how much energy does a gallon of diesel contain? Assume that input heat is equal to all the energy from the fuel.

493. A certain heat engine operates at 54 percent efficiency burning fuel that provides 2.53×10^7 joules of energy per kilogram. The engine uses a 5.2-kilogram heat sink with a specific heat capacity of 3,820 joules per kilogram per kelvin. By how much does the temperature of the heat sink increase per kilogram of burned fuel? (Hint: $Q = mc\Delta T$.)

Evaluating Efficiency of Heat Engines

494–497

494. What is the definitive attribute of a Carnot engine?

495. What is the maximum theoretical efficiency of a heat engine based on a heat source of 250 kelvins and heat sink at 100 kelvins? (Hint: efficiency $= 1 - \left(T_c / T_h \right)$.)

496. The heat sink of a Carnot engine with an efficiency of 43 percent is maintained at 87 kelvins. At what temperature is the heat source?

497. A 65-percent-efficient Carnot engine provides 2.38×10^8 joules of heat energy to power a certain process. If the heat source of the engine is maintained at 90 degrees Celsius, at what temperature is the heat sink in degrees Celsius? (Hints: efficiency $= 1 - \left(T_c / T_h \right)$, and $Q_h = W$/efficiency.)

Going against the Flow with Heat Pumps

498–501

498. A normal heater requires you to input more energy than is returned to you as heat, whereas a heat pump actually outputs more heat than you put into it as fuel. How is this possible if there's no such thing as a 100-percent-efficient engine?

499. A certain Carnot process uses 750 joules of work to pump 2,500 joules of heat from a 210-kelvin heat sink. What is the temperature of the heat source?

Hint: $W = Q_h \left(1 - \left(\dfrac{T_c}{T_h} \right) \right)$

500. A certain heat pump with a COP of 14 is used to transfer 12,350 joules of heat. How much work is required?

Hint: $COP = \dfrac{Q_h}{W}$.

501. How much heat can you theoretically pump from a −35 degrees-Celsius source to a 33 degrees-Celsius target with 7,250 joules of work and a Carnot engine?

Hints: 0 °C = 273.15 K; $COP = \dfrac{1}{1 - \dfrac{T_c}{T_h}}$

$COP = \dfrac{Q_h}{W}$.

2 The Answers

Here's where you can find the answers and explanations for all the problems in this book. As you read through the explanations, if you find that you need a little more help with certain concepts, *For Dummies* has your back. The following titles (all published by John Wiley & Sons) are great additional resources to shore up your physics knowledge:

Physics I For Dummies, 3rd Edition

Physics I Workbook For Dummies, 3rd Edition

Physics Essentials For Dummies

Visit www.dummies.com for more information.

Chapter 16

Answers

1. $m = \dfrac{y-3}{2}$

First you have to isolate the variable m on one side of the equals sign by performing the same operations on each side of the equation. Start by subtracting 3 from each side of the equation. This gives

$$\begin{aligned} y - 3 &= 2m + 3 - 3 \\ &= 2m \end{aligned}$$

Now divide each side of the equation by 2, which gives

$$\begin{aligned} \frac{y-3}{2} &= \frac{\cancel{2}m}{\cancel{2}} \\ &= m \end{aligned}$$

Normally, you write this as $m = \dfrac{y-3}{2}$, although it's not mandatory.

2. $m_0 + m_1 = \dfrac{2I}{r^2}$

Isolate the sum $m_0 + m_1$ on one side of the equation by performing the same operations on each side of the equation. Solve for m first; then use the second equation to replace m with $m_0 + m_1$.

Start by multiplying both sides of the first equation by 2:

$$\begin{aligned} 2I &= \cancel{2}\left(\frac{1}{\cancel{2}}mr^2\right) \\ &= mr^2 \end{aligned}$$

Now divide both sides by r^2 to get

$$\begin{aligned} \frac{2I}{r^2} &= \frac{mr^{\cancel{2}}}{r^{\cancel{2}}} \\ &= m \end{aligned}$$

Finally, replace m with $m_0 + m_1$:

$$\frac{2I}{r^2} = m$$

$$= m_0 + m_1$$

Write your answer as $m_0 + m_1 = \frac{2I}{r^2}$.

3. $v = \pm c\sqrt{1 - \frac{1}{\gamma^2}}$

Isolate the variable v on one side of the equation by performing the same operations on each side of the equation. Start by squaring both sides of the equation:

$$\gamma^2 = \left(\frac{1}{\sqrt{1 - \frac{v^2}{c^2}}}\right)^2$$

$$= \frac{1}{1 - \frac{v^2}{c^2}}$$

Next, multiply both sides by $1 - \frac{v^2}{c^2}$ to find

$$\gamma^2\left(1 - \frac{v^2}{c^2}\right) = \frac{1}{1 - \frac{v^2}{c^2}}\left(1 - \frac{v^2}{c^2}\right)$$

$$= 1$$

Now divide both sides by γ^2 to get

$$\frac{\gamma^2\left(1 - \frac{v^2}{c^2}\right)}{\gamma^2} = \frac{1}{\gamma^2}$$

$$\left(1 - \frac{v^2}{c^2}\right) =$$

Subtract 1 from each side:

$$\left(1 - \frac{v^2}{c^2}\right) - 1 = \frac{1}{\gamma^2} - 1$$

$$-\frac{v^2}{c^2} =$$

Multiply both sides by -1:

$$(-1)\left(-\frac{v^2}{c^2}\right) = (-1)\left(\frac{1}{\gamma^2} - 1\right)$$

$$\frac{v^2}{c^2} = 1 - \frac{1}{\gamma^2}$$

Multiply both sides by c^2:

$$\left(\frac{v^2}{c^2}\right)c^2 = \left(1 - \frac{1}{\gamma^2}\right)c^2$$
$$v^2 =$$

Finally, take the square root of both sides to find

$$\sqrt{v^2} = \pm\sqrt{\left(1 - \frac{1}{\gamma^2}\right)c^2}$$
$$v = \pm c\sqrt{1 - \frac{1}{\gamma^2}}$$

You don't know whether v is positive or negative, so insert the plus–minus sign.

4. 6.4 m

$\cos\theta$ is defined as the adjacent side divided by the hypotenuse, so the following is true:

$$\cos\theta = \frac{\text{adjacent}}{\text{hypotenuse}} = \frac{x}{r}$$

Insert $\cos\theta = 0.8$ and hypotenuse $= 8$ m, and solve for x to find

$$\cos\theta = \frac{x}{r}$$
$$0.8 = \frac{x}{8\,\text{m}}$$
$$0.8\,(8\,\text{m}) = x$$
$$6.4\,\text{m} = x$$

5. 2.9 m

The line you dive through the water forms the hypotenuse of a right triangle. The bottom of the pool forms the side adjacent the angle θ, and the wall of the pool forms the side opposite the angle θ. Because the bottom of the pool is parallel to the surface, you have $\theta = 35°$. The pool is 2 meters deep, so $y = 2.0$ m. Using the tan function, you can find the distance x along the bottom of the pool from the wall to where you hit the bottom:

$$\tan\theta = \frac{y}{x}$$
$$x = \frac{y}{\tan\theta}$$
$$= \frac{2.0\,\text{m}}{\tan 35°}$$
$$= \frac{2.0\,\text{m}}{0.70}$$
$$= 2.9\,\text{m}$$

6. 33,000 ft

One meter is the same as 3.3 feet, so the conversion factor is $1 = \dfrac{3.3\,\text{ft}}{1\,\text{m}}$.

Multiply the height in meters by the conversion factor to find the height in feet:

$$10,000\,\cancel{\text{m}} \times \frac{3.3\,\text{ft}}{1\,\cancel{\text{m}}} = 10,000 \times 3.3\,\text{ft}$$
$$= 33,000\,\text{ft}$$

7. 63 mi/hr

One mile equals 1.6 kilometers, so the conversion factor is $1 = \dfrac{1\,\text{mi}}{1.6\,\text{km}}$.

Multiply the conversion factor by the speed in kilometers per hour to get the speed in miles per hour: $100\,\cancel{\text{km}}/\text{hr} \times \dfrac{1.0\,\text{mi}}{1.6\,\cancel{\text{km}}} = 63\,\text{mi/hr}$.

8. 4 hours

The painting rate per painter is $\dfrac{10\,\text{m}^2}{1\,\text{hr}}$. You have 5 painters, so the total rate of painting is $5 \times \dfrac{10\,\text{m}^2}{1\,\text{hr}}$. To find the time it takes to paint the house, divide the total surface area by the total painting rate:

$$t = \frac{200\,\cancel{\text{m}^2}}{5 \times 10\,\cancel{\text{m}^2}/\text{hr}} = 4.0\,\text{hr}$$

9. 40 days

The rate at which your dog eats is $\text{rate} = \dfrac{1/4\,\text{lb}}{1\,\text{d}} = \dfrac{1}{4}\,\text{lb/d}$.

Divide the weight of a dog food bag by the rate at which your dog eats to find how long a bag of food lasts: $t = \dfrac{10\,\cancel{\text{lb}}}{1/4\,\cancel{\text{lb}}/\text{d}} = 40\,\text{d}$.

10. 120 yards

First write the ratios given in the problem. The knitting rate in bobbins per hour is $\dfrac{1\,\text{bobbin}}{2\,\text{h}}$. The yards per bobbin is $\dfrac{10\,\text{yd}}{1\,\text{bobbin}}$. The total number of hours to knit a sweater is $3\,\cancel{\text{d}} \times \dfrac{8\,\text{h}}{1\,\cancel{\text{d}}} = 24\,\text{h}$.

To find the length ℓ in yards of the yarn in the sweater, multiply or divide these three quantities so that, after cancelling the units, you're left with yards. This gives

$$\ell = 24\,\cancel{\text{h}} \times \frac{1\,\cancel{\text{bobbin}}}{2\,\cancel{\text{h}}} \times \frac{10\,\text{yd}}{1\,\cancel{\text{bobbin}}} = 120\,\text{yd}.$$

11. $3 \times 10^8 \, \text{m/s}$

You can write this speed as $3 \times 100{,}000{,}000 \, \text{m/s}$.

But $100{,}000{,}000$ has 8 zeroes behind the 1, so it's the same as 10^8. Thus, you can write the speed as $3 \times 10^8 \, \text{m/s}$.

12. $1 \times 10^{-15} \, \text{s}$

One million has 6 zeroes behind the 1, so 1 million is 10^6. One millionth is 1 divided by 1 million, so it is $1 \, \text{millionth} = \dfrac{1}{10^6} = 10^{-6}$.

Likewise, 1 billion is $1 \, \text{billionth} = \dfrac{1}{10^9} = 10^{-9}$.

A femtosecond (fs) is 1 millionth of 1 billionth of a second, so you can write it as follows:

$$
\begin{aligned}
1 \, \text{fs} &= 10^{-6} \times 10^{-9} \times 1 \, \text{s} \\
&= 1 \times 10^{-15} \, \text{s}
\end{aligned}
$$

13. $1.412 \times 10^{27} \, \text{m}^3$

Plug the given value of the sun's radius into the equation for the volume of a sphere to find

$$
\begin{aligned}
V &= \frac{4}{3} \pi r^3 \\
&= \frac{4}{3} \pi \left(6.960 \times 10^8 \, \text{m} \right)^3 \\
&= \frac{4}{3} \pi (6.960)^3 \times \left(10^8 \, \text{m} \right)^3 \\
&= 1412 \times 10^{24} \, \text{m}^3
\end{aligned}
$$

You can write the prefactor $1{,}412$ as 1.412×10^3, so you can write the volume as follows:

$$
\begin{aligned}
V &= 1{,}412 \times 10^{24} \, \text{m}^3 \\
&= 1.412 \times 10^3 \times 10^{24} \, \text{m}^3 \\
&= 1.412 \times 10^{27} \, \text{m}^3
\end{aligned}
$$

14. $8 \times 10^{-14} \, \text{kg}$

A billion is 10^9, so 7 billion is 7×10^9. The total mass of the human population is the number of humans (7×10^9) multiplied by the mass of each person (70 kg). This gives

$$
\begin{aligned}
m_{\text{humans}} &= 7 \times 10^9 \times 70 \, \text{kg} \\
&= 490 \times 10^9 \, \text{kg}
\end{aligned}
$$

The fraction of the mass of Earth that is due to humans is

$$
\begin{aligned}
f &= \frac{m_{\text{humans}}}{m_{\text{Earth}}} \\
&= \frac{490 \times 10^9 \, \text{kg}}{6 \times 10^{24} \, \text{kg}} \\
&= 8 \times 10^{-14} \, \text{kg}
\end{aligned}
$$

15. **1.5 g**

A gram contains 1,000 milligrams, so 500 milligrams is half of a gram. Three times $\frac{1}{2}$ is $1\frac{1}{2}$, so you need $1\frac{1}{2}$ grams of salt.

16. **10^9**

A watt has 1,000 milliwatts, and a megawatt consists of 1 million watts, so there are 1 thousand million milliwatts in a megawatt, or:

$$1{,}000 \times 1{,}000{,}000 \text{ mW} = 1 \text{ MW}$$
$$10^3 \times 10^6 \text{ mW} = 1 \text{ MW}$$
$$10^9 \text{ mW} = 1 \text{ MW}$$

17. **4**

The zero is significant because it's neither leading nor trailing. So the total number of significant digits is 4.

18. **28.4**

Adding the numbers gives

$$\begin{array}{r} 21.21 \\ 4.8 \\ +2.35 \\ \hline 28.36 \end{array}$$

The rightmost column in which all the numbers contain significant digits is the tenths column, so you have to write the result of the addition with three significant digits. You thus have to round up your answer to 28.4.

19. **4**

Because the zeroes are neither trailing nor leading, they're significant. Thus, the number 5,003 has four significant digits.

20. **5**

Both terms in the sum in the square root have the same number of significant digits after the decimal point (three), so the sum also has three significant digits after the decimal point. The sum in the square root thus has five significant digits overall. Taking a square root is like multiplication, so you retain five significant digits in the answer.

21. **126.93**

The sum is

$$
\begin{array}{r}
98.374 \\
+28.56 \\
\hline
126.934
\end{array}
$$

The rightmost column in which all the digits are significant is the hundredth column. Thus, you must round the answer to the nearest hundredth, to get 126.93.

22. **10 gal**

The sum is

$$
\begin{array}{r}
10 \text{ gal} \\
+0.25 \text{ gal} \\
\hline
10.25 \text{ gal}
\end{array}
$$

The rightmost column in which all the digits are significant is the ones column, so you have to round the answer to the nearest unity. The result is 10 gallons.

23. $\mathbf{8.0 \times 10^1}$

The first term gives $5.01 \times 4.4 = 22.044$. You must round this to two significant digits because 4.4 has two significant digits: $5.01 \times 4.4 = 22$.

The second term gives $3.2 \times 18 = 57.6$. You must also round to two significant digits, which gives $3.2 \times 18 = 58$.

Add these two results: $22 + 58 = 80$.

To show that the trailing zero is significant, you have to use scientific notation and write the answer as 8.0×10^1.

24. **21 m**

The building has 8 floors, each of which is 2.6 meters high, so the total height h is

$$
\begin{aligned}
h &= 8 \times 2.6 \text{ m} \\
&= 20.8 \text{ m}
\end{aligned}
$$

Because you know the height of your floor in a measurement with only two significant digits, you can determine the height of the building to only two significant digits. Therefore, you have to round the height from 20.8 meters to 21 meters.

25. 3×10^3

Begin by calculating the fraction in the second factor. You must round the result to two significant figures because the numerator has two significant figures: $\frac{65}{4} = 16.25$.

You have to round up this result to 20 (where the trailing zero in 20 is not significant). Doing the addition now gives

$$\begin{array}{r} 18.54 \\ +20 \\ \hline 38.54 \end{array}$$

The rightmost column in which all the digits are significant is the tens column (recall that the 0 in 20 isn't significant, in this case). Therefore, you have to round up your result to 40 (again, the 0 in 40 isn't significant). Now do the multiplication to get $63.005 \times 40 = 2,520.2$.

Because the 40 has only one significant digit, you have to round up the final result to 3,000. To show that this result has only one significant figure, write it in scientific notation as 3×10^3.

26. 0 blocks

The initial position and the final position are the same, so $S_i = S_f$. The displacement s is just the difference between the initial and final position, which is

$$\begin{aligned} s &= s_i - s_f \\ &= 0 \text{ blocks} \end{aligned}$$

So the displacement is 0 blocks.

27. 4 miles north

You can find the overall displacement by adding together all the intermediate displacements. Consider north the positive direction and south the negative direction. The first displacement is 5 miles north, so $s_1 = +5$ mi.

The second displacement is 3 miles south, so $s_2 = -3$ mi.

The third displacement is 2 miles north, so $s_3 = +2$ mi.

Add these to find the total displacement:

$$\begin{aligned} s &= s_1 + s_2 + s_3 \\ &= +5 \text{ mi} - 3 \text{ mi} + 2 \text{ mi} \\ &= +4 \text{ mi} \end{aligned}$$

The + sign means the total displacement is to the north, so the answer is 4 miles north.

28. 5 m

Call the displacement of the two skaters together on the ice $s_{ice} = 10\,\text{m}$. The displacement of the female skater in the air is S_{air}. Add these displacements to find the total displacement of the female skater, which is $s = 15\,\text{m}$. Thus:

$$s = s_{ice} + s_{air}$$
$$15\,\text{m} = 10\,\text{m} + s_{air}$$
$$5\,\text{m} = s_{air}$$

29. 3 floors

The initial position of the elevator is $s_i = 0\,\text{floor}$. The final position of the elevator is when Ms. Smith gets on at the third floor, so $s_f = 3\,\text{floor}$. The displacement is the final position minus the initial position:

$$s = s_f - s_i$$
$$= 3\,\text{floors} - 0\,\text{floors}$$
$$= 3\,\text{floors}$$

30. 5 cm

The change in the x coordinate is the final x coordinate minus the initial x coordinate:

$$\Delta x = x_f - x_i$$
$$= 5\,\text{cm} - 2\,\text{cm}$$
$$= 3\,\text{cm}$$

Likewise, the change in the y coordinate is

$$\Delta y = y_f - y_i$$
$$= 8\,\text{cm} - 4\,\text{cm}$$
$$= 4\,\text{cm}$$

The magnitude s of the displacement is given by the distance formula:

$$s = \sqrt{\Delta x^2 + \Delta y^2}$$
$$= \sqrt{(3\,\text{cm})^2 + (4\,\text{cm})^2}$$
$$= 5\,\text{cm}$$

31. 76°

Treat the direction east as the positive x-axis and the direction north as the positive y-axis. The change in the x coordinate is $\Delta x = 1$ block. The change in the y coordinate is $\Delta y = 4$ blocks.

With respect to the x-axis (the direction east), the angle of the displacement is

$$
\tan\theta = \frac{\Delta y}{\Delta x}
$$
$$
\theta = \tan^{-1}\left(\frac{\Delta y}{\Delta x}\right)
$$
$$
= \tan^{-1}\left(\frac{4\text{ block}}{1\text{ block}}\right)
$$
$$
= 76°
$$

32. 5.4 ft

Let the positive y-axis represent the upward vertical direction, and let the x-axis represent the horizontal direction from the player to the basket. The change in the y position is the final y position (10 feet) minus the initial y position (8 feet):

$$
\Delta y = y_f - y_i
$$
$$
= 10\text{ ft} - 8\text{ ft}
$$
$$
= 2\text{ ft}
$$

If you call the player's position $x_i = 0$ ft, then the basket is at $x_i = 5$ ft. The change in the x position is then

$$
\Delta x = x_f - x_i
$$
$$
= 5\text{ ft} - 0\text{ ft}
$$
$$
= 5\text{ ft}
$$

Using the distance formula, the magnitude of the displacement is

$$
s = \sqrt{\Delta x^2 + \Delta y^2}
$$
$$
= \sqrt{(5\text{ ft})^2 + (2\text{ ft})^2}
$$
$$
= 5.4\text{ ft}
$$

33. 5.7 squares at 45° from the positive x-axis

Let the first coordinate be the x coordinate and the second coordinate be the y coordinate. The change in the x position is

$$
\Delta x = x_f - x_i
$$
$$
= 7\text{ squares} - 3\text{ squares}
$$
$$
= 4\text{ squares}
$$

Likewise, the change in the y position is

$$\Delta y = y_f - y_i$$
$$= 5 \text{ squares} - 1 \text{ squares}$$
$$= 4 \text{ squares}$$

Using the distance formula, the magnitude of the displacement is

$$s = \sqrt{\Delta x^2 + \Delta y^2}$$
$$= \sqrt{(4 \text{ squares})^2 + (4 \text{ squares})^2}$$
$$= 5.7 \text{ squares}$$

The angle θ of the bishop's displacement is

$$\tan\theta = \frac{\Delta y}{\Delta x}$$
$$\theta = \tan^{-1}\left(\frac{\Delta y}{\Delta x}\right)$$
$$= \tan^{-1}\left(\frac{4 \text{ squares}}{4 \text{ squares}}\right)$$
$$= 45°$$

So the displacement is 5.7 squares at 45 degrees from the positive x-axis.

34.   4.2 m

This problem gives you changes in position, so you have to start by adding the changes in position in the x and y directions to find the total change in position in these two directions. Let the y direction be the direction down the hall. The child makes two displacements in the y direction. The first is 5 meters down the hall, so $\Delta y_1 = 5$ meters. The second is the displacement after the child's second right turn, in which case the child is heading in the direction opposite to the child's initial direction, so $\Delta y_2 = -2$ meters. The total displacement in the y direction is

$$\Delta y = \Delta y_1 + \Delta y_2$$
$$= 5 \text{ m} - 2 \text{ m}$$
$$= 3 \text{ m}$$

The displacement of the child in the x direction is just the single displacement of 3 meters to the right, so $\Delta x = 3$ m.

Using the distance formula, the magnitude of the displacement is

$$s = \sqrt{\Delta x^2 + \Delta y^2}$$
$$= \sqrt{(3 \text{ m})^2 + (3 \text{ m})^2}$$
$$= 4.2 \text{ m}$$

35. $22°$

Let the vertical upward direction be the positive y direction and the horizontal direction from the laser toward the mirror ball be the positive x direction. The difference in the x position between the laser and the mirror ball is $\Delta x = 5$ meters. The distance from the floor to the ceiling is 4 meters, the laser is 1 meter above the floor, and the mirror ball is 1 meter below the ceiling, so the difference in the y position between the laser and the disco ball is

$$\Delta y = 4\,\text{m} - 1\,\text{m} - 1\,\text{m}$$
$$= 2\,\text{m}$$

The angle θ at which to aim the laser is

$$\tan\theta = \frac{\Delta y}{\Delta x}$$
$$\theta = \tan^{-1}\left(\frac{\Delta y}{\Delta x}\right)$$
$$= \tan^{-1}\left(\frac{2.0\,\text{m}}{5.0\,\text{m}}\right)$$
$$= 22°$$

36. 5.0 m/s

Average speed is the distance covered divided by the time it takes to cover that distance. In this case, the distance covered is $\Delta x = 300$ meters.

In seconds, the time it takes to cover this distance is

$$\Delta t = 1.0\,\text{min}$$
$$= 60\,\text{s}$$

Your average speed is

$$v = \frac{\Delta x}{\Delta t}$$
$$= \frac{300\,\text{m}}{60\,\text{s}}$$
$$= 5.0\,\text{m/s}$$

37. 30 mph

Instantaneous speed is your speed at any given point in time. It doesn't matter how long you maintain that speed. The maximum instantaneous speed in the traffic jam is 30 miles per hour.

38. instantaneous speed

The speed is measured over a relatively short time interval, so it's the instantaneous speed. This instantaneous speed can't be the same as the average speed because it's measured at the end of the ball's trajectory, where the ball is moving at its maximum speed.

39. 1.4 hr

You're given the average speed of $v = 11$ miles per hour and the distance traveled of $\Delta x = 15$ miles. To find the time such a trip takes, solve the equation for average speed for the time:

$$v = \frac{\Delta x}{\Delta t}$$

$$\Delta t = \frac{\Delta x}{v}$$

$$= \frac{15 \text{ mi}}{11 \text{ mi/hr}}$$

$$= 1.4 \text{ hr}$$

40. Runner B finishes first.

Calculate the times it takes each runner to finish the race. The runner with the shortest time wins. Call the average speed of runner A v_A; the distance the runner must travel is $\Delta x_A = 100$ meters. The time for runner A to finish the race is

$$v_A = \frac{\Delta x_A}{\Delta t_A}$$

$$\Delta t_A = \frac{\Delta x_A}{v_A}$$

Likewise, the time for runner B to finish the race is $\Delta t_B = \frac{\Delta x_B}{v_B}$.

Here, $\Delta x_B = 90$ meters. Take the ratio of these times and use the information that $v_A = 1.1 v_B$ to find

$$\frac{\Delta t_A}{\Delta t_B} = \frac{\Delta x_A}{v_A} \frac{v_B}{\Delta x_B}$$

$$= \frac{\Delta x_A}{\Delta x_B} \frac{v_B}{v_A}$$

$$= \frac{100 \text{ m}}{90 \text{ m}} \times \frac{1}{1.1}$$

$$= 1.01$$

Because this ratio is greater than 1, Δt_B must be less than Δt_A, so runner B finishes the race first.

41. **21 mph**

Let the x-axis represent the direction east and the y-axis represent the direction north. Your change in position in the x direction is $\Delta x = 30$ miles. Your change in position in the y direction is $\Delta y = 80$ miles. The total displacement is

$$s = \sqrt{\Delta x^2 + \Delta y^2}$$
$$= \sqrt{(30\,\text{mi})^2 + (80\,\text{mi})^2}$$
$$= 85\,\text{mi}$$

Divide this displacement by the time the trip takes to find the magnitude of the average velocity:

$$v = \frac{\Delta x}{\Delta t}$$
$$= \frac{85\,\text{mi}}{4\,\text{hr}}$$
$$= 21\,\text{mph}$$

42. **28 mph**

Let the x-axis represent the direction east and the y-axis represent the direction north. Your change in position in the x direction is $\Delta x = 30$ miles. Your change in position in the y direction is $\Delta y = 80$ miles. The total distance traveled is

$$d = \Delta x + \Delta y$$
$$= 30\,\text{mi} + 80\,\text{mi}$$
$$= 110\,\text{mi}$$

Divide this distance by the time the trip takes to get the average speed:

$$v = \frac{d}{\Delta t}$$
$$= \frac{110\,\text{mi}}{4\,\text{hr}}$$
$$= 28\,\text{mph}$$

43. **80 mph at 60° north of east**

The average velocity is the total displacement divided by the total time of the trip. Let the y-axis represent north and the x-axis represent east. The displacement in the x direction is $\Delta x = 20$ miles. The displacement in the y direction is $\Delta y = 35$ miles.

The total displacement is

$$s = \sqrt{\Delta x^2 + \Delta y^2}$$
$$= \sqrt{(20\,\text{mi})^2 + (35\,\text{mi})^2}$$
$$= 40\,\text{mi}$$

The total time is $\Delta t = 30$ minutes. To convert minutes to hours, use the conversion factor $1 = \dfrac{1\,\text{hr}}{60\,\text{min}}$, which gives

$$\Delta t = 30\,\text{min} \times \frac{1\,\text{hr}}{60\,\text{min}}$$

$$= 0.50\,\text{hr}$$

The magnitude of the average velocity is thus

$$v = \frac{s}{\Delta t}$$

$$= \frac{40\,\text{mi}}{0.50\,\text{hr}}$$

$$= 80\,\text{mph}$$

The direction of the average velocity is

$$\tan\theta = \frac{\Delta y}{\Delta x}$$

$$\theta = \tan^{-1}\left(\frac{\Delta y}{\Delta x}\right)$$

$$= \tan^{-1}\left(\frac{35\,\text{mi}}{20\,\text{mi}}\right)$$

$$= 60°$$

Thus, the average velocity is 80 miles per hour at 60 degrees from the x-axis, which is 60 degrees north of east.

---

44. 0.37 m/s

First use the average speed to determine the total distance traveled. Subtract the known distances from this to find the distance walked in the south direction. Use this to find the total displacement and the average velocity.

The total distance d traveled is

$$d = v\Delta t$$

$$= (1.0\,\text{m/s})(1.0\,\text{hr}) \times \frac{60\,\text{min}}{1\,\text{hr}} \times \frac{60\,\text{s}}{1\,\text{min}}$$

$$= 3600\,\text{m}$$

Subtracting the known distances gives the distance d_{south} walked in the south direction:

$$d_{\text{south}} = d - \left(10\,\text{block} \times \frac{100\,\text{m}}{1\,\text{block}}\right) - \left(3\,\text{block} \times \frac{100\,\text{m}}{1\,\text{block}}\right)$$

$$= 3{,}600\,\text{m} - 1{,}000\,\text{m} - 300\,\text{m}$$

$$= 2{,}300\,\text{m}$$

Now let the north direction be the positive y-axis and the south direction be the negative y-axis. The east direction is the positive x-axis. The displacement in the y direction is then the displacement north minus the displacement south:

$$\Delta y = 1{,}000\,\text{m} - 2{,}300\,\text{m}$$
$$= -1{,}300\,\text{m}$$

The displacement in the x direction is $\Delta x = 300\,\text{m}$. The total displacement is

$$s = \sqrt{\Delta x^2 + \Delta y^2}$$
$$= \sqrt{(300\,\text{m})^2 + (-1{,}300\,\text{m})^2}$$
$$= 1334\,\text{m}$$

Finally, the average velocity is

$$v = \frac{s}{\Delta t}$$
$$= \frac{1334\,\text{m}}{1\,\text{hr}} \times \frac{1\,\text{hr}}{60\,\text{min}} \times \frac{1\,\text{min}}{60\,\text{s}}$$
$$= 0.37\,\text{m/s}$$

45.   50 mph

Average speed is the total distance divided by the time it takes to travel that distance. The total distance is

$$d = 40\,\text{mi} + 30\,\text{mi} + 20\,\text{mi} + 10\,\text{mi}$$
$$= 100\,\text{mi}$$

The time for the trip is $\Delta t = 2\,\text{hr}$, so the average speed is

$$v = \frac{d}{\Delta t}$$
$$= \frac{100\,\text{mi}}{2\,\text{hr}}$$
$$= 50\,\text{mph}$$

46. $3.5\,\text{m/s}^2$

Acceleration is the change in velocity divided by the time it takes to make the change. The change in velocity is the final velocity ($v_f = 7.0$ meters per second) minus the initial velocity ($v_i = 0$ meters per second):

$$\Delta v = v_f - v_i$$
$$= 7.0\,\text{m/s} - 0\,\text{m/s}$$
$$= 7.0\,\text{m/s}$$

The time it takes to change velocity is $\Delta t = 2.0$ seconds. The acceleration is therefore

$$a = \frac{\Delta v}{\Delta t}$$
$$= \frac{7.0 \text{ m/s}}{2.0 \text{ s}}$$
$$= 3.5 \text{ m/s}^2$$

47. **4.9 m/s**

You're given the acceleration of the ball ($a = 9.8$ meters per second per second) and the time during which it accelerates ($\Delta t = 0.5$ seconds). The definition of acceleration is

$$a = \frac{\Delta v}{\Delta t}$$
$$= \frac{v_f - v_i}{\Delta t}$$

The initial velocity of the ball is 0 meters per second, so this equation reduces to $a = \frac{v_f}{\Delta t}$.
You can solve for v_f to get this result:

$$a = \frac{v_f}{\Delta t}$$
$$a\Delta t = v_f$$
$$9.8 \frac{\text{m}}{\text{s}^2} \times 0.5 \text{ s} =$$
$$4.9 \text{ m/s} =$$

48. **0.15 s**

You're given the acceleration of the child ($a = 2$ meters per second per second) and the initial and final velocities ($v_i = 0$ meters per second and $v_f = 0.3$ meters per second). Note that the acceleration and final velocity have the same sign because they're in the same direction.

The definition of acceleration is

$$a = \frac{\Delta v}{\Delta t}$$
$$= \frac{v_f - v_i}{\Delta t}$$

You can solve for Δt. The result is

$$a = \frac{v_f - v_i}{\Delta t}$$
$$\Delta t = \frac{v_f - v_i}{a}$$
$$= \frac{0.3 \text{ m/s} - 0 \text{ m/s}}{2 \text{ m/s}^2}$$
$$= 0.15 \text{ s}$$

49. **29 s**

You're given the final velocity of the plane in kilometers per hour, so begin by converting this to meters per second. Using the conversion factors

$$1 = \frac{1\,\text{hr}}{3600\,\text{s}} \text{ and } 1 = \frac{1{,}000\,\text{m}}{1\,\text{km}} \text{ gives:}$$

$$v_f = 300\,\frac{\text{km}}{\text{hr}} \times \frac{1\,\text{hr}}{3600\,\text{s}} \times \frac{1{,}000\,\text{m}}{1\,\text{km}}$$

$$= 83.33\,\text{m/s}$$

The initial velocity of the plane is 0 meters per second. The acceleration of the plane is $a = 2.9$ meters per second per second. Note that the acceleration has the same sign as the velocity because they're both in the same direction.

The definition of acceleration is

$$a = \frac{\Delta v}{\Delta t}$$

$$= \frac{v_f - v_i}{\Delta t}$$

You can solve for Δt to get this result:

$$a = \frac{v_f - v_i}{\Delta t}$$

$$\Delta t = \frac{v_f - v_i}{a}$$

$$= \frac{83.33\,\text{m/s} - 0\,\text{m/s}}{2.9\,\text{m/s}^2}$$

$$= 29\,\text{s}$$

50. **13 m/s in the opposite direction as the initial velocity**

The problem involves two opposite directions. Call the direction of the initial velocity positive and the direction of the final velocity negative. The initial velocity is then $v_i = 10$ meters per second, the acceleration is $a = -2.3$ meters per second per second, and the time during which you accelerate is $\Delta t = 10$ seconds. You can solve the equation for acceleration for the final velocity:

$$a = \frac{\Delta v}{\Delta t}$$

$$= \frac{v_f - v_i}{\Delta t}$$

$$a\Delta t = v_f - v_i$$

$$v_i + a\Delta t = v_f$$

$$10\,\text{m/s} + \left(-2.3\,\text{m/s}^2\right)\left(10\,\text{s}\right) =$$

$$-13\,\text{m/s} =$$

The minus sign means you're moving in the opposite direction of the initial velocity.

51. 4 m

The equation relating acceleration a, time t, and distance s is $s = v_i t + \frac{1}{2} a t^2$.

v_i is the initial speed, which is 0 meters per second in this problem because you start from rest. The acceleration is $a = 2$ meters per second per second, and the time during which you accelerate is $t = 2$ seconds. Plugging these values into the equation for distance gives

$$s = v_i t + \frac{1}{2} a t^2$$
$$= (0\text{ m/s})(2\text{ s}) + \frac{1}{2}\left(2\text{ m/s}^2\right)(2\text{ s})^2$$
$$= 4\text{ m}$$

52. 3.9 s

The initial velocity of the car is $v_i = 0$ meters per second. In this case, the equation relating distance, acceleration, and time reduces to

$$s = v_i t + \frac{1}{2} a t^2$$
$$= \frac{1}{2} a t^2$$

The car's acceleration is $a = 4.0$ meters per second per second, and the distance it travels during its acceleration is $s = 30$ meters. Solve the previous equation for time and insert these values:

$$s = \frac{1}{2} a t^2$$
$$\frac{2s}{a} = t^2$$
$$\pm\sqrt{\frac{2s}{a}} = t$$
$$t = \pm\sqrt{\frac{2 \times 30\text{ m}}{4.0x\text{ m/s}^2}}$$
$$= \pm 3.9\text{ s}$$

Because you want the length of time the car accelerates, you use the positive sign in the equation. Thus, the final answer is 3.9 seconds.

53. 4.0 m/s²

The initial velocity of the motorcycle is $v_i = 0$ meters per second. In this case, the equation relating distance, acceleration, and time reduces to

$$s = v_i t + \frac{1}{2} a t^2$$
$$= \frac{1}{2} a t^2$$

Solving this equation for the acceleration gives

$$s = \frac{1}{2}at^2$$

$$\frac{2s}{t^2} = a$$

Plug in the given values of $t = 10$ seconds and $s = 200$ meters to find

$$a = \frac{2s}{t^2}$$

$$= \frac{2 \times 200 \text{ m}}{(10 \text{ s})^2}$$

$$= 4.0 \text{ m/s}^2$$

---

54. 1 m

If the ball starts at a standstill, then its initial velocity is $v_i = 0$ meters per second. Its final speed is $v_f = 100$ miles per hour. Use the conversion factors $1 = \frac{1,609 \text{ m}}{1 \text{ mi}}$ and $1 = \frac{1 \text{ hr}}{3,600 \text{ s}}$ to convert from miles per hour to meters per second. The result is

$$v_f = 100 \frac{\text{mi}}{\text{hr}} \times \frac{1,609 \text{ m}}{\text{mi}} \times \frac{1 \text{ hr}}{3,600 \text{ s}}$$

$$= 44.7 \text{ m/s}$$

It takes $\Delta t = 0.05$ seconds to reach the final speed, so the ball's acceleration is

$$a = \frac{\Delta v}{t}$$

$$= \frac{v_f - v_i}{\Delta t}$$

$$= \frac{44.7 \text{ m/s} - 0 \text{ m/s}}{0.05 \text{ s}}$$

$$= 894 \text{ m/s}^2$$

Use this result and the same time period to find the distance traveled:

$$s = v_i t + \frac{1}{2}at^2$$

$$= (0 \text{ m/s})(0.05 \text{ s}) + \frac{1}{2}\left(894 \text{ m/s}^2\right)(0.05 \text{ s})^2$$

$$= 1 \text{ m}$$

---

55. 3 m/s² south

Your initial velocity is $v_i = 20$ meters per second, and your final velocity is $v_f = 10$ meters per second. During the acceleration, you cover a distance of $s = 50$ meters. Solve the equation relating initial and final velocities, acceleration, and distance for acceleration; then insert these values to find the answer. The result is

$$v_f^2 - v_i^2 = 2as$$

$$\frac{v_f^2 - v_i^2}{2s} = a$$

$$a = \frac{(10\,\text{m/s})^2 - (20\,\text{m/s})^2}{2 \times 50\,\text{m}}$$

$$= -3\,\text{m/s}^2$$

The minus sign indicates that the acceleration is in the direction opposite the initial velocity, so the final answer is $a = 3\,\text{m/s}^2$ south.

56. 9.4 m/s

Your initial velocity is $v_i = 3$ meters per second. You accelerate at $a = 2$ meters per second per second over a distance of $s = 20$ meters. Solve the equation relating initial and final velocities, acceleration, and distance for the final velocity; then insert these values to find the answer. The result is

$$v_f^2 - v_i^2 = 2as$$

$$v_f^2 = 2as + v_i^2$$

$$v_f = \pm\sqrt{2as + v_i^2}$$

$$= \pm\sqrt{2 \times 2.0\,\text{m/s}^2 \times 20\,\text{m} + (3.0\,\text{m/s})^2}$$

$$= \pm 9.4\,\text{m/s}$$

The plus and minus signs indicate that you don't know the direction of the final velocity, but that's okay because you just need the speed, which is 9.4 meters per second.

57. 400 m/s²

The initial velocity of the ball is $v_i = 0$ meters per second. The final velocity of the ball is $v_f = 90$ miles per hour. To convert this to meters per second, use the conversion factors $1 = \frac{1{,}609\,\text{m}}{1\,\text{mi}}$ and $1 = \frac{1\,\text{hr}}{3{,}600\,\text{s}}$. Thus, in meters per second, the final velocity of the ball is

$$v_f = 90\,\frac{\cancel{\text{mi}}}{\cancel{\text{hr}}} \times \frac{1{,}609\,\text{m}}{1\,\cancel{\text{mi}}} \times \frac{1\,\cancel{\text{hr}}}{3{,}600\,\text{s}}$$

$$= 40.2\,\text{m/s}$$

The distance over which the acceleration occurs is $s = 2.0$ meters. Solve the equation relating initial and final velocities, acceleration, and distance for acceleration; then insert these values to find the answer. The result is

$$v_f^2 - v_i^2 = 2as$$

$$\frac{v_f^2 - v_i^2}{2s} = a$$

$$a = \frac{(40.2\,\text{m/s})^2 - (0\,\text{m/s})^2}{2 \times 2.0\,\text{m}}$$

$$= 400\,\text{m/s}^2$$

58. 11 km

Your initial velocity is $v_i = 200$ meters per second, and your final velocity is $v_f = 500$ meters per second. Your acceleration is $a = 10$ meters per second per second. Solve the equation relating initial and final velocities, acceleration, and distance for the distance covered; then insert these values to find the answer. The result is

$$v_f^2 - v_i^2 = 2as$$

$$\frac{v_f^2 - v_i^2}{2a} = s$$

$$s = \frac{(500 \text{ m/s})^2 - (200 \text{ m/s})^2}{2 \times 10 \text{ m/s}^2}$$

$$= 10,500 \text{ m}$$

Use the conversion factor $1 = \dfrac{1 \text{ km}}{1,000 \text{ m}}$ to convert this answer to kilometers. The result is

$$s = 10,500 \text{ m} \times \frac{1 \text{ km}}{1,000 \text{ m}}$$

$$= 11 \text{ km}$$

59. 3.6 m/s²

The speed skater's initial velocity is $v_i = 0$ meters per second, covering a distance of $s = 12$ meters in a time of $t = 2.6$ seconds. Solve the equation relating distance to initial velocity, acceleration, and time for acceleration; then insert these values to find the answer:

$$s = v_i t + \frac{1}{2}at^2$$

$$s - v_i t = \frac{1}{2}at^2$$

$$2(s - v_i t) = at^2$$

$$\frac{2(s - v_i t)}{t^2} = a$$

$$a = \frac{2(12 \text{ m} - 0 \text{ m} \times 2.6 \text{ s})}{(2.6 \text{ s})^2}$$

$$= 3.6 \text{ m/s}^2$$

60. 3.4 s

Your initial velocity is $v_i = 0$ meters per second. You then accelerate at $a = 3.4$ meters per second per second over a distance of $s = 20$ meters. Because $v_i = 0$ meters per second, the equation relating distance to initial velocity, acceleration, and time reduces to

$$s = v_i t + \frac{1}{2}at^2$$

$$= \frac{1}{2}at^2$$

Solve this equation for the time t:

$$s = \frac{1}{2}at^2$$

$$2s = at^2$$

$$\frac{2s}{a} = t^2$$

$$\pm\sqrt{\frac{2s}{a}} = t$$

$$t = \pm\sqrt{\frac{2 \times 20 \text{ m}}{\left(3.4 \text{ m/s}^2\right)}}$$

$$= \pm 3.4 \text{ s}$$

The negative time isn't possible, so choose the positive time. The result is 3.4 seconds.

---

61. 50 m

The initial velocity is $v_i = 18$ meters per second. The acceleration must be in the direction opposite the velocity to make the car stop, so the sign of the acceleration must be negative. Thus, $a = -2.8$ meters per second per second. The time it takes you to stop is $t = 4.0$ seconds. Insert these values into the equation relating distance to initial velocity, acceleration, and time to find

$$s = v_i t + \frac{1}{2}at^2$$

$$= 18 \text{ m/s} \times 4.0 \text{ s} + \frac{1}{2}\left(-2.8 \text{ m/s}^2\right)(4.0 \text{ s})^2$$

$$= 50 \text{ m}$$

---

62. 14 m west

The boat's initial velocity is $v_i = 1.3$ meters per second to the east, which you can call the positive direction. The acceleration is to the west, so it's negative: $a = -0.20$ meters per second per second. The boat accelerates for $t = 20$ seconds. Plug these values into the equation relating distance to initial velocity, acceleration, and time to find the displacement:

$$s = v_i t + \frac{1}{2}at^2$$

$$= 1.3 \text{ m/s} \times 20 \text{ s} + \frac{1}{2}\left(-0.20 \text{ m/s}^2\right)(20 \text{ s})^2$$

$$= -14 \text{ m}$$

The negative sign means the displacement is to the west of the original position, so the answer is that the boat is displaced a distance of 14 meters west of its original position.

63. $18 \, \text{m/s}^2$

Solve the equation relating initial and final velocity to displacement and acceleration for the acceleration:

$$v_f^2 - v_i^2 = 2as$$

$$\frac{v_f^2 - v_i^2}{2s} = a$$

The initial velocity of the cheetah is $v_i = 0$ miles per hour. The final velocity of the cheetah is $v_f = 60$ miles per hour. Use the conversion factors $1 = \dfrac{1{,}609 \, \text{m}}{1 \, \text{mi}}$ and $1 = \dfrac{1 \, \text{hr}}{3{,}600 \, \text{s}}$ to convert the final speed to meters per second. This gives

$$s = 60 \, \frac{\text{mi}}{\text{hr}} \times \frac{1{,}609 \, \text{m}}{1 \, \text{mi}} \times \frac{1 \, \text{hr}}{3{,}600 \, \text{s}}$$

$$= 26.8 \, \text{m/s}$$

The displacement of the cheetah during its acceleration is $s = 20$ meters. Insert these values into the previous equation for acceleration to find

$$a = \frac{v_f^2 - v_i^2}{2s}$$

$$= \frac{(26.8 \, \text{m/s})^2 - (0 \, \text{m/s})^2}{2 \times 20 \, \text{m}}$$

$$= 18 \, \text{m/s}^2$$

64. $0.86 \, \text{m/s}^2$

Solve the equation relating initial and final velocity to displacement and acceleration for the acceleration:

$$v_f^2 - v_i^2 = 2as$$

$$\frac{v_f^2 - v_i^2}{2s} = a$$

The boat's initial velocity is $v_i = 3.0$ meters per second and your final velocity is

$$v_f = 3v_i$$

$$= 3 \times 3.0 \, \text{m/s}$$

$$= 9.0 \, \text{m/s}$$

The boat's displacement during the acceleration is $s = 42$ meters. Insert these values into the previous equation for acceleration to find

$$a = \frac{v_f^2 - v_i^2}{2s}$$

$$= \frac{(9.0 \, \text{m/s})^2 - (3.0 \, \text{m/s})^2}{2 \times 42 \, \text{m}}$$

$$= 0.86 \, \text{m/s}^2$$

65. 0.23 m/s²

Solve the equation relating initial and final velocity to displacement and acceleration for the acceleration:

$$v_f^2 - v_i^2 = 2as$$
$$\frac{v_f^2 - v_i^2}{2s} = a$$

The initial speed of the feather is $v_i = 0$ meters per second, and the final speed is $v_f = 0.30$ meters per second. The distance it moves during its acceleration is $s = 0.20$ meters. Insert these values into the previous equation for acceleration to find

$$a = \frac{v_f^2 - v_i^2}{2s}$$
$$= \frac{(0.30\ \text{m/s})^2 - (0\ \text{m/s})^2}{2 \times 0.20\ \text{m}}$$
$$= 0.23\ \text{m/s}^2$$

66. 0.0096 m/s² southward

Solve the equation relating initial and final velocity to displacement and acceleration for the acceleration:

$$v_f^2 - v_i^2 = 2as$$
$$\frac{v_f^2 - v_i^2}{2s} = a$$

Call north the positive direction and south the negative direction. The initial velocity is $v_i = 2.3$ meters per second, and the final velocity is $v_f = 1.2$ meters per second. The displacement during the acceleration is $s = 200$ meters. Insert these values into the equation for the acceleration:

$$a = \frac{v_f^2 - v_i^2}{2s}$$
$$= \frac{(1.2\ \text{m/s})^2 - (2.3\ \text{m/s})^2}{2(200\ \text{m})}$$
$$= -0.0096\ \text{m/s}^2$$

The minus sign means that the acceleration is southward, so the final answer is 0.0096 m/s² southward.

67. 8.5 m/s northward

Solve the equation relating initial and final velocity to displacement and acceleration for the final velocity:

$$v_f^2 - v_i^2 = 2as$$
$$v_f^2 = 2as + v_i^2$$
$$v_f = \pm\sqrt{2as + v_i^2}$$

The initial velocity is $v_i = 2.0$ meters per second, its acceleration is $a = 0.34$ meters per second per second, and the distance over which it accelerates is $s = 100$ meters. Insert these values into the previous equation for the final velocity to find

$$v_f = \pm\sqrt{2as + v_i^2}$$
$$= \pm\sqrt{2 \times 0.34 \text{ m/s}^2 \times 100 \text{ m} + (2.0 \text{ m/s})^2}$$
$$= \pm 8.5 \text{ m/s}$$

Because the acceleration is in the same direction as the initial velocity, the final velocity must be in the same direction (north). Therefore, you use the plus sign and your final answer is $v_f = 8.5$ m/s northward.

68. 24 m/s

Solve the equation relating initial and final velocity to displacement and acceleration for the initial velocity:

$$v_f^2 - v_i^2 = 2as$$
$$v_i^2 = v_f^2 - 2as$$
$$v_i = \pm\sqrt{v_f^2 - 2as}$$

The final velocity is $v_f = 0$ meters per second, so this equation reduces to $v_i = \pm\sqrt{-2as}$

The acceleration is $a = -0.10$ meters per second per second. The minus sign indicates that the train is slowing down, so the acceleration is in the direction opposite the initial velocity, and the two need to have opposite signs. The distance over which it accelerates is $s = 3,000$ meters. Insert these values into the previous equation for the final velocity to find

$$v_i = \pm\sqrt{-2as}$$
$$= \pm\sqrt{-2(-0.10 \text{ m/s}^2)(3,000 \text{ m})}$$
$$= \pm 24 \text{ m/s}$$

The initial speed is therefore 24 meters per second.

69.   12 m/s, 24 m/s

You know that $v_f = 2v_i$, so the equation relating initial and final velocity to displacement and acceleration becomes

$$v_f^2 - v_i^2 = 2as$$
$$(2v_i)^2 - v_i^2 =$$
$$4v_i^2 - v_i^2 =$$
$$3v_i^2 =$$
$$v_i = \pm\sqrt{\frac{2as}{3}}$$

The acceleration is $a = 4.5$ meters per second per second, and the distance over which it accelerates is $s = 50$ meters. Insert these values into the previous equation for the initial velocity to find

$$v_i = \pm\sqrt{\frac{2as}{3}}$$
$$= \pm\sqrt{\frac{2 \times 4.5 \text{ m/s}^2 \times 50 \text{ m}}{3}}$$
$$= \pm 12 \text{ m/s}$$

Using $v_f = 2v_i$, the final velocity is

$$v_f = 2v_i$$
$$= 2 \times (\pm 12 \text{ m/s})$$
$$= \pm 24 \text{ m/s}$$

The initial and final speeds are the magnitudes of the velocities, which are positive by definition. The initial and final speeds are therefore $v_i = 12 \text{ m/s}$ and $v_f = 24 \text{ m/s}$.

70.   21 m/s

Solve the equation relating initial and final velocity to displacement and acceleration for the final velocity:

$$v_f^2 - v_i^2 = 2as$$
$$v_f^2 = 2as + v_i^2$$
$$v_f = \pm\sqrt{2as + v_i^2}$$

The initial velocity is $v_i = 0$ meters per second, the acceleration is $a = 9.1$ meters per second per second, and the distance over which it accelerates is $s = 25$ meters. Insert these values into the previous equation for the final velocity to find

$$v_f = \pm\sqrt{2as + v_i^2}$$
$$= \pm\sqrt{2as}$$
$$= \pm\sqrt{2 \times 9.1\,\text{m/s}^2 \times 25\,\text{m}}$$
$$= \pm 21\,\text{m/s}$$

Because the falcon doesn't accelerate after the initial 25 meters, this speed is the speed at which it hits the pigeon. The final speed must be positive, so the answer is 21 meters per second.

71. 2

One number is required for each dimension of a vector. In two dimensions, you need a value for the x-direction and one for the y-direction.

72. 45 kilometers

A vector has two components: a magnitude and a direction. The *magnitude* is the vector's length. Marcus's trip is 45 kilometers long.

73. northwest

A *resultant vector* is the sum of two or more vectors. Add vectors by placing the tail of a vector at the tip of the preceding vector in the path. If you walk in a westerly direction, stop, and then walk in a northerly direction, you'll be northwest of your original position. Thus, the resultant vector — the "result" of following the vectors — points to the northwest.

74. 10 cm

A *resultant vector* is the sum of two or more vectors. Graphically, this means the tail of one vector touches the tip of another vector.

Illustration by Thomson Digital

Proceeding 3 centimeters to the right, then 5 more centimeters to the right, and then 2 more to the right results in a final position of 10 centimeters from the start $3 + 5 + 2 = 10$.

75. 0 m

Whether Jake walks 12 meters to the left and then 14 meters back or vice versa, He ends up 2 meters to the right of the initial starting point. This is an example of the commutative property of vector addition. The order of addition isn't important.

76. (5, 12)

To add vectors, sum their x-coordinates and then sum their y-coordinates:

$$\begin{aligned}\mathbf{A}+\mathbf{B} &= (2,4)+(3,8)\\ &= (2+3,\ 4+8)\\ &= (5,12)\end{aligned}$$

77. (3, −2)

A number that immediately precedes a vector is called a *scalar*. Multiply the scalar's value by each component of a vector to determine the new result:

$$\begin{aligned}\tfrac{1}{2}\mathbf{v} &= \tfrac{1}{2}(6,-4)\\ &= \left[\tfrac{1}{2}(6),\ \tfrac{1}{2}(-4)\right]\\ &= (3,-2)\end{aligned}$$

78. (9, 11)

To multiply a vector times a *scalar* (a number, such as the 3 before **A**), simply multiply that scalar by each of the vector's coordinates. To add vectors, sum their x-coordinates and then sum their y-coordinates.

$$\begin{aligned}\mathbf{3A}+\mathbf{5B} &= 3(-2,2)+5(3,1)\\ &= [3(-2),3(2)]+[5(3),5(1)]\\ &= (-6,6)+(15,5)\\ &= (-6+15,6+5)\\ &= (9,11)\end{aligned}$$

79. (5, −27)

Substitute the given vector values into the expression and solve:

$$\begin{aligned}\mathbf{2A}-\mathbf{3B} &= \mathbf{D}-\mathbf{3C}\\ 2(7,-3)-3(0,4) &= \mathbf{D}-3(-3,-3)\\ (14,-6)-(0,12) &= \mathbf{D}-(-9,-9)\\ (14,-18) &= \mathbf{D}-(-9,-9)\\ (14,-18)+(-9,-9) &= \mathbf{D}\\ (5,-27) &= \mathbf{D}\end{aligned}$$

80. 27.6 cm

To calculate A_y — the y-component of vector **A** — use the relationship $A_y = A\sin\theta$, where A is the magnitude of **A** and θ is the angle relative to the x-axis.

$$A_y = A\sin\theta$$
$$= (28\text{ cm})\sin 80°$$
$$= (28)(0.98)$$
$$= 27.6\text{ cm}$$

81. −5.1 m

Calculate a vector's vertical component by using the formula $C_y = C\sin\theta$, where C is the vector's magnitude and θ is the angle it makes with the positive x-axis. Because the given angle is below the x-axis, you need to use a negative sign:

$$C_y = C\sin\theta$$
$$= (8\text{ m})\sin\left(-40°\right)$$
$$= -5.1\text{ m}$$

82. 1,840 N·m

Jeffrey drags the box, meaning the box is moving in the horizontal direction. So you need to calculate the force's horizontal component to use the relationship between work and force.

Calculate a vector's horizontal component (in this case, the force) by using the trigonometric relationship $V_x = V\cos\theta$, where V is the vector's magnitude and θ is the angle the vector makes with the horizontal.

$$V_x = V\cos\theta$$
$$= (150\text{ N})\cos 35°$$
$$= 122.9\text{ N}$$

(N is the abbreviation for newtons.)

The work is equal to this force times the distance that Jeffrey pulled the box, so you have one final calculation:

$$W = Fd$$
$$= (122.9\text{ N})(15\text{ m})$$
$$= 1,843\text{ N·m}$$

Ignoring the units and rounding, you end up with 1,840.

83. $(-17, 163)$

A *resultant vector* is the sum of all the individual vectors. The easiest way to add vectors is when they're in component form.

First, convert each of the three vectors into its component form, using the formulas $F_x = F\cos\theta$ and $F_y = F\sin\theta$, where F_x and F_y are the individual components, F is the magnitude of the force, and θ is the angle the path makes with the positive x-axis.

$$
\begin{aligned}
F_{x1} &= F_1\cos\theta \\
&= 100\cos 20° \\
&= 94 \\
F_{y1} &= F_1\sin\theta \\
&= 100\sin 20° \\
&= 34.2 \\
F_{x2} &= F_2\cos\theta \\
&= 60\cos 80° \\
&= 10.4 \\
F_{y2} &= F_2\sin\theta \\
&= 60\sin 80° \\
&= 59.1 \\
F_{x3} &= F_3\cos\theta \\
&= 140\cos 150° \\
&= -121.2 \\
F_{y3} &= F_3\sin\theta \\
&= 140\sin 150° \\
&= 70
\end{aligned}
$$

Finally, add all the horizontal (x) components together, and then add all the vertical (y) components together:

$$
\begin{aligned}
F_{\text{resultant}} &= \left(F_{x1} + F_{x2} + F_{x3}, F_{y1} + F_{y2} + F_{y3}\right) \\
&= [94 + 10.4 + (-121.2), 34.2 + 59.1 + 70] \\
&= (-17, 163)
\end{aligned}
$$

84. $6.7; 26.6°$

You can use the Pythagorean theorem to calculate the magnitude (W) of a vector ($\mathbf{w}$) given in component form $\left(W_x, W_y\right)$.

$$
\begin{aligned}
W^2 &= W_x^2 + W_y^2 \\
W &= \sqrt{W_x^2 + W_y^2} \\
&= \sqrt{6^2 + 3^2} \\
&= \sqrt{45} = 6.7
\end{aligned}
$$

And you can use the following trigonometric relationship to calculate the angle that a vector makes with respect to the positive x-axis:

$$\tan \theta = \frac{W_y}{W_x}$$

$$\tan \theta = \frac{3}{6} = 0.5$$

$$\theta = \tan^{-1}(0.5) = 26.6°$$

---

85.   **45.0°**

To find an angle when you're given vectors in component notation, first combine the vectors into a single vector:

$$\begin{aligned} \mathbf{C} &= \mathbf{A} + \mathbf{W} \\ &= (3, -3) + (-2, 4) \\ &= [3 + (-2), -3 + 4] \\ &= (1, 1) \end{aligned}$$

Solve for the angle that $\mathbf{C}$ makes with the x-axis by using the formula:

$$\tan \theta = \frac{C_y}{C_x}$$

$$\tan \theta = \frac{1}{1} = 1$$

$$\theta = \tan^{-1}(1) = 45°$$

---

86.   **11; 146°**

Summarize your four vectors in component form:

$$\mathbf{n} = (0, 12)$$
$$\mathbf{e} = (11, 0)$$
$$\mathbf{s} = (0, -6)$$
$$\mathbf{w} = (-20, 0)$$

where $\mathbf{n}$ represents the northern path, $\mathbf{e}$ represents the eastern path, $\mathbf{s}$ represents the southern path, and $\mathbf{w}$ represents the western path.

Now sum the vectors to obtain the resultant vector in component form:

$$\begin{aligned} \mathbf{v}_{\text{resultant}} &= \mathbf{n} + \mathbf{e} + \mathbf{s} + \mathbf{w} \\ &= (0, 12) + (11, 0) + (0, -6) + (-20, 0) \\ &= [0 + 11 + 0 + (-20), 12 + 0 + (-6) + 0] \\ &= (11 - 20, 12 - 6) \\ &= (-9, 6) \end{aligned}$$

Use the Pythagorean theorem to calculate the magnitude of the resultant:

$$v^2 = v_x^2 + v_y^2$$
$$v = \sqrt{v_x^2 + v_y^2}$$
$$= \sqrt{(-9)^2 + 6^2}$$
$$= \sqrt{81 + 36}$$
$$= \sqrt{117} = 11$$

Finally, use the following trigonometric relationship to calculate the angle that the resultant vector makes with respect to an easterly direction:

$$\tan\theta = \frac{v_y}{v_x}$$
$$\tan\theta = \frac{6}{-9} = -0.67$$
$$\theta = \tan^{-1}(-0.67) = -34°$$

However, this isn't the value you want because this is a fourth quadrant result — the calculator only gives inverse-tangent values in the first and fourth quadrants. $\mathbf{V}_{\text{resultant}}$ terminates in the second quadrant, where the x-value is negative and the y-value is positive. To switch your result to the correct quadrant, add or subtract 180 degrees to obtain a number between −180 and 180.

$$-34° + 180° = 146°$$
$$-34° - 180° = -214°$$

−214° is too small, but 146° is within the acceptable range of values.

87. 256 km

Because Candace starts and finishes her journey at the same place after traveling along the three paths, her net displacement is 0 kilometers. In vector notation that means $\mathbf{s}_{\text{resultant}} = 0 = (0, 0)$

Write a formula showing that the resultant of the three vectors sum to 0, $\mathbf{s}_{\text{resultant}} = \mathbf{s}_1 + \mathbf{s}_2 + \mathbf{s}_3$, where $\mathbf{s}_1$ is the car ride, $\mathbf{s}_2$ is the flight from Seneca to Westsmith, and $\mathbf{s}_3$ is the flight from Westsmith to the airfield.

The question gives you the components of the car ride (100 kilometers west and 250 kilometers north), so $\mathbf{s}_1 = (-100 \text{ km}, 250 \text{ km})$. You don't know any values for $\mathbf{s}_2$. To obtain component values for $\mathbf{s}_3$, convert the magnitude s_3 (300 kilometers) and direction θ (15 degrees south of east), recalling that $s_{3x} = s_3 \cos\theta$, and $s_{3y} = s_3 \sin\theta$.

$$\mathbf{s}_3 = (s_{3x}, \mathbf{s}_{3y})$$
$$= (s_3 \cos\theta, s_3 \sin\theta)$$
$$= \left[300 \cos(-15°) \text{ km}, 300 \sin(-15°) \text{ km} \right]$$
$$= (289.8 \text{ km}, -77.6 \text{ km})$$

You now have all vectors in the correct form to solve for the component form of $\mathbf{s}_2$, the flight from Seneca to Westsmith:

$$\mathbf{s}_{resultant} = \mathbf{s}_1 + \mathbf{s}_2 + \mathbf{s}_3$$
$$(0,0) = (-100\text{ km}, 250\text{ km}) + \mathbf{s}_2 + (289.8\text{ km}, -77.6\text{ km})$$
$$(0,0) = \mathbf{s}_2 + [-100\text{ km} + 289.8\text{ km}, 250\text{ km} + (-77.6\text{ km})]$$
$$(0,0) = \mathbf{s}_2 + (189.8\text{ km}, 172.4\text{ km})$$
$$\mathbf{s}_2 = (0 - 189.8\text{ km}, 0 - 172.4\text{ km})$$
$$\mathbf{s}_2 = (-189.8\text{ km}, -172.4\text{ km})$$

To find the length — or magnitude — of the flight, use the Pythagorean theorem:

$$s_2^2 = s_{2x}^2 + s_{2y}^2$$
$$s_2 = \sqrt{s_{2x}^2 + s_{2y}^2}$$
$$= \sqrt{(-189.8\text{ km})^2 + (-172.4\text{ km})^2}$$
$$= \sqrt{36,024 + 29,722}\text{ km}$$
$$= \sqrt{65,746}\text{ km} = 256\text{ km}$$

88.   47 m/s

To solve for the northward component of the velocity vector, use the formula $v_y = v\sin\theta$, where v_y is the component of the velocity in the vertical/northerly/y-direction, v is the magnitude of the car's velocity, and θ is the angle the car's velocity makes with respect to the positive x-axis (east).

$$v_y = v\sin\theta$$
$$= (50\text{ m/s})\sin 70°$$
$$= 47\text{ m/s}$$

89.   20 m

Traditionally, the north-south axis is synonymous with the y-axis, so you need to calculate y-components to solve this question.

In the first part of the trip, the y-component is

$$s_y = s\sin\theta$$
$$= (25\text{ m})\sin 30°$$
$$= (25\text{ m})(0.5)$$
$$= 12.5\text{ m}$$

In the second part of the trip, the y-component is

$$s_y = s\sin\theta$$
$$= (15\text{ m})\sin 30°$$
$$= (15\text{ m})(0.5)$$
$$= 7.5\text{ m}$$

The question asks for distance, not displacement, so add the absolute values of the two y-components to get the solution:

$$|12.5 \text{ m}| + |7.5 \text{ m}| = 12.5 \text{ m} + 7.5 \text{ m} = 20 \text{ m}$$

90. 44 m

Because Jake can't take the direct route, he has to take the "component" route. Jake first has to walk north and then east (or vice versa) to reach the bin that's positioned north of east of his current location. You need to calculate each component and then add them to find the total number of meters Jake actually walks to reach his destination. Use the expressions $s_x = s \cos \theta$ and $s_y = s \sin \theta$ to find the x- and y-components of Jake's motion, respectively:

$$s_x = s \cos \theta$$
$$= (34 \text{ m}) \cos 70°$$
$$= (34 \text{ m})(0.34)$$
$$= 11.63 \text{ m}$$

$$s_y = s \sin \theta$$
$$= (34 \text{ m}) \sin 70°$$
$$= (34 \text{ m})(0.94)$$
$$= 31.95 \text{ m}$$

Add these two distances to find Jake's total walking distance: $11.63 \text{ m} + 31.95 \text{ m} = 43.58 \text{ m}$.

91. (80.0 m, 20.8 m)

Given their magnitudes and directions, the easiest way to add vectors is to convert them to component form using the formula:

$$s_x = s \cos \theta$$
$$s_y = s \sin \theta$$

Then you can add all the x-components together and all the y-components together. So start by computing $A_x, A_y, B_x, B_y, C_x,$ and C_y.

$$A_x = (45 \text{ m}) \cos 20°$$
$$= (45 \text{ m})(0.94)$$
$$= 42.28 \text{ m}$$
$$A_y = (45 \text{ m}) \sin 20°$$
$$= (45 \text{ m})(0.34)$$
$$= 15.39 \text{ m}$$
$$B_x = (18 \text{ m}) \cos 65°$$
$$= (18 \text{ m})(0.42)$$
$$= 7.61 \text{ m}$$
$$B_y = (18 \text{ m}) \sin 65°$$
$$= (18 \text{ m})(0.91)$$
$$= 16.31 \text{ m}$$

$$C_x = (32 \, \text{m}) \cos(-20°)$$
$$= (32 \, \text{m})(0.94)$$
$$= 30.07 \, \text{m}$$
$$C_y = (32 \, \text{m}) \sin(-20°)$$
$$= (32 \, \text{m})(-0.34)$$
$$= -10.94 \, \text{m}$$

Finally, calculate the x- and y-components of $\mathbf{D}$ (D_x and D_y, respectively) by summing the individual components:

$$D_x = A_x + B_x + C_x$$
$$= 42.28 \, \text{m} + 7.61 \, \text{m} + 30.07 \, \text{m}$$
$$= 79.96 \, \text{m}$$
$$D_y = A_y + B_y + C_y$$
$$= 15.39 \, \text{m} + 16.31 \, \text{m} + (-10.94 \, \text{m})$$
$$= 20.76 \, \text{m}$$

92.   42 m

Use the Pythagorean theorem to solve for the length of a resultant vector when given its component displacement.

$$s^2 = s_x^2 + s_y^2$$
$$s = \sqrt{s_x^2 + s_y^2}$$
$$= \sqrt{(50 \, \text{m})^2 + (150 \, \text{m})^2}$$
$$= \sqrt{2{,}500 + 22{,}500} \, \text{m} = \sqrt{25{,}000} \, \text{m}$$
$$= 158.1 \, \text{m}$$

Walking west and then south requires a person to walk 200 meters. The direct route is $200 - 158.1 = 41.9$ meters shorter.

93.   −45°

The easiest way to add vectors is to write them in component form. First, for the easterly path:

$$\mathbf{s}_{\text{west}} = (s_x, s_y)$$
$$= (s \cos \theta, s \sin \theta)$$
$$= (5 \cos 0° \, \text{m}, 5 \sin 0° \, \text{m})$$
$$= (5 \, \text{m}, 0 \, \text{m})$$

Likewise, for the southerly path:

$$\mathbf{s}_{\text{north}} = (s_x, s_y)$$
$$= (s \cos \theta, s \sin \theta)$$
$$= \left[5 \cos(-90°) \, \text{m}, 5 \sin(-90°) \, \text{m} \right]$$
$$= (0 \, \text{m}, -5 \, \text{m})$$

Now, add the two vectors to determine the component values of the resultant vector:

$$\mathbf{s}_{resultant} = \mathbf{s}_{west} + \mathbf{s}_{north}$$
$$= (5\,\text{m}, 0\,\text{m}) + (0\,\text{m}, -5\,\text{m})$$
$$= [5\,\text{m} + 0\,\text{m}, 0\,\text{m} + (-5)\,\text{m}]$$
$$= (5\,\text{m}, -5\,\text{m})$$

Solve for the angle relative to east by using the formula:

$$\tan\theta = \frac{s_y}{s_x}$$
$$\tan\theta = \frac{-5}{5} = -1$$
$$\theta = \tan^{-1}(-1) = -45°$$

94. 2.1 km

To get the magnitude of the resultant vector's displacement, you need the individual vectors to be *displacement* vectors — currently, you have *velocity* vectors. Recall that $s = vt$, where s is displacement, v is velocity, and t is time. If the river were completely still, the time required to swim 500 meters at 2 meters per second would be

$$s = vt$$
$$500\,\text{m} = (2\,\text{m/s})\,t$$
$$250\,\text{s} = t$$

Therefore, the swimmer will be in the water for 250 seconds. You can now calculate how far downriver the current will move the swimmer by the time the swimmer reaches the opposite shore:

$$s = vt$$
$$= (8\,\text{m/s})(250\,\text{s})$$
$$= 2,000\,\text{m}$$

The component form of the swimmer's displacement is therefore $(500\,\text{m}, -2,000\,\text{m})$.

Use the Pythagorean theorem to solve for the length of the resultant:

$$s^2 = s_x^2 + s_y^2$$
$$s = \sqrt{s_x^2 + s_y^2}$$
$$= \sqrt{(500\,\text{m})^2 + (2,000\,\text{m})^2}$$
$$= \sqrt{250,000 + 4,000,000} = \sqrt{4,250,000}$$
$$= 2,062\,\text{m}$$

Finally, convert this figure into the requested units:

$$(2,062\,\text{m})\left(\frac{1\,\text{km}}{1,000\,\text{m}}\right) = 2.062\,\text{km}$$

95. yes; 8 m/s

All the values requested for the solution are in terms of the east-west direction, or x-axis. Start this problem by turning the velocity and acceleration vectors into their component forms:

$$\mathbf{v} = \left(v_x, v_y\right)$$
$$= (v\cos\theta, v\sin\theta)$$
$$= \left(10\cos 15° \text{ m/s}, 10\sin 15° \text{ m/s}\right)$$
$$= (9.65 \text{ m/s}, 2.59 \text{ m/s})$$
$$\mathbf{a} = \left(a_x, a_y\right)$$
$$= (a\cos\theta, a\sin\theta)$$
$$= \left(-3\cos 15° \text{ m/s}^2, -3\sin 15° \text{ m/s}^2\right)$$
$$= \left(-2.9 \text{ m/s}^2, -0.78 \text{ m/s}^2\right)$$

To figure out whether the ball makes it out of the gravel patch, you need to find out how far it will travel before its final velocity is 0 meters per second. Use the velocity-displacement formula, $\mathbf{v}_f^2 = \mathbf{v}_i^2 + 2\mathbf{as}$, where $\mathbf{v}_i$ is the initial velocity, $\mathbf{a}$ is the acceleration, $\mathbf{s}$ is the displacement, and $\mathbf{v}_f$ is the final velocity — which, in this case, you want to equal 0 meters per second, 0 meters per second.

$$\mathbf{v}_f^2 = \mathbf{v}_i^2 + 2\mathbf{as}$$
$$(0,0)^2 = (9.65 \text{ m/s}, 2.59 \text{ m/s})^2 + 2\left(-2.9 \text{ m/s}^2, -0.78 \text{ m/s}^2\right)\left(s_x, s_y\right)$$
$$(0,0) = (93.1 \text{ m}, 6.71 \text{ m}) + \left(-5.8s_x, -1.56s_y\right)$$
$$(0,0) = \left(93.1 \text{ m} - 5.8s_x, 6.71 \text{ m} - 1.56s_y\right)$$

You're only concerned with the horizontal (east-west) direction, so solve the x-component set of values:

$$0 = 93.1 \text{ m} - 5.8s_x$$
$$-93.1 \text{ m} = -5.8s_x$$
$$16.1 \text{ m} = s_x$$

So, even if the gravel were 16 meters wide, the basketball would have still kept rolling. It definitely travels the necessary 5 meters to escape the patch. Now, all that remains for you to do is to calculate the x-component of the velocity after 5 meters (when the basketball will be just breaching the eastern edge of the gravel patch). Use the velocity-displacement formula one more time, this time solving for the final horizontal velocity:

$$v_{fx}^2 = v_{ix}^2 + 2a_x s_x$$
$$v_{fx} = \sqrt{v_{ix}^2 + 2a_x s_x}$$
$$= \sqrt{(9.65 \text{ m/s})^2 + 2\left(-2.9 \text{ m/s}^2\right)(5 \text{ m})}$$
$$= \sqrt{93.1 - 29} \text{ m/s} = \sqrt{64.1} \text{ m/s}$$
$$= 8 \text{ m/s}$$

96.   147 km

The easiest way to add vectors is to write them in component form. For the first leg, be sure to accurately describe the angle relative to the positive x-axis, or, in compass terms, relative to due east:

$$\mathbf{s}_1 = \left(s_x, s_y \right)$$
$$= \left(s \cos \theta, s \sin \theta \right)$$
$$= \left[180 \cos 110^\circ \text{ km}, 180 \sin 110^\circ \text{ km} \right]$$
$$= \left(-61.6 \text{ km}, 169.1 \text{ km} \right)$$

Likewise, compute the values for the second and third legs:

$$\mathbf{s}_2 = \left(s_x, s_y \right)$$
$$= \left(s \cos \theta, s \sin \theta \right)$$
$$= \left[45 \cos \left(-90^\circ \right) \text{ km}, 45 \sin \left(-90^\circ \right) \text{ km} \right]$$
$$= \left(0 \text{ km}, -45 \text{ km} \right)$$
$$\mathbf{s}_3 = \left(s_x, s_y \right)$$
$$= \left(s \cos \theta, s \sin \theta \right)$$
$$= \left[18 \cos 180^\circ \text{ km}, \ 18 \sin 180^\circ \text{ km} \right]$$
$$= \left(-18 \text{ km}, 0 \text{ km} \right)$$

Now, add the three vectors to determine the component values of the resultant vector:

$$\mathbf{s}_{\text{resultant}} = \mathbf{s}_1 + \mathbf{s}_2 + \mathbf{s}_3$$
$$= \left(-61.6 \text{ km}, 169.1 \text{ km} \right) + \left(0 \text{ km}, -45 \text{ km} \right) + \left(-18 \text{ km}, 0 \text{ km} \right)$$
$$= \left[-61.6 \text{ km} + 0 \text{ km} + \left(-18 \text{ km} \right), 169.1 \text{ km} + \left(-45 \text{ km} \right) + 0 \text{ km} \right]$$
$$= \left(-79.6 \text{ km}, 124.1 \text{ km} \right)$$

Finally, use the Pythagorean theorem to solve for the length of a resultant vector when given its component displacement:

$$s^2 = s_x^2 + s_y^2$$
$$s = \sqrt{s_x^2 + s_y^2}$$
$$= \sqrt{\left(-79.6 \text{ km} \right)^2 + \left(124.1 \text{ km} \right)^2}$$
$$= \sqrt{6,336.2 + 15,400.8} \text{ km} = \sqrt{21,737} \text{ km}$$
$$= 147 \text{ km}$$

97.   0.6 s

The acceleration due to gravity is 9.8 m/s^2 downward. Summarize the given information about the displacement, velocity, and acceleration vectors:

$$\mathbf{s} = \left(s_x, -2 \text{ m} \right)$$
$$\mathbf{v}_i = \left(v_x, 0 \text{ m/s} \right)$$
$$\mathbf{a} = \left(0 \text{ m/s}^2, -9.8 \text{ m/s}^2 \right)$$

To solve for the time the marble spends in the air, use the vertical form of the displacement equation $\mathbf{s} = \mathbf{v}_i t + \frac{1}{2} \mathbf{a} t^2$:

$$s_y = v_y t + \frac{1}{2} a_y t^2$$
$$-2\,\text{m} = (0\,\text{m/s})t + \frac{1}{2}\left(-9.8\,\text{m/s}^2\right)t^2$$
$$-2 = -4.9 t^2$$
$$0.41 = t^2$$
$$0.64\,\text{s} = t$$

98.   1.4 m

First, summarize the information given in the question:

$$\mathbf{s} = \left(s_x, s_y\right) = \left(s_x, -22\,\text{m}\right)$$
$$\mathbf{v}_i = \left(v_x, v_y\right) = (0.65\,\text{m/s}, 0\,\text{m/s})$$
$$\mathbf{a} = \left(a_x, a_y\right) = \left(0\,\text{m/s}^2, -9.8\,\text{m/s}^2\right)$$

where $\mathbf{s}$, $\mathbf{v}$, and $\mathbf{a}$ are the position, velocity, and acceleration vectors, respectively, and the x- and y-subscripts indicate the horizontal and vertical components of those vectors.

Next, substitute the y-values of the vectors into the vertical–component displacement formula $s_y = v_y t + \frac{1}{2} a_y t^2$:

$$-22\,\text{m} = (0\,\text{m/s})t + \frac{1}{2}\left(-9.8\,\text{m/s}^2\right)t^2$$
$$-22 = -4.9 t^2$$
$$2.12 = t$$

Finally, solve for the horizontal displacement:

$$s_x = v_x t + \frac{1}{2} a_x t^2$$
$$= (0.65\,\text{m/s})(2.12\,\text{s}) + \frac{1}{2}\left(0\,\text{m/s}^2\right)(2.12\,\text{s})^2$$
$$= 1.4\,\text{m}$$

99.   47 m/s

First, convert the units of the velocity so that they match those of the vertical displacement (or vice versa):

$$\left(\frac{132\,\text{km}}{\text{h}}\right)\left(\frac{1{,}000\,\text{m}}{1\,\text{km}}\right)\left(\frac{1\,\text{h}}{60\,\text{min}}\right)\left(\frac{1\,\text{min}}{60\,\text{s}}\right) = \frac{132{,}000\,\text{m}}{3{,}600\,\text{s}} = 36.7\,\text{m/s}$$

Then, summarize the given information about the displacement, velocity, and acceleration vectors:

$$\mathbf{s} = \left(s_x, -45\,\text{m} \right)$$
$$\mathbf{v}_i = (36.7\,\text{m/s}, 0\,\text{m/s})$$
$$\mathbf{a} = \left(0\,\text{m/s}^2, -9.8\,\text{m/s}^2 \right)$$

The velocity's horizontal component doesn't change during flight, but the vertical component does because there's a force in that direction (gravity) accelerating (changing the velocity of) the car. To find the final vertical velocity without knowing the time the car spends in the air, use the velocity-displacement equation, $\mathbf{v}_f^2 = \mathbf{v}_i^2 + 2\mathbf{as}$, where $\mathbf{v}_f$ is the final vertical velocity at the end of the displacement $\mathbf{s}$, and $\mathbf{v}_i$ is the initial vertical velocity at launch. Writing out the vertical form of the velocity-displacement equation:

$$v_{fy}^2 = v_{iy}^2 + 2a_y s_y$$
$$= (0\,\text{m/s})^2 + 2\left(-9.8\,\text{m/s}^2 \right)(-45\,\text{m})$$
$$= 0 + 882\,\text{m/s}$$
$$v_{fy} = \pm\sqrt{882}\,\text{m/s}$$
$$= \pm 29.7\,\text{m/s}$$

Because the car is falling (moving in the conventional negative direction), use –29.7 m/s as the solution.

You now have the component form of the final velocity vector: $\mathbf{v} = \left(v_x, v_y \right)$ (36.7 m/s, – 29.7 m/s).

To find the magnitude of a vector given its components, use the Pythagorean theorem:

$$v^2 = v_x^2 + v_y^2$$
$$v = \sqrt{v_x^2 + v_y^2}$$
$$= \sqrt{(36.7\,\text{m/s})^2 + (-29.7\,\text{m/s})^2}$$
$$= \sqrt{1{,}346.9 + 882.1} = \sqrt{2{,}229}$$
$$= 47.2\,\text{m/s}$$

100.   5 m/s

When you want to calculate the horizontal component of an object's known velocity, use the equation $v_x = v\cos\theta$, where v is the initial speed and θ is the angle of the object relative to the ground.

$$v_x = v\cos\theta$$
$$= (10\,\text{m/s})\cos 60°$$
$$= 5.0\,\text{m/s}$$

101. **0 m/s**

A projectile's vertical component is always 0 meters per second at its zenith because it's transitioning from traveling upward (positive velocity) to traveling downward (negative velocity).

102. **41 m**

First, summarize the information given in the question, including the fact that the cannonball will land at the same height it started (0 meters):

$$\mathbf{s} = (s_x, s_y) = (s_x, 0\,\text{m})$$
$$\mathbf{v} = (v_x, v_y) = (v\cos\theta, v\sin\theta)$$
$$\mathbf{a} = (a_x, a_y) = (0\,\text{m/s}^2, -9.8\,\text{m/s}^2)$$

s, **v**, and **a** are the position, velocity, and acceleration vectors, respectively, and the *x*- and *y*-subscripts indicate the horizontal and vertical components of those vectors.

The horizontal (v_x) and vertical (v_y) components of the initial velocity are:

$$
\begin{aligned}
v_x &= v\cos\theta \\
&= (25\,\text{m/s})\cos 20° \\
&= 23.5\,\text{m/s}
\end{aligned}
$$

and

$$
\begin{aligned}
v_y &= v\sin\theta \\
&= (25\,\text{m/s})\sin 20° \\
&= 8.55\,\text{m/s}
\end{aligned}
$$

Second, substitute the six vector values you've compiled (two each for displacement, velocity, and acceleration) into the displacement formula, $\mathbf{s} = \mathbf{v}_i t + \frac{1}{2}\mathbf{a}t^2$.

In the horizontal direction,

$$
\begin{aligned}
s_x &= v_x t + \frac{1}{2}a_x t^2 \\
&= (23.5\,\text{m/s})t + \frac{1}{2}(0\,\text{m/s}^2)t^2 \\
&= 23.5t
\end{aligned}
$$

And in the vertical direction,

$$
\begin{aligned}
s_y &= v_y t + \frac{1}{2}a_y t^2 \\
0 &= (8.55\,\text{m/s})\,t + \frac{1}{2}(-9.8\,\text{m/s}^2)t^2 \\
0 &= 8.55t - 4.9t^2
\end{aligned}
$$

Solving the vertical direction equation, $t = 0$ or 1.75 s. This means that after 0 seconds (when the cannonball was fired) and after 1.75 seconds, the cannonball is at its starting height. Therefore, it hits the ground after 1.75 seconds. You can now use this value in the result you obtained from the horizontal direction equation:

$$s_x = (23.5 \text{ m/s})(1.75 \text{ s})$$
$$= 41 \text{ m}$$

103.   234 m

First, summarize the information given in the question, including the time before impact, $t = 8.2$ seconds:

$$\mathbf{s} = (s_x, s_y) = (s_x, -100 \text{ m})$$
$$\mathbf{v} = (v_x, v_y) = (40 \cos \theta \text{ m/s}, 40 \sin \theta \text{ m/s})$$
$$\mathbf{a} = (a_x, a_y) = (0 \text{ m/s}^2, -9.8 \text{ m/s}^2)$$

s, **v**, and **a** are the position, initial velocity, and acceleration vectors, respectively, and the x- and y-subscripts indicate the horizontal and vertical components of those vectors.

Substitute these values into the displacement equation, $\mathbf{s} = \mathbf{v}_i t + \frac{1}{2}\mathbf{a}t^2$, first for the vertical components:

$$s_y = v_y t + \frac{1}{2}a_y t^2$$
$$-100 \text{ m} = (40 \sin \theta \text{ m/s})(8.2 \text{ s}) + \frac{1}{2}(-9.8 \text{ m/s}^2)(8.2 \text{ s})^2$$
$$-100 = 328 \sin \theta - 329.5$$
$$0.70 = \sin \theta$$
$$44.4° = \theta$$

And then for the horizontal components:

$$s_x = v_x t + \frac{1}{2}a_x t^2$$
$$= [40 \cos 44.4° \text{ m/s}](8.2 \text{ s}) + \frac{1}{2}(0 \text{ m/s}^2)(8.2 \text{ s})^2$$
$$= 234 \text{ m}$$

104. 63°

Whenever a projectile launches and lands at the same height, the zenith (highest point) of its path — where the vertical velocity is 0 meters per second — occurs halfway through the trip. Using this information, summarize the information given in the question:

$$\mathbf{s} = (s_x, s_y) = (150\,\text{m}, s_y)$$
$$\mathbf{v}_i = (v_{ix}, v_{iy}) = (v_i \cos\theta, v_i \sin\theta)$$
$$\mathbf{v}_f = (v_{fx}, v_{fy}) = (v_f \cos\theta, 0\,\text{m/s})$$
$$\mathbf{a} = (a_x, a_y) = (0\,\text{m/s}^2, -9.8\,\text{m/s}^2)$$

$\mathbf{s}$, $\mathbf{v}$, and $\mathbf{a}$ are the position, velocity, and acceleration vectors, respectively, and the x- and y-subscripts indicate the horizontal and vertical components of those vectors. $\mathbf{v}_i$ represents the projectile's initial ($t = 0$) velocity, and $\mathbf{v}_f$ represents the velocity at the zenith — when $t = 5.5$ seconds.

This problem has two unknowns: the speed v and the angle θ at which the cannonball was fired. So you need two equations to solve the problem.

Use the vertical-component form of the velocity formula $\mathbf{v}_f = \mathbf{v}_i + \mathbf{a}t$ and the horizontal-component form of the displacement formula $\mathbf{s} = \mathbf{v}_i t + \frac{1}{2}\mathbf{a}t^2$ (the only equation that's useful in a direction where $a = 0$ m/s^2):

$$v_{fy} = v_{iy} + a_y t$$
$$0\,\text{m/s} = v\sin\theta + (-9.8\,\text{m/s}^2)(5.5\,\text{s})$$
$$0 = v\sin\theta - 53.9\,\text{m/s}$$
$$53.9\,\text{m/s} = v\sin\theta$$

$$s_x = v_x t + \frac{1}{2}a_x t^2$$
$$150\,\text{m} = (v\cos\theta)(5.5\,\text{s}) + \frac{1}{2}(0\,\text{m/s}^2)(5.5\,\text{s})^2$$
$$150\,\text{m} = 5.5\,v\cos\theta$$
$$27.3\,\text{m/s} = v\cos\theta$$

To eliminate one of the two unknown variables, divide the vertical result by the horizontal one:

$$\frac{53.9 = v\sin\theta}{27.3 = v\cos\theta} \Rightarrow$$

$$\frac{53.9}{27.3} = \frac{v\sin\theta}{v\cos\theta}$$
$$1.98 = \frac{\sin\theta}{\cos\theta} = \tan\theta$$
$$63° = \theta$$

105.   yes; 150 cm

First, summarize the information given in the question:

$$\mathbf{s} = \left(s_x, s_y\right) = \left(120\,\text{m}, s_y\right)$$
$$\mathbf{v} = \left(v_x, v_y\right) = \left(v\cos\theta, v\sin\theta\right)$$
$$\mathbf{a} = \left(a_x, a_y\right) = \left(0\,\text{m/s}^2, -9.8\,\text{m/s}^2\right)$$

$\mathbf{s}$, $\mathbf{v}$, and $\mathbf{a}$ are the position, velocity, and acceleration vectors, respectively, and the x- and y-subscripts indicate the horizontal and vertical components of those vectors.

Then, calculate the horizontal and vertical components of the velocity:

$$v_x = (35\,\text{m/s})\cos 40° = 26.8\,\text{m/s}$$
$$v_y = (35\,\text{m/s})\sin 40° = 22.5\,\text{m/s}$$

If the vertical displacement 120 meters away (from where the baseball was struck by the bat) is greater than 1 meter — the difference between the starting height and the height of the fence — the result will be a home run.

Use the formula for calculating displacement, $\mathbf{s} = \mathbf{v}_i t + \frac{1}{2}\mathbf{a}t^2$, to write out each component's equation. First, horizontally:

$$s_x = v_x t + \frac{1}{2}a_x t^2$$
$$120\,\text{m} = (26.8\,\text{m/s})t + \frac{1}{2}\left(0\,\text{m/s}^2\right)t^2$$
$$120 = 26.8t$$
$$4.48\,\text{s} = t$$

And then, vertically:

$$s_y = v_y t + \frac{1}{2}a_y t^2$$
$$= (22.5\,\text{m/s})t + \frac{1}{2}\left(-9.8\,\text{m/s}^2\right)t^2$$
$$= 22.5t\,\text{m/s} - 4.9t^2\,\text{m/s}^2$$

Then, substitute the time required for the baseball to reach the fence that you calculated from the horizontal equation:

$$s_y = (22.5\,\text{m/s})(4.48\,\text{s}) - \left(4.9\,\text{m/s}^2\right)(4.48\,\text{s})^2$$
$$= 2.46\,\text{m}$$

This value is 1.46 meters above the displacement required for a home run. Convert this into centimeters:

$$1.46\,\text{m}\left(\frac{100\,\text{cm}}{1\,\text{m}}\right) = 146\,\text{cm}$$

and then round to the nearest 10 centimeters (150 centimeters).

106. kilogram (kg)

In physics, kilograms are the standard units of mass (inertia). (SI is the abbreviation for *Le Syst ème Internationale d'Unités*, the International System of Units.)

107. inertia

Although inertia is a property of mass, mass is the measurement by which an object's inertia is quantified.

108. 40 m/s², east

Use Newton's second law, $F_{net} = ma_{net}$, with the given force (F_{net}) and mass (m) values to solve for the particle's acceleration.

$$F_{net} = ma_{net}$$
$$10\,N = (0.25\,kg)a_{net}$$
$$40\,m/s^2 = a_{net}$$

109. 13.5 m/s

Although you would expect a block sliding on a floor to slow down in the real world, the question stipulates that there are no forces — including friction — acting along the block's path of motion. Therefore, because there is no net force on the block, its speed does not change. (If $F_{net} = 0$, then $a_{net} = 0$, and that means that $\Delta v = 0$.)

110. D

The force of gravity between Earth and an object always points straight down in a force diagram.

111. A and C

Regardless of direction, tension forces always connect one object — upon which the force is exerted — to a second object that exerts the force. Anytime you see a taut rope in a physics question, tension forces exist along the rope. (In the given diagram, the direction of **A** indicates that the rope happens to be pulling up a second, unseen object with the help of the pulley.)

112. A and C

If the mass doesn't accelerate vertically, then C and D must be equal; likewise, if the mass doesn't accelerate horizontally, then B and E must be equal. However, neither of those situations is stipulated by any information you've been given. The only thing you know for sure is that the magnitudes of the two tension forces — A and C — must be equal, because they are on the same rope. They point in opposite directions to satisfy Newton's third law of equal-and-opposite forces (or else the rope would rip itself apart).

113. $D > C$

For an object to accelerate in a certain direction, there must be a net force on the object *in that direction*. So, for the mass to accelerate downward, the net force must also point down. Only **C** and **D** point in the up–down direction, so **B** and **E** can both be ignored — and **D** must be larger than **C** for the downward force to "win."

114. A

The normal force points perpendicularly away from a surface, starting from the point of contact between the object in question and the surface itself.

115. $A = D$

Completely supported by the table, the block never accelerates downward (or upward), so the net force on the block in the vertical direction must equal 0:

$$F_{y,\,net} = \sum F_y$$
$$0 = A + (-D)$$
$$0 = A - D$$
$$D = A$$

Note that there is a negative sign on the magnitude of **D** because that vector is pointing down (the negative direction, conventionally). The magnitudes of **A** and **D** are equal, but their directions are opposite.

116. $K = B - C$

Because the box moves at a constant velocity (after the initial push), it doesn't experience net acceleration in either the horizontal or vertical directions. Therefore, the box "feels" no net force in any direction. You're interested in K, so focus on the forces that are in **K**'s direction: **B** and **C**. You take the right side to be the positive direction, as is the convention:

$$F_{x,\,net} = \sum F_x$$
$$0 = K + C + (-B)$$
$$0 = K + C - B$$
$$-K = C - B$$
$$K = -C + B = B - C$$

117. 4 N

If the two forces involved were 8 newtons westerly and 8 newtons easterly, they would exactly cancel. However, an extra 4 newtons are pulling west in this situation, making that the net force. (Direction is not required in the answer because the question only asks for the magnitude of the force vector.)

Mathematically, if east is the positive direction, you can calculate the net force like this:

$$F_{net} = \sum F$$
$$= (-12\,N) + (8\,N)$$
$$= -4\,N$$

The net force is 4 newtons in the negative (westerly) direction.

118.   70 N

When adding vectors, first convert them into component form. The question only asks for the net force in the east/west direction, so you only have half of the usual computations to deal with. As with all components, the value of the x (east-west) component is the vector's magnitude multiplied by the cosine of the angle relative to the positive x (eastern) axis.

For the 58-newton force, the angle of 12 degrees north of east is a *positive* 12 degrees relative to the positive x-axis, because it's *above* the axis:

$$F_{x1} = F_1 \cos \theta_1$$
$$= (58\,N) \cos \left(12°\right)$$
$$= (58\,N)(0.978)$$
$$= 56.7\,N$$

For the 30-newton force, the angle is 64 degrees south of east, which is *negative* 64 degrees relative to the positive x-axis because it's *below* the axis.

$$F_{x2} = F_2 \cos \theta_2$$
$$= (30\,N) \cos \left(-64°\right)$$
$$= (30\,N)(0.438)$$
$$= 13.1\,N$$

Now, add up the two x components:

$$F_x = F_{x1} + F_{x2}$$
$$= (56.7\,N) + (13.1\,N)$$
$$= 69.8\,N \approx 70\,N$$

119.   322 N, 29° north of east

When adding vectors, convert them into component form first. As with all components, the value of the x (east-west) component is the vector's magnitude multiplied by the cosine of the angle relative to the positive x (eastern) axis. Similarly, the y (north-south) component is the vector's magnitude multiplied by the sine of the angle.

For the 340-newton force, the angle of 25 degrees south of east is *negative* 25 degrees relative to the positive x-axis because it's *below* the axis. Therefore, the x and y components are

$$F_{x1} = F_1 \cos\theta_1$$
$$= (340\,\text{N})\cos\left(-25°\right)$$
$$= (340\,\text{N})(0.906)$$
$$= 308\,\text{N}$$

and

$$F_{y1} = F_1 \sin\theta_1$$
$$= (340\,\text{N})\sin\left(-25°\right)$$
$$= (340\,\text{N})(-0.423)$$
$$= -144\,\text{N}$$

For the 300-newton force, the angle is 85 degrees north of west, or 95 degrees "away from" the positive x-axis. It's *positive* because it's *above* the axis.

$$F_{x2} = F_2 \cos\theta_2$$
$$= (300\,\text{N})\cos\left(95°\right)$$
$$= (300\,\text{N})(-0.08716)$$
$$= -26.1\,\text{N}$$

and

$$F_{y2} = F_2 \sin\theta_2$$
$$= (300\,\text{N})\sin\left(95°\right)$$
$$= (300\,\text{N})(0.996)$$
$$= 299\,\text{N}$$

Separately add the x components and the y components to find the components of the resultant vector:

$$F_x = F_{x1} + F_{x2}$$
$$= (308\,\text{N}) + (-26.1\,\text{N})$$
$$= 281.9\,\text{N}$$
$$F_y = F_{y1} + F_{y2}$$
$$= (-144\,\text{N}) + (299\,\text{N})$$
$$= 155\,\text{N}$$

To find the magnitude of a vector from its components, use the Pythagorean theorem.

$$F^2 = F_x^2 + F_y^2$$
$$F = \sqrt{F_x^2 + F_y^2}$$
$$= \sqrt{(281.9\,\text{N})^2 + (155\,\text{N})^2}$$
$$= \sqrt{79,468\,\text{N}^2 + 24,025\,\text{N}^2}$$
$$= \sqrt{103,493\,\text{N}^2}$$
$$= 322\,\text{N}$$

And to find the angle relative to the x-axis, use the following trigonometric relationship:

$$\tan \theta = \frac{F_y}{F_x}$$

$$\theta = \tan^{-1}\left(\frac{F_y}{F_x}\right)$$

$$= \tan^{-1}\left(\frac{155 \text{ N}}{281.9 \text{ N}}\right)$$

$$= \tan^{-1}\left(\frac{155}{281.9}\right)$$

$$= \tan^{-1}(0.55)$$

$$= 29°$$

120. 9.0 m

Start with the force version of Newton's second law — $F_{net} = ma_{net}$ — where the net force exerted on an object is equal to the product of the object's mass and its acceleration in the direction of the force. Solve for the acceleration of the crate:

$$F_{net} = ma_{net}$$
$$(50 \text{ N}) = (25 \text{ kg})a_{net}$$
$$2 \text{ m/s}^2 = a_{net}$$

Then use the displacement formula $s = vt + \frac{1}{2}at^2$ — where v is the initial velocity, a is the acceleration, and t is time — along with the given data to solve for the displacement, s:

$$s = vt + \frac{1}{2}at^2$$

$$= (0 \text{ m/s})(3 \text{ s}) + \frac{1}{2}\left(2 \text{ m/s}^2\right)(3 \text{ s})^2$$

$$= (0)(3) \text{ m} + \frac{1}{2}(2)(9) \text{ m}$$

$$= 0 \text{ m} + 9 \text{ m}$$

$$= 9 \text{ m}$$

121. $\frac{Mv^2}{2F}$

First, find an expression for the box's acceleration using the given variables by making use of Newton's second law:

$$F_{net} = ma_{net}$$
$$\frac{F_{net}}{m} = a_{net}$$
$$\frac{F}{M} = a_{net}$$

Then use the velocity-displacement formula, $v_f^2 = v_i^2 + 2as$ – where v_i and v_f are an object's initial and final velocities, respectively, and s is the object's displacement — to solve for s:

$$v_f^2 = v_i^2 + 2as$$

$$v_f^2 - v_i^2 = 2as$$

$$\frac{v_f^2 - v_i^2}{2a} = s$$

$$\frac{v^2 - (0)^2}{2v\left(\dfrac{F}{M}\right)} = s$$

$$\frac{v^2}{\dfrac{2F}{M}} = s$$

$$v^2 \cdot \frac{M}{2F} = s$$

$$\frac{Mv^2}{2F} = s$$

122.   1.1 km

First, convert the velocity into the "correct" units:

$$\left(200\,\frac{\text{km}}{\text{h}}\right)\left(\frac{1{,}000\,\text{m}}{1\,\text{km}}\right)\left(\frac{1\,\text{h}}{60\,\text{min}}\right)\left(\frac{1\,\text{min}}{60\,\text{s}}\right) = 55.6\,\text{m/s}$$

Zero kilometers per hour equals 0 meters per second, because any number(s) multiplied by 0 equals 0.

For the stage where the speed increases, use Newton's second law, $F_{\text{net}} = ma_{\text{net}}$, with the given force of the engine (F_{net}) and mass (m) to solve for the sports car's acceleration.

$$F_{\text{net}} = ma_{\text{net}}$$

$$40{,}600\,\text{N} = (5{,}850\,\text{kg})a_{\text{net}}$$

$$6.94\,\text{m/s}^2 = a_{\text{net}}$$

Then use the velocity-displacement equation to solve for the car's displacement during this stage:

$$v_f^2 = v_i^2 + 2as$$

$$(55.6\,\text{m/s})^2 = (0\,\text{m/s})^2 + 2\left(6.94\,\text{m/s}^2\right)s$$

$$55.6^2\,\text{m}^2/\text{s}^2 = 0^2\,\text{m}^2/\text{s}^2 + (2)\left(6.94\,\text{m/s}^2\right)s$$

$$3{,}091.4\,\text{m}^2/\text{s}^2 = \left(13.88\,\text{m/s}^2\right)s$$

$$222.7\,\text{m} = s$$

No forces are involved in the direction of the sports car's motion in the constant-velocity stage, and so — as expected when velocity doesn't change — the acceleration then is 0 meters per second squared. Use the displacement formula to find the car's displacement during this stage, remembering that it now has an "initial" velocity of 55.6 meters per second:

$$s = vt + \frac{1}{2}at^2$$
$$= (55.6 \text{ m/s})(10 \text{ s}) + \frac{1}{2}\left(0 \text{ m/s}^2\right)(10 \text{ s})^2$$
$$= (55.6)(10) \text{ m} + \frac{1}{2}(0)(100) \text{ m}$$
$$= 556 \text{ m} + 0 \text{ m}$$
$$= 556 \text{ m}$$

For the final stage — where the speed decreases — use Newton's second law and the velocity-displacement equation as you did in the first go-round.

$$F_{net} = ma_{net}$$
$$-31,800 \text{ N} = (5,850 \text{ kg})a_{net}$$
$$-5.436 \text{ m/s}^2 = a_{net}$$

$$v_f^2 = v_i^2 + 2as$$
$$(0 \text{ m/s})^2 = (55.6 \text{ m/s})^2 + 2\left(-5.436 \text{ m/s}^2\right)s$$
$$0^2 \text{ m}^2/\text{s}^2 = 55.6^2 \text{ m}^2/\text{s}^2 + (2)\left(-5.436 \text{ m/s}^2\right)s$$
$$0 \text{ m}^2/\text{s}^2 = 3,091.4 \text{ m}^2/\text{s}^2 - \left(10.872 \text{ m/s}^2\right)s$$
$$-3,091.4 \text{ m}^2/\text{s}^2 = -\left(10.872 \text{ m/s}^2\right)s$$
$$284.3 \text{ m} = s$$

Add the three displacement values you've obtained to tally the total displacement of the sports car during its trip through this question: $222.7 \text{ m} + 556 \text{ m} + 284.3 \text{ m} = 1,063 \text{ m}$.

Finally, convert into the requested units — kilometers — and round: (1,063 m)

$$\left(\frac{1 \text{ km}}{1,000 \text{ m}}\right) = 1.063 \text{ km} \approx 1.1 \text{ km}.$$

123.   6 m/s

Before calculating any results, make sure to convert the given velocity into the "correct" units:

$$\left(-10.8 \frac{\text{km}}{\text{h}}\right)\left(\frac{1,000 \text{ m}}{1 \text{ km}}\right)\left(\frac{1 \text{ h}}{60 \text{ min}}\right)\left(\frac{1 \text{ min}}{60 \text{ s}}\right) = -3 \frac{\text{m}}{\text{s}}$$

(The negative sign symbolizes the "backward" part of the velocity value.)

You need to use Newton's second law, $F_{net} = ma_{net}$, to calculate the vehicle's acceleration. The engine exerts 450 newtons of force (F_{engine}), but 10 percent of that amount is exerted by friction (F_F) in the opposite direction, making the net force

$$F_{net} = \sum F$$
$$= F_{engine} - F_F$$
$$= F_{engine} - 0.1F_{engine}$$
$$= 0.9F_{engine}$$
$$= 0.9(450 \text{ N})$$
$$= 405 \text{ N}$$

Therefore, the acceleration of the vehicle is

$$F_{net} = ma_{net}$$
$$405 \text{ N} = (180 \text{ kg})a_{net}$$
$$2.25 \text{ m/s}^2 = a_{net}$$

Finally, use the velocity equation $v_f = v_i + at$ to solve for the final velocity v_f given the initial velocity, acceleration, and time $(v_i, a, \text{and } t)$:

$$v_f = v_i + at$$
$$= (-3 \text{ m/s}) + (2.25 \text{ m/s}^2)(4 \text{ s})$$
$$= (-3 \text{ m/s}) + (2.25)(4) \text{ m/s}$$
$$= -3 \text{ m/s} + 9 \text{ m/s}$$
$$= 6 \text{ m/s}$$

124.   360 km/h

Although kilonewtons are technically correct units (they're simply a power-of-ten multiple of newtons), you'll find the math more convenient if you convert them into "regular" newtons: $(96 \text{ kN})\left(\dfrac{1 \times 10^3 \text{N}}{1 \text{ kN}}\right) = 9.6 \times 10^4 \text{ N}$.

With that out of the way, use Newton's second law to find the magnitude of the acceleration provided by the engine using the given force F_{net} and the car's mass m:

$$F_{net} = ma_{net}$$
$$9.6 \times 10^4 \text{ N} = (7.200 \text{ kg})a_{net}$$
$$13.3 \text{ m/s}^2 = a_{net}$$

You know the initial velocity ("from rest" means "0 meters per second") and the time of the acceleration, so you can use the velocity formula to find the final velocity of the car:

$$v_f = v_i + at$$
$$= (0 \text{ m/s}) + (13.3 \text{ m/s}^2)(7.5 \text{ s})$$
$$= 0 \text{ m/s} + (13.3)(7.5) \text{ m/s}$$
$$= 0 \text{ m/s} + 99.75 \text{ m/s}$$
$$= 99.75 \text{ m/s}$$

Finish the solution by converting this into the requested units and rounding to two significant digits:

$$\left(99.75 \, \frac{m}{s}\right)\left(\frac{1 \, km}{1,000 \, m}\right)\left(\frac{60 \, s}{1 \, min}\right)\left(\frac{60 \, min}{1 \, h}\right) = 359.1 \, km/h \approx 360 \, \frac{km}{h}$$

125. $4.32 \times 10^3 \, N$

Break (no pun intended) the problem down into the two sections of the trip. For the section from the tenth floor down to the point halfway between the sixth and seventh floors, when the elevator is increasing in speed, two forces are involved: tension in the cable $\left(F_{T,\text{brakes off}}\right)$ and gravity $\left(F_G\right)$. The net force is therefore

$$
\begin{aligned}
F_{1,\text{net}} &= \sum F \\
&= F_{T,\text{brakes off}} - F_G \\
&= F_{T,\text{brakes off}} - ma_G \\
&= F_{T,\text{brakes off}} - mg \\
&= F_{T,\text{brakes off}} - (800 \, \text{kg})\left(9.8 \, \text{m/s}^2\right) \\
&= F_{T,\text{brakes off}} - 7,840 \, \text{N}
\end{aligned}
$$

Relate the cable's tension to the cab's acceleration by using Newton's second law:

$$
\begin{aligned}
F_{1,\text{net}} &= ma_{1,\text{net}} \\
F_{T,\text{brakes off}} - 7,840 \, \text{N} &= (800 \, \text{kg})a_{1,\text{net}} \\
F_{T,\text{brakes off}} &= (800 \, \text{kg})a_{1,\text{net}} + 7,840 \, \text{N}
\end{aligned}
$$

Follow the same procedure for the braking section:

$$
\begin{aligned}
F_{2,\text{net}} &= \sum F \\
&= F_{T,\text{brakes on}} + F_B - F_G \\
&= F_{T,\text{brakes on}} + F_B - ma_G \\
&= F_{T,\text{brakes on}} + F_B - mg \\
&= F_{T,\text{brakes on}} + F_B - (800 \, \text{kg})\left(9.8 \, \text{m/s}^2\right) \\
&= F_{T,\text{brakes on}} + F_B - 7,840 \, \text{N}
\end{aligned}
$$

where F_B is the force of the brakes. Applying this to Newton's second law, you obtain

$$
\begin{aligned}
F_{2,\text{net}} &= ma_{2,\text{net}} \\
F_{T,\text{brakes on}} + F_B - 7,840 \, \text{N} &= (800 \, \text{kg})a_{2,\text{net}} \\
F_{T,\text{brakes off}} &= (800 \, \text{kg})a_{2,\text{net}} + 7,840 \, \text{N} - F_B \\
\frac{F_{T,\text{brakes off}}}{2} &= (800 \, \text{kg})a_{2,\text{net}} + 7,840 \, \text{N} - F_B \\
F_{T,\text{brakes off}} &= (1600 \, \text{kg})a_{2,\text{net}} + 15,680 \, \text{N} - 2F_B
\end{aligned}
$$

Set the two equations equal to each other to eliminate $F_{T,\text{brakes off}}$ for the moment:

$$F_{T,\text{brakes off}} = (800\text{ kg})a_{1,\text{net}} + 7{,}840\text{ N} = (1600\text{ kg})a_{2,\text{net}} + 15{,}680\text{ N} - 2F_B$$
$$(800\text{ kg})a_{1,\text{net}} + 7{,}840\text{ N} = (1600\text{ kg})a_{2,\text{net}} + 15{,}680\text{ N} - 2F_B$$
$$(800\text{ kg})a_{1,\text{net}} = (1600\text{ kg})a_{2,\text{net}} - 2F_B + 7{,}840\text{ N}$$
$$a_{1,\text{net}} = 2a_{2,\text{net}} - \frac{F_B}{400\text{ kg}} + 9.8\text{ m/s}^2$$

At this point, it may seem as if you have too many unknowns to solve the problem, but the question gives you still more data. If each floor is 20 meters high, then traveling the 3.5 floors from the tenth to halfway between the sixth and seventh floors is equivalent to descending $(3.5)(20\text{ meters}) = 70$ meters. Similarly, the braking section involves a descent of $(1.5)(20\text{ meters}) = 30$ meters. You also know the initial and final velocities in each section. Use the velocity–displacement equation $\left(v_f^2 = v_i^2 + 2as\right)$ to solve for the accelerations.

For the "speed–up" section, $v_i = 0$ and $v_f = -5$, so

$$v_f^2 = v_i^2 + 2a_1 s$$
$$(-5\text{ m/s})^2 = (0\text{ m/s})^2 + 2a_1(-70\text{ m})$$
$$(-5)^2\text{ m}^2/\text{s}^2 = 0^2\text{m}^2/\text{s}^2 + (2)(-70\text{ m})a_1$$
$$25\text{ m}^2/\text{s}^2 = 0\text{ m}^2/\text{s}^2 - (140\text{ m})a_1$$
$$25\text{ m}^2/\text{s}^2 = -(140\text{ m})a_1$$
$$-0.1786\text{ m/s}^2 = a_1 = a_{1,\text{net}}$$

For the braking section, $v_i = -5$ and $v_f = 0$, so

$$v_f^2 = v_i^2 + 2a_2 s$$
$$(0\text{ m/s})^2 = (-5\text{ m/s})^2 + 2a_2(-30\text{ m})$$
$$0^2\text{ m}^2/\text{s}^2 = (-5)^2\text{ m}^2/\text{s}^2 + (2)(-30\text{ m})a_2$$
$$0\text{ m}^2/\text{s}^2 = 25\text{ m}^2/\text{s}^2 - (60\text{ m})a_2$$
$$-25\text{ m}^2/\text{s}^2 = -(60\text{ m})a_2$$
$$0.4167\text{ m/s}^2 = a_2 = a_{2,\text{net}}$$

Now, substitute these two accelerations into the equation you developed earlier, relating them to the braking force, F_B:

$$a_{1,\text{net}} = 2a_{2,\text{net}} - \frac{F_B}{400\text{ kg}} + 9.8\text{ m/s}^2$$

$$-0.1786\text{ m/s}^2 = 2\left(0.4167\text{ m/s}^2\right) - \frac{F_B}{400\text{ kg}} + 9.8\text{ m/s}^2$$

$$-10.812\text{ m/s}^2 = -\frac{F_B}{400\text{ kg}}$$

$$10.812\text{ m/s}^2 = \frac{F_B}{400\text{ kg}}$$

$$4{,}324.8\text{ N} = F_B = 4.32\times10^3\text{ N}$$

126. 12 N

Whenever object A (Joe) exerts a force on object B (brick wall), object B exerts a force of *equal magnitude* (12 newtons) on object A. The only difference in the two forces is that they point in opposite directions.

127. 3.8×10^{-1} m/s²

By Newton's third law, if the astronaut exerts 45 newtons of force on the shuttle, the shuttle exerts 45 newtons of force on the astronaut. Therefore, the net force felt by the astronaut is 45 newtons. Use the equation form of Newton's second law to solve for the astronaut's acceleration:

$$F_{net} = ma_{net}$$
$$45 \text{ N} = (120 \text{ kg})a_{net}$$
$$0.38 \text{ m/s}^2 = a_{net} = 3.8 \times 10^{-1} \text{m/s}^2$$

128. 4

Four boxes require four free-body diagrams:

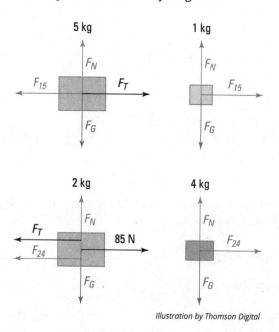

Illustration by Thomson Digital

F_{15} is the force of the 1-kilogram box pushing on the 5-kilogram box, and vice versa — the same magnitude of force applies in each direction by Newton's third law. Not surprisingly, F_{24} is the force of the 2-kilogram box pushing on the 4-kilogram box. There is no vertical acceleration — or friction — involved, so F_N and F_G can be ignored. F_{15} denotes the tension force in the rope connecting the 2-kilogram box to the 5-kilogram box.

The net force and the second-law relationship for the 1-kilogram box are

$$F_{net} = \sum F$$
$$= F_{15}$$

and

$$F_{net} = ma_{net}$$
$$F_{15} = ma$$
$$F_{15} = (1\,kg)a$$

And, similarly, for the 4-kilogram box:

$$F_{net} = \sum F$$
$$= F_{24}$$

and

$$F_{net} = ma_{net}$$
$$F_{24} = ma$$
$$F_{24} = (4\,kg)a$$

So the contact force between the 2- and 4-kilogram boxes is $(4\,kg)a$, and the contact force between the 5- and 1-kilogram boxes is $(1\,kg)a$. Therefore, the F_{24} contact force is $\frac{(4\,kg)a}{(1\,kg)a} = 4$ times more powerful than the F_{15} contact force.

129. 0.5z

The force of friction acting on a carton of milk is proportional to the carton's mass. The less massive carton (0.6 kilograms) has half the mass of the more massive carton (1.2 kilograms), and therefore it experiences half the force of friction that the more massive carton experiences.

130. 77 m/s²

One form of Newton's second law of motion states that $\sum F = ma_{net}$: The sum of all the forces on an object equals the product of the object's mass and its net acceleration. Use this to set up an equation relating the force with which Neil strikes the puck (F_{Neil}) to the frictional force exerted by the street/concrete (F_F).

$$\sum F = ma_{net}$$
$$F_{Neil} + F_F = ma_{net}$$
$$(44\,N) + (-3\,N) = ma_{net}$$
$$41\,N = ma_{net}$$

Friction is negative because it acts in the opposite direction of motion. To solve for the acceleration, you need the mass, which you can calculate from the puck's weight of 5.2 newtons because weight equals the force of gravity:

$$W = F_G$$
$$= ma_G$$
$$= mg$$
$$5.2\,\text{N} = m\left(9.8\,\text{m/s}^2\right)$$
$$0.53\,\text{kg} = m$$

Substitute this into the earlier equation:

$$41\,\text{N} = ma_{\text{net}}$$
$$41\,\text{N} = (0.53\,\text{kg})a_{\text{net}}$$
$$77\,\text{m/s}^2 = a_{\text{net}}$$

131.   140 N

Use Newton's second law to relate the forces involved with the dummy's acceleration. If F is the force with which Hans pushes the dummy and F_F is the force with which friction pushes against the tackling dummy, then

$$\sum F = ma_{\text{net}}$$
$$F_{\text{Hans}} + F_F = ma_{\text{net}}$$
$$F + F_F = (312\,\text{kg})\left(0.15\,\text{m/s}^2\right)$$
$$F + F_F = 46.8\,\text{N}$$
$$F = 46.8 - F_F$$

and, in the second instance,

$$\sum F = ma_{\text{net}}$$
$$F_{\text{Hans}} + F_F = ma_{\text{net}}$$
$$2F + F_F = (312\,\text{kg})\left(0.75\,\text{m/s}^2\right)$$
$$2F + F_F = 234\,\text{N}$$
$$2F = 234\,\text{N} - F_F$$
$$F = \frac{234\,\text{N} - F_F}{2}$$
$$F = 117\,\text{N} - 0.5F_F$$

Set the two equations equal to each other to solve for F_F:

$$46.8\,\text{N} - F_F = F = 117\,\text{N} - 0.5F_F$$
$$46.8\,\text{N} - F_F = 117\,\text{N} - 0.5F_F$$
$$46.8\,\text{N} = 117\,\text{N} + 0.5F_F$$
$$-70.2\,\text{N} = 0.5F_F$$
$$-140.4\,\text{N} = F_F$$

The negative sign indicates that the frictional force operates in the opposite direction from the motion of the dummy, as expected.

132. T

Because both masses are connected by the same rope, the same force of tension acts throughout, regardless of the varying masses on each side of the pulley. "Same rope" means "same tension."

133. 147.0 N

Use Newton's second law, $F_{net} = ma_{net}$, where F_{net} is the net force on the hay bale (the sum of all the forces acting on the bale), m is the bale's mass, and a_{net} is its acceleration. When you examine a force diagram of the hay bale, only two forces appear: F_G (the force due to gravity) and the tension force, F_T (the force exerted by the rope on the hay bale). The former points down, and the latter points up, so a sum of the forces on the hay bale looks like this (using the convention that "up" is "positive"):

$$\begin{aligned}F_{net} &= \sum F \\ &= F_T - F_G \\ &= F_T - ma_G \\ &= F_T - mg\end{aligned}$$

For the bale of hay to travel upward with constant velocity, it must have a net acceleration of 0 meters per second squared. Because $a_{net} = 0$,

$$\begin{aligned}F_{net} &= ma_{net} \\ &= m\left(0\ \text{m/s}^2\right) \\ &= 0\ \text{N}\end{aligned}$$

Substitute this into the earlier force-summation equation.

$$\begin{aligned}F_{net} &= F_T - mg \\ 0\ \text{N} &= F_T - (15\ \text{kg})\left(9.8\ \text{m/s}^2\right) \\ 0 &= F_T - 147\ \text{N} \\ 147\ \text{N} &= F_T\end{aligned}$$

This is the same force throughout the entire rope, so it's also the force with which Farmer Dell is pulling on the rope.

134. **4.2 m/s²**

First, draw a free-body diagram for each mass:

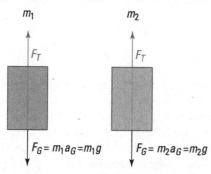

Illustration by Thomson Digital

Whenever you do calculations for a pulley problem involving multiple masses, you must choose a direction of positive acceleration. For the solution given here, "positive" means that mass 2 will fall, pulling mass 1 up as it "follows" in the linked train. So that means "down" is positive for the m_2 diagram, and "up" is positive for the m_1 diagram. The value of each mass's acceleration must be the same, because they're connected by ropes; they're a single entity in terms of their motion.

Sum all the forces on each mass and write Newton's second law for each. Remember to convert the masses into the "correct" units before inserting them into equations:

$$(300\text{g})\left(\frac{1\,\text{kg}}{1,000\,\text{g}}\right) = 0.3\,\text{kg} \text{ and } (750\,\text{g})\left(\frac{1\,\text{kg}}{1,000\,\text{g}}\right) = 0.75\,\text{kg}$$

For m_1:

$$F_T - m_1 g = m_1 a$$
$$F_T - (0.3\,\text{kg})\left(9.8\text{m/s}^2\right) = (0.3\,\text{kg})a$$
$$F_T - 2.94\,\text{N} = (0.3\,\text{kg})a$$
$$F_T = (0.3\,\text{kg})a + 2.94\,\text{N}$$

where F_T is the force of tension in the rope.

For m_2:

$$m_2 g - F_T = m_2 a$$
$$(0.75\,\text{kg})\left(9.8\,\text{m/s}^2\right) - F_T = (0.75\,\text{kg})a$$
$$7.35\,\text{N} - F_T = (0.75\,\text{kg})a$$
$$-F_T = (0.75\,\text{kg})a - 7.35\,\text{N}$$
$$F_T = -(0.75\,\text{kg})a + 7.35\,\text{N}$$

Set the two expressions for F_T equal to each other and solve for the acceleration of the system.

$$(0.3\ \text{kg})a + 2.94\ \text{N} = (-0.75\ \text{kg})a + 7.35\ \text{N}$$
$$(1.05\ \text{kg})a + 2.94\ \text{N} = 7.35\ \text{N}$$
$$(1.05\ \text{kg})a = 4.41\ \text{N}$$
$$a = 4.2\ \text{m/s}^2$$

135. 18.4 N

First, draw three free–body diagrams, one for each of the three masses:

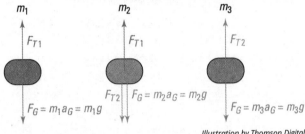

Illustration by Thomson Digital

Whenever you do calculations for a pulley problem involving multiple masses, you must choose a direction of positive acceleration. For the solution given here, "positive" means that masses 2 and 3 will fall, pulling mass 1 up as it follows the other two in the linked train. So that means "down" is positive for the m_2 and m_3 diagrams, and "up" is positive for the m_1 diagram. The value of each mass's acceleration must be the same, because all three are connected by ropes; they're a single entity in terms of their motion.

Sum all the forces on each mass and write Newton's second law for each.

For m_1:

$$F_{1,\text{net}} = m_1 a_{\text{net}}$$
$$F_{T1} - m_1 g = m_1 a$$
$$F_{T1} - (3\ \text{kg})\left(9.8\ \text{m/s}^2\right) = (3\ \text{kg})a$$
$$F_{T1} - 29.4\ \text{N} = (3\ \text{kg})a$$
$$F_{T1} = (3\ \text{kg})a + 29.4\ \text{N}$$

where F_{T1} is the force of tension in rope 1.

For m_2:

$$F_{2,\text{net}} = m_2 a_{\text{net}}$$
$$F_{T2} + m_2 g - F_{T1} = m_2 a$$
$$F_{T2} + (8\ \text{kg})\left(9.8\ \text{m/s}^2\right) - F_{T1} = (8\ \text{kg})a$$
$$F_{T2} + 78.4\ \text{N} - F_{T1} = (8\ \text{kg})a$$
$$F_{T2} - F_{T1} = (8\ \text{kg})a - 78.4\ \text{N}$$

where F_{T2} is the force of tension in rope 2.

For m_3:

$$F_{3,\text{net}} = m_3 a_{\text{net}}$$
$$m_3 g - F_{T2} = m_3 a$$
$$(5\text{ kg})\left(9.8\text{ m/s}^2\right) - F_{T2} = (5\text{ kg})a$$
$$49\text{ N} - F_{T2} = (5\text{ kg})a$$
$$-F_{T2} = (5\text{ kg})a - 49\text{ N}$$
$$F_{T2} = -(5\text{ kg})a + 49\text{ N}$$

Substitute your results from the first and third masses into your result from the second mass to leave a as the only remaining variable.

$$F_{T2} - F_{T1} = (8\text{ kg})a - 78.4\text{ N}$$
$$(-(5\text{ kg})a + 49\text{ N}) - ((3\text{ kg})a + 29.4\text{ N}) = (8\text{ kg})a - 78.4\text{ N}$$
$$-(5\text{ kg})a + 49\text{ N} - (3\text{ kg})a - 29.4\text{ N} = (8\text{ kg})a - 78.4\text{ N}$$
$$-(8\text{ kg})a + 19.6\text{ N} = (8\text{ kg})a - 78.4\text{ N}$$
$$19.6\text{ N} = (16\text{ kg})a - 78.4\text{ N}$$
$$98\text{ N} = (16\text{ kg})a$$
$$6.125\text{ m/s}^2 = a$$

To solve for the tension in rope 2 $\left(F_{T2}\right)$, use this acceleration in the result from the third mass:

$$F_{T2} = -(5\text{ kg})a + 49\text{ N}$$
$$= -(5\text{ kg})\left(6.125\text{ m/s}^2\right) + 49\text{ N}$$
$$= -30.625\text{ N} + 49\text{ N}$$
$$= 18.4\text{ N}$$

136. $\frac{4}{3}m_1 g$

First, draw a free-body diagram for each mass:

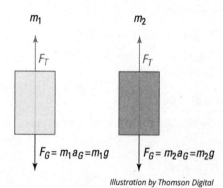

Illustration by Thomson Digital

Whenever you do calculations for a pulley problem involving multiple masses, you must choose a direction of positive acceleration. For the solution given here, "positive" means that mass 2 will fall, pulling mass 1 up as it follows in the linked train. So that means "down" is positive for the m_2 diagram, and "up" is positive for the m_1 diagram. The value of each mass's acceleration must be the same, because they're connected by ropes; they're a single entity in terms of their motion.

Sum all the forces on each mass and write Newton's second law for each.

For m_1:

$$F_T - m_1 g = m_1 a$$
$$F_T = m_1 a + m_1 g$$

where F_T is the force of tension in the rope.

For m_2:

$$m_2 g - F_T = m_2 a$$
$$(2m_1)g - F_T = (2m_1)a$$
$$2m_1 g - F_T = 2m_1 a$$
$$-F_T = 2m_1 a - 2m_1 g$$
$$F_T = -2m_1 a + 2m_1 g$$

Set the two expressions for F_T equal to each other and solve for the acceleration of the system:

$$m_1 a + m_1 g = -2m_1 a + 2m_1 g$$
$$2m_1 a + m_1 a + m_1 g = 2m_1 g$$
$$2m_1 a + m_1 a = 2m_1 g - m_1 g$$
$$3m_1 a = m_1 g$$
$$3a = g$$
$$a = \frac{g}{3}$$

The question asks for the rope's tension, so you must make a final substitution using either equation you obtained for F_T (this solution uses the result from m_1's free-body diagram):

$$F_T = m_1 a + m_1 g$$
$$= m_1 \left(\frac{g}{3} \right) + m_1 g$$
$$= \frac{1}{3} m_1 g + m_1 g$$
$$= \frac{4}{3} m_1 g$$

137. acceleration

Although the net force *on* an object must equal 0, *acceleration* is the only word that can correctly fill in the gap in the sentence.

138. velocity

The object's acceleration must remain constant (0 meters per second squared), but, as long as its velocity remains constant (the requirement to have $a = 0$), that velocity can be a nonzero value.

139. 690 N

Start with the banner's free-body diagram:

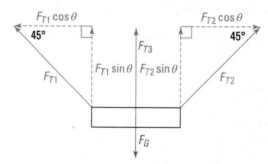

Illustration by Thomson Digital

For a system to be in equilibrium, its acceleration must be 0 in all directions: $a_{x,\text{net}} = 0\ \text{m/s}^2$ and $a_{y,\text{net}} = 0\ \text{m/s}^2$. The force form of Newton's second law states that $F_{\text{net}} = ma_{\text{net}}$, so:

$$F_{\text{net}} = ma_{\text{net}}$$
$$F_{x,\text{net}} = ma_{x,\text{net}} \qquad\qquad F_{y,\text{net}} = ma_{y,\text{net}}$$
$$= m\left(0\ \text{m/s}^2\right) \text{ and } \qquad = m\left(0\ \text{m/s}^2\right)$$
$$= 0\ \text{N} \qquad\qquad\qquad = 0\ \text{N}$$

Sum up the forces in the y direction and set them equal to $F_{y,\text{net}}$:

$$F_{y,\text{net}} = \sum F_y$$
$$F_{y,\text{net}} = F_{T1}\sin\theta + F_{T2}\sin\theta + F_{T3} - F_G$$
$$F_{y,\text{net}} = F_T\sin\theta + F_{T2}\sin\theta + F_{T3} - mg$$
$$0\ \text{N} = (100\ \text{N})\sin 45° + (100\ \text{N})\sin 45° + F_{T3} - (85\ \text{kg})\left(9.8\ \text{m/s}^2\right)$$
$$0 = (100\ \text{N})(0.707) + (100\ \text{N})(0.707) + F_{T3} - 85(9.8)\ \text{N}$$
$$0 = 70.7\ \text{N} + 70.7\ \text{N} + F_{T3} - 833\ \text{N}$$
$$0 = -691.6\ \text{N} + F_{T3}$$
$$691.6\ \text{N} = F_{T3}$$
$$690\ \text{N} \approx F_{T3}$$

140. 180 N

Start with the crate's free-body diagram:

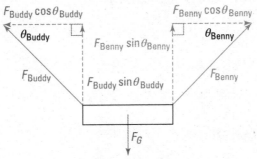

Illustration by Thomson Digital

For a system to be in equilibrium, its acceleration must be 0 in all directions: $a_{x,\text{net}} = 0 \text{ m/s}^2$ and $a_{y,\text{net}} = 0 \text{ m/s}^2$. The force form of Newton's second law states that $F_{\text{net}} = ma_{\text{net}}$, so:

$$F_{x,\text{net}} = ma_{x,\text{net}} \qquad F_{y,\text{net}} = ma_{y,\text{net}}$$
$$= m\left(0 \text{ m/s}^2\right) \text{ and} \qquad = m\left(0 \text{ m/s}^2\right)$$
$$= 0 \text{ N} \qquad\qquad = 0 \text{ N}$$

Sum up the forces in the x direction and set them equal to $F_{x,\text{net}}$:

$$F_{x,\text{net}} = \sum F_x$$

$$F_{x,\text{net}} = F_{\text{Benny}} \cos \theta_{\text{Benny}} - F_{\text{Buddy}} \cos \theta_{\text{Buddy}}$$
$$0 \text{ N} = F_{\text{Benny}} \cos 40^\circ - F_{\text{Buddy}} \cos 25^\circ$$
$$0 = F_{\text{Benny}}(0.766) - F_{\text{Buddy}}(0.906)$$
$$0 = 0.766 F_{\text{Benny}} - 0.906 F_{\text{Buddy}}$$
$$0.906 F_{\text{Buddy}} = 0.766 F_{\text{Benny}}$$
$$F_{\text{Buddy}} = 0.8455 F_{\text{Benny}}$$

Then sum the forces in the y direction and set them equal to $F_{y,\text{net}}$:

$$F_{y,\text{net}} = \sum F_y$$
$$F_{y,\text{net}} = F_{\text{Buddy}} \sin \theta_{\text{Buddy}} + F_{\text{Benny}} \sin \theta_{\text{Benny}} - F_G$$
$$F_{y,\text{net}} = F_{\text{Buddy}} \sin \theta_{\text{Buddy}} + F_{\text{Benny}} \sin \theta_{\text{Benny}} - mg$$
$$0 \text{ N} = F_{\text{Buddy}} \sin 25^\circ + F_{\text{Benny}} \sin 40^\circ - \left[(8.4 \text{ kg}) + 106(1.05 \text{ kg})\right]\left(9.8 \text{ m/s}^2\right)$$
$$0 = F_{\text{Buddy}}(0.423) + F_{\text{Benny}}(0.643) - (119.7 \text{ kg})\left(9.8 \text{ m/s}^2\right)$$
$$0 = 0.423 F_{\text{Buddy}} + 0.643 F_{\text{Benny}} - 1,173.1 \text{ N}$$

Substitute your earlier result for F_{Buddy} into this one to solve for the amount of force Benny is exerting to hold up the crate of gold.

$$0 = 0.423\left(0.8455F_{\text{Benny}}\right) + 0.643F_{\text{Benny}} - 1{,}173.1\,\text{N}$$
$$0 = 0.358F_{\text{Benny}} + 0.643F_{\text{Benny}} - 1{,}173.1\,\text{N}$$
$$1{,}173.1\,\text{N} = 1.001F_{\text{Benny}}$$
$$1{,}172\,\text{N} = F_{\text{Benny}}$$

And because $F_{\text{Buddy}} = 0.8455F_{\text{Benny}}$,

$$F_{\text{Buddy}} = 0.8455(1{,}172\,\text{N})$$
$$= 991\,\text{N}$$

which makes a difference of

$$F = F_{\text{Benny}} - F_{\text{Buddy}}$$
$$= (1{,}172\,\text{N}) - (991\,\text{N})$$
$$= 181\,\text{N} \approx 180\,\text{N}$$

141.   980 N

The question isn't asking for a component of the gravitational force relative to the ramp, so you only need to calculate F_G. Because $F = ma$, $F_G = ma_G$, and, because the ramp is (presumably) on the surface of Earth, $a_G = g$. Therefore,

$$F_G = mg$$
$$= (100\,\text{kg})\left(9.8\,\text{m/s}^2\right)$$
$$= 980\,\text{kg}\cdot\text{m/s}^2 = 980\,\text{N}$$

142.   310 N

Only two forces act on the motorcycle, so draw a free-body diagram, breaking any necessary vectors into their components.

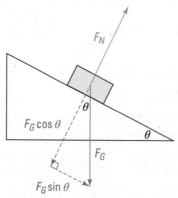

Illustration by Thomson Digital

In the "vertical" direction — perpendicular to the surface of the hill — the net acceleration is 0 meters per second squared (the motorcycle isn't accelerating upward or downward). Therefore, the net force in the vertical direction is 0 newtons. By definition, the net force is the sum of the forces in a particular direction, so you can write the following relationship:

$$F_{y,net} = \sum F_y$$
$$0\,N = F_N - F_G \cos\theta$$
$$0 = F_N - F_G \cos\theta$$
$$F_G \cos\theta = F_N$$

Weight is the colloquial term for the force of Earth's gravity, so substitute the 400 newtons as well as the 40-degree angle to solve for the normal force:

$$F_G \cos\theta = F_N$$
$$(400\,N)\cos 40° = F_N$$
$$(400\,N)(0.766) = F_N$$
$$306.4\,N = F_N \approx 310\,N$$

---

143. $4.9\,m/s^2$

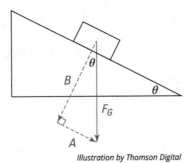

Illustration by Thomson Digital

The force of gravity, F_G, always points straight down. You're interested in the axis of the ramp's surface, so you need to solve for that component of the gravitational force. The component of the force that points along the surface axis is the side of the triangle labeled *A*. By the definitions of the sine function:

$$\sin\theta = \frac{A}{F_G}$$
$$F_G \sin\theta = A$$
$$ma_G \sin\theta = A$$
$$mg \sin\theta = A$$

where m is the mass of the box.

$F = ma$, or $\dfrac{F}{m} = a$, so the acceleration along the axis is the force along vector A divided by the mass of the box.

$$a = \frac{F}{m}$$
$$= \frac{mg \sin \theta}{m}$$
$$= g \sin \theta$$
$$= \left(9.8 \text{ m/s}^2\right) \sin 30^\circ$$
$$= \left(9.8 \text{ m/s}^2\right)(0.5)$$
$$= 4.9 \text{ m/s}^2$$

144. 29°

Examining a force diagram on Audrey as she stands on the hill,

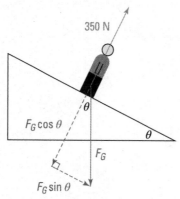

Illustration by Thomson Digital

you notice that the 350 newtons of force she feels from the ground is in the exact opposite direction as $F_G \cos \theta$. No acceleration is occurring in this direction (Audrey isn't plummeting through the ground or rocketing into the air), and therefore, the net force in that direction must be 0:

$$F_{y,\text{net}} = \sum F_y$$
$$= 350 \text{ N} + \left(-F_G \cos \theta\right)$$
$$0 = 350 \text{ N} - F_G \cos \theta$$
$$0 = 350 \text{ N} - (400 \text{ N})\cos \theta$$
$$-350 \text{ N} = -(400 \text{ N})\cos \theta$$
$$0.875 = \cos \theta$$
$$\cos^{-1}(0.875) = \theta$$
$$29^\circ = \theta$$

145. 29°

First draw the free-body diagram:

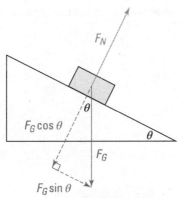

Illustration by Thomson Digital

Only one force is pointing along the axis of motion — $F_G \sin \theta$ — so that's the only contributor to the net acceleration along the hill's surface:

$$F_{x,\text{net}} = \sum F_x$$
$$ma_{x,\text{net}} = mg \sin \theta$$
$$a_{x,\text{net}} = g \sin \theta$$
$$4.8 \text{ m/s}^2 = \left(9.8 \text{ m/s}^2\right) \sin \theta$$
$$\frac{4.8 \text{ m/s}^2}{9.8 \text{ m/s}^2} = \sin \theta$$
$$0.49 = \sin \theta$$
$$\sin^{-1}(0.49) = \theta$$
$$29.3° = \theta$$

Three forces are involved (F_N, the normal force; F_G, the force due to Earth's gravity; and F_P, the force from Samantha's push). Draw a free-body diagram showing the vectors and any necessary components to align the forces into two perpendicular axes:

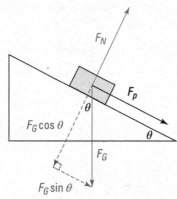

Illustration by Thomson Digital

The sled moves along the axis parallel to the ramp's surface, so that's the direction of the acceleration you need to examine. Adding up the forces in that direction:

$$F_{x,\text{net}} = \sum F_x$$
$$= F_G \sin\theta + F_P$$
$$= ma_G \sin\theta + F_P$$
$$= mg \sin\theta + F_P$$

(This setup uses "down the slope" as the positive direction.) By Newton's second law, $F_{\text{net}} = ma_{\text{nev}}$, so

$$F_{x,\text{net}} = mg\sin\theta + F_P$$
$$ma_{x,\text{net}} = mg\sin\theta + F_P$$
$$(40\,\text{kg})(1.9\,\text{m/s}^2) = (40\,\text{kg})(9.8\,\text{m/s}^2)\sin\theta + (50\,\text{N})$$
$$76\,\text{N} = (392\,\text{N})\sin\theta + 50\,\text{N}$$
$$26\,\text{N} = (392\,\text{N})\sin\theta$$
$$0.0663 = \sin\theta$$
$$\sin^{-1}(0.0663) = \theta$$
$$3.8° = \theta \approx 4°$$

147.　340 N

Three forces are involved (F_N, the normal force; F_G, the force due to Earth's gravity; and F_P, the force from the official's push). Draw a free-body diagram showing the vectors and any necessary components to align the forces into two perpendicular axes:

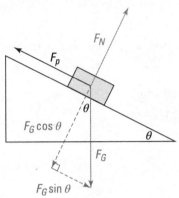

Illustration by Thomson Digital

The wheelchair and its occupant move along the axis parallel to the ramp's surface, so that's the direction of the 0.7-meters-per-second-squared acceleration. Adding up the forces in that direction:

$$F_{x,\,net} = \sum F_x$$
$$= F_G \sin \theta - F_P$$
$$= m a_G \sin \theta - F_P$$
$$= mg \sin \theta - F_P$$

(This setup uses "down the slope" as the positive direction.) By Newton's second law, $F_{net} = m a_{net}$, so

$$F_{x,\,net} = mg \sin \theta - F_P$$
$$m a_{x,\,net} = mg \sin \theta - F_P$$
$$m\left(-0.7\,\text{m/s}^2\right) = m\left(9.8\,\text{m/s}^2\right)\sin \theta - (75\,\text{N})$$

You can solve for θ by using trigonometry: If the ramp's length is 12 meters, and it rises 1.8 meters from start to finish, the triangle looks like this:

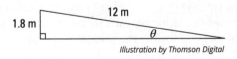

Illustration by Thomson Digital

Therefore, $\sin\theta = \dfrac{1.8\,\text{m}}{12\,\text{m}}$. Solving for θ,

$$\sin\theta = \dfrac{1.8\,\text{m}}{12\,\text{m}}$$
$$\sin\theta = 0.15$$
$$\theta = \sin^{-1}(0.15)$$
$$= 8.63°$$

Substitute this back into your result from $F_{x,\text{net}}$ to determine the combined mass of the patient and the wheelchair:

$$m\left(-0.7\,\text{m/s}^2\right) = m\left(9.8\,\text{m/s}^2\right)\sin\theta - (75\,\text{N})$$
$$m\left(-0.7\,\text{m/s}^2\right) = m\left(9.8\,\text{m/s}^2\right)(0.15) - 75\,\text{N}$$
$$\left(-0.7\,\text{m/s}^2\right)m = \left(1.47\,\text{m/s}^2\right)m - 75\,\text{N}$$
$$\left(-2.17\,\text{m/s}^2\right)m = -75\,\text{N}$$
$$m = 34.6\,\text{kg}$$

The force that the ramp exerts on the chair is called the normal force; from the free-body diagram, you can see that its value is related to the vertical acceleration (which must be 0 meters per second squared if the wheelchair isn't leaping off the ground or plummeting into it). Sum up the forces in the vertical direction and use Newton's second law once more to solve for the normal force:

$$F_{y,\text{net}} = \sum F_y$$
$$ma_y = F_N - F_G \cos\theta$$
$$ma_y = F_N - ma_G \cos\theta$$
$$ma_y = F_N - mg \cos\theta$$
$$(34.6\,\text{kg})\left(0\,\text{m/s}^2\right) = F_N - (34.6\,\text{kg})\left(9.8\,\text{m/s}^2\right)\cos 8.63°$$
$$0\,\text{N} = F_N - (34.6\,\text{kg})\left(9.8\,\text{m/s}^2\right)(0.989)$$
$$0\,\text{N} = F_N - 335\,\text{N}$$
$$335\,\text{N} = F_N$$

Rounded to two significant digits, this becomes 340 newtons.

148. $F_F < F_N$

The equation for a frictional force is $F_F = \mu F_N$, where μ is the coefficient of either static or kinetic friction — whichever the situation calls for. μ is always a positive number less than 1; therefore, no matter what value the normal force takes on, the frictional force must be smaller.

149. 274 N

To use the formula for friction, you need a value for the normal force on the chair. The only vertical forces acting on the chair are the gravitational force of attraction to Earth and the normal force. The latter points "up" (conventionally, the positive direction), and the former points "down." Therefore, the net force on the chair in the y direction is

$$
\begin{aligned}
F_{y,\text{net}} &= \sum F_y \\
&= F_N - F_G \\
&= F_N - ma_G \\
&= F_N - mg
\end{aligned}
$$

The acceleration of the chair in the y direction is 0 meters per second squared, so the net force in the vertical direction must also equal 0:

$$
\begin{aligned}
F_{y,\text{net}} &= ma_{y,\text{net}} \\
&= m\left(0\,\text{m/s}^2\right) \\
&= 0
\end{aligned}
$$

Equating these two results produces the relationship

$$
\begin{aligned}
F_N - mg &= F_{y,\text{net}} = 0 \\
F_N - mg &= 0 \\
F_N &= mg
\end{aligned}
$$

which, for a mass of 140 kilograms, results in a normal force equal to

$$
\begin{aligned}
F_N &= mg \\
&= (140\,\text{kg})\left(9.8\,\text{m/s}^2\right) \\
&= 1{,}372\,\text{N}
\end{aligned}
$$

Now use the friction formula to calculate the amount of friction exerted on the chair by the floor:

$$
\begin{aligned}
F_F &= \mu F_N \\
&= (0.2)(1{,}372\,\text{N}) \\
&= 274.4\,\text{N}
\end{aligned}
$$

yes; 0.8 s

First draw a free-body diagram showing the forces exerted on the trunk (this answer presumes that the trunk is being pulled toward the right):

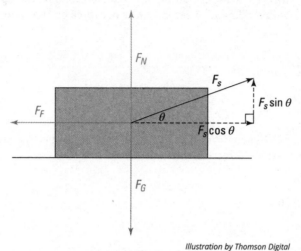

Illustration by Thomson Digital

(F_s denotes the force exerted by the porter on the trunk's strap.)

The sum of the vertical (y-direction) forces — and therefore the net force in the y direction — is

$$
\begin{aligned}
F_{y,\text{net}} &= \sum F_y \\
&= F_N + F_s \sin\theta - F_G \\
&= F_N + F_s \sin\theta - ma_G \\
&= F_N + F_s \sin\theta - mg
\end{aligned}
$$

By Newton's second law, this is equal to the mass of the trunk times its net acceleration in the y direction, which itself is equal to 0 meters per second squared, as the trunk is not accelerating off the floor (up) or into the floor (down):

$$
\begin{aligned}
F_{y,\text{net}} &= ma_{y,\text{net}} \\
F_N + F_s \sin\theta - mg &= ma_{y,\text{net}} \\
F_N + (5{,}000\,\text{N}) \sin 20° - (795\,\text{kg})\left(9.8\,\text{m/s}^2\right) &= (795\,\text{kg})\left(0\,\text{m/s}^2\right) \\
F_N + (5{,}000\,\text{N})(0.342) - (795\,\text{kg})\left(9.8\,\text{m/s}^2\right) &= (795\,\text{kg})\left(0\,\text{m/s}^2\right) \\
F_N + 1{,}710\,\text{N} - 7{,}791\,\text{N} &= 0\,\text{N} \\
F_N - 6{,}081\,\text{N} &= 0\,\text{N} \\
F_N &= 6{,}081\,\text{N}
\end{aligned}
$$

The sum of the horizontal (x-direction) forces — and therefore the net force in the x direction — is

$$
\begin{aligned}
F_{x,\text{net}} &= \sum F_x \\
&= F_s \cos\theta - F_F \\
&= F_s \cos\theta - \mu F_N
\end{aligned}
$$

using the friction formula $F_F = \mu F_N$.

Once again use Newton's second law — this time in the x direction — to solve for the trunk's acceleration in the x direction.

$$
\begin{aligned}
F_{x,\text{net}} &= ma_{x,\text{net}} \\
F_s \cos\theta - \mu F_N &= ma_{x,\text{net}} \\
(5,000\,\text{N})\cos 20^\circ - (0.65)(6,081\,\text{N}) &= (795\,\text{kg})a_{x,\text{net}} \\
(5,000\,\text{N})(0.9397) - (0.65)(6,081\,\text{N}) &= (795\,\text{kg})a_{x,\text{net}} \\
4{,}698.5\,\text{N} - 3{,}952.7\,\text{N} &= (795\,\text{kg})a_{x,\text{net}} \\
745.8\,\text{N} &= (795\,\text{kg})a_{x,\text{net}} \\
0.938\,\text{m/s}^2 &= a_{x,\text{net}}
\end{aligned}
$$

With that acceleration, how long will the porter need to travel 40 meters? Use the displacement formula ($s = v_i t + \frac{1}{2}at^2$, where s is displacement, v_i is initial velocity, and t is time) to find:

$$
\begin{aligned}
s &= v_i t + \frac{1}{2}at^2 \\
40\,\text{m} &= (0\,\text{m/s})\,t + \frac{1}{2}\left(0.938\,\text{m/s}^2\right)t^2 \\
40\,\text{m} &= 0\,\text{m} + \left(0.469\,\text{m/s}^2\right)t^2 \\
40\,\text{m} &= \left(0.469\,\text{m/s}^2\right)t^2 \\
85.29\text{s}^2 &= t^2 \\
\sqrt{85.29\,\text{s}^2} &= t \\
9.24\,\text{s} &= t
\end{aligned}
$$

That's $10 - 9.24 = 0.76$ seconds short of the cutoff time for the tip, so the porter earns the tip. Rounding to the nearest tenth, the porter's spare time is 0.8 seconds.

151. static

Static friction is the frictional force that must be overcome for one object to *start* moving relative to another object.

152.   0.43

Use the formula for friction, $F_F = \mu F_N$, where F_F is the magnitude of the frictional force, μ is the coefficient of friction, and F_N is the magnitude of the normal force. To calculate the latter, examine the vertical forces at work on the cabinet: Gravity pulls down, and the normal force pushes up. Added together, they equal the net force on the cabinet in the vertical direction, which, because the cabinet has no acceleration in the vertical direction, must equal 0 newtons. Therefore,

$$F_{net} = \sum F$$
$$0\,N = -F_G + F_N$$
$$0\,N = -ma_G + F_N$$
$$0\,N = -mg + F_N$$
$$mg = F_N$$

Substitute this into the formula for friction, which here contains the subscript s standing for static:

$$F_{F,s} = \mu_s F_N$$
$$= \mu_s mg$$
$$= \mu_s (800\,\text{kg})\left(9.8\,\text{m/s}^2\right)$$
$$= (7{,}840\,\text{N})\,\mu_s$$

A force of 3,400 newtons is required to start the cabinet moving; therefore, that is the amount of static friction involved. Substitute that value into your last equation to solve for μ:

$$F_{F,s} = (7{,}840\,\text{N})\mu_s$$
$$3{,}400\,\text{N} = (7{,}840\,\text{N})\mu_s$$
$$\frac{3{,}400\,\text{N}}{7{,}840\,\text{N}} = \mu_s$$
$$0.43 = \mu_s$$

153.   5.4 m/s²

Whenever you're asked for an acceleration when dealing with the force of static friction, check to make sure enough pushing/pulling force is present to counteract the friction and actually *move* the object. Use the formula for friction, $F_F = \mu F_N$, where F_F is the magnitude of the frictional force, μ is the coefficient of friction, and F_N is the magnitude of the normal force. To calculate the latter, examine the vertical forces at work on the table: Gravity pulls down, and the normal force pushes up. Added together, they equal the net force on the table in the vertical direction, which, because the table has no acceleration in the vertical direction, must equal 0 newtons. Therefore,

$$F_{net} = \sum F$$
$$0\,N = -F_G + F_N$$
$$0 = -ma_G + F_N$$
$$0 = -mg + F_N$$
$$mg = F_N$$

Substitute this into the friction formula, which here contains the subscript s standing for static:

$$F_{F,s} = \mu_s F_N$$
$$= \mu_s mg$$
$$= (0.3)(60\text{ kg})\left(9.8\text{ m/s}^2\right)$$
$$= 176.4\text{ N}$$

Franz's 500 newtons is more than enough to compensate for the force of friction; therefore, the table indeed starts to move. To calculate its acceleration (immediately after it moves), sum the horizontal forces and use Newton's second law to solve for a:

$$F_{x,\text{net}} = \sum F_x$$
$$ma_x = F_{\text{Franz}} - F_{F,x}$$
$$(60\text{ kg})\,a_x = (500\text{ N}) - (176.4\text{ N})$$
$$(60\text{ kg})\,a_x = 323.6\text{ N}$$
$$a_x = 5.4\text{ m/s}^2$$

154. 488 kg

First draw a free-body diagram:

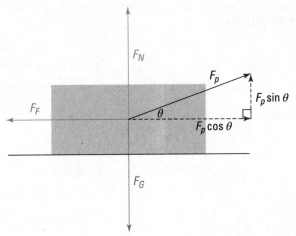

Illustration by Thomson Digital

F_p represents the pulling force, and, because you're interested in getting the crate initially moving, you need to deal with the force of *static* friction, represented by F_F.

With that in mind, the sum of the vertical (y-direction) forces — and therefore the net force in the y direction — is

$$F_{y,\text{net}} = \sum F_y$$
$$= F_N + F_p \sin\theta - F_G$$
$$= F_N + F_p \sin\theta - ma_G$$
$$= F_N + F_p \sin\theta - mg$$

By Newton's second law, this is equal to the mass of the crate times its net acceleration in the y direction, which itself is equal to 0 meters per second squared, as the crate is not accelerating off the floor (up) or into the floor (down):

$$F_{y,\text{net}} = ma_{y,\text{net}}$$

$$F_N + F_p \sin\theta - mg = ma_{y,\text{net}}$$

$$F_N + (890\text{ N})\sin 28° - m\left(9.8\text{ m/s}^2\right) = m\left(0\text{ m/s}^2\right)$$

$$F_N + (890\text{ N})(0.469) - m\left(9.8\text{ m/s}^2\right) = 0\text{ N}$$

$$F_N + 417.4\text{ N} - \left(9.8\text{ m/s}^2\right)m = 0\text{ N}$$

$$F_N = \left(9.8\text{ m/s}^2\right)m - 417.4\text{ N}$$

And the sum of the horizontal (x-direction) forces — and therefore the net force in the x direction — is

$$F_{x,\text{net}} = \sum F_x$$
$$= F_p \cos\theta - F_F$$
$$= F_p \cos\theta - \mu_s F_N$$

using the formula for static friction, $F_{F,s} = \mu_s F_N$.

Just before the crate starts moving, its acceleration is 0 meters per second in the horizontal direction, so use Newton's second law once more — this time in the x direction — to solve for the unknown mass. You need to substitute your previous result for the normal force in the process:

$$F_{x,\text{net}} = ma_{x,\text{net}}$$

$$F_p \cos\theta - \mu_s F_N = ma_{x,\text{net}}$$

$$(890\text{ N})\cos 28° - (0.18)\left(\left(9.8\text{ m/s}^2\right)m - 417.4\text{ N}\right) = m\left(0\text{ m/s}^2\right)$$

$$(890\text{ N})(0.883) - \left[(0.18)\left(\left(9.8\text{ m/s}^2\right)m\right) + (0.18)(-417.4\text{ N})\right] = 0\text{ N}$$

$$785.87\text{ N} - \left(\left(1.764\text{ m/s}^2\right)m - 75.13\text{ N}\right) = 0\text{ N}$$

$$785.87\text{ N} - \left(1.764\text{ m/s}^2\right)m + 75.13\text{ N} = 0\text{ N}$$

$$861\text{ N} - \left(1.764\text{ m/s}^2\right)m = 0\text{ N}$$

$$-\left(1.764\text{ m/s}^2\right)m = -861\text{ N}$$

$$m = 488\text{ kg}$$

155. 1.2×10^4 N; west

Friction always acts in the opposite direction as the motion, so it must be pointing west if the car is traveling east. For the car to move at a constant speed, its acceleration must be 0; therefore, by Newton's second law, the net force in the direction of travel — the x or horizontal direction — must also equal 0:

$$
\begin{aligned}
F_{x,\text{net}} &= \sum F_x \\
F_{x,\text{net}} &= F_{\text{engine}} - F_F \\
0\,\text{N} &= (12{,}000\,\text{N}) - F_F \\
-12{,}000\,\text{N} &= -F_F \\
12{,}000\,\text{N} &= F_F
\end{aligned}
$$

156. 0.08

First sketch a free-body diagram for the skater.

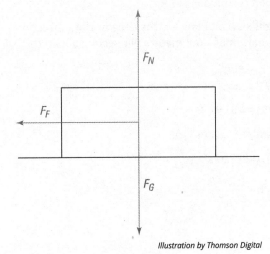

Illustration by Thomson Digital

The diagram designates the motion of the skater as being to the right, which is also the positive direction in the calculations below.

The sum of the vertical (y-direction) forces — and therefore the net force in the y direction — is

$$
\begin{aligned}
F_{y,\text{net}} &= \sum F_y \\
&= F_N - F_G \\
&= F_N - ma_G \\
&= F_N - mg
\end{aligned}
$$

By Newton's second law, this is equal to the mass of the skater times the skater's net acceleration in the y direction — which itself is equal to 0 meters per second squared, as the skater is not accelerating off of the ground (up) or into it (down):

$$F_{y,\text{net}} = ma_{y,\text{net}}$$
$$F_N - mg = ma_{y,\text{net}}$$
$$F_N - m\left(9.8 \text{ m/s}^2\right) = m\left(0 \text{ m/s}^2\right)$$
$$F_N - \left(9.8 \text{ m/s}^2\right)m = 0 \text{ N}$$
$$F_N = \left(9.8 \text{ m/s}^2\right)m$$

The sum of the horizontal (x-direction) forces — and therefore the net force in the x direction — is

$$F_{x,\text{net}} = \sum F_x$$
$$= -F_F$$
$$= -\mu F_N$$

substituting the friction formula $F_F = \mu F_N$.

Once again use Newton's second law — this time in the x direction — to solve for μ using the given acceleration of -0.8 meters per second squared.

$$F_{x,\text{net}} = ma_{x,\text{net}}$$
$$-\mu F_N = ma_x$$
$$-\mu\left(\left(9.8 \text{ m/s}^2\right)m\right) = m\left(-0.8 \text{ m/s}^2\right)$$
$$-\left(9.8 \text{ m/s}^2\right)\mu m = -\left(0.8 \text{ m/s}^2\right)m$$
$$\left(9.8 \text{ m/s}^2\right)\mu = 0.8 \text{ m/s}^2$$
$$\mu = \frac{0.8 \text{ m/s}^2}{9.8 \text{ m/s}^2} = 0.082$$

157.   4.7 m

To calculate a distance, you need to know acceleration first. So start by analyzing the forces on the roller-skater by sketching a free-body diagram.

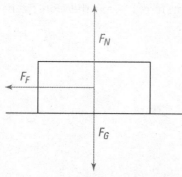

Illustration by Thomson Digital

The diagram designates the motion of the roller-skater as being to the right, which is also the positive direction in the calculations below.

The sum of the vertical (y-direction) forces — and therefore the net force in the y direction — is

$$
\begin{aligned}
F_{y,\text{net}} &= \sum F_y \\
&= F_N - F_G \\
&= F_N - ma_G \\
&= F_N - mg
\end{aligned}
$$

By Newton's second law, this is equal to the mass of the roller-skater times her net acceleration in the y direction — which itself is equal to 0 meters per second squared, as she is not accelerating off of the ground (up) or into it (down):

$$
\begin{aligned}
F_{y,\text{net}} &= ma_{y,\text{net}} \\
F_N - mg &= ma_{y,\text{net}} \\
F_N - (95\ \text{kg})\left(9.8\ \text{m/s}^2\right) &= (95\ \text{kg})\left(0\ \text{m/s}^2\right) \\
F_N - 931\ \text{N} &= 0\ \text{N} \\
F_N &= 931\ \text{N}
\end{aligned}
$$

The sum of the horizontal (x-direction) forces — and therefore the net force in the x direction — is

$$
\begin{aligned}
F_{x,\text{net}} &= \sum F_x \\
&= -F_F \\
&= -\mu F_N
\end{aligned}
$$

substituting the friction formula $F_F = \mu F_N$.

Once again use Newton's second law — this time in the x direction — to solve for the roller-skater's acceleration.

$$
\begin{aligned}
F_{x,\text{net}} &= ma_{x,\text{net}} \\
-\mu F_N &= ma_x \\
-(0.7)(931\ \text{N}) &= (95\ \text{kg})a_x \\
-651.7\ \text{N} &= (95\ \text{kg})a_x \\
-6.86\ \text{m/s}^2 &= a_x
\end{aligned}
$$

Finally, use this acceleration in the velocity-displacement formula ($v_f^2 = v_i^2 + 2as$, where v_i and v_f are the initial and final velocities, respectively, a is the acceleration, and s is the displacement) to find the distance traveled while changing from a speed of 8 meters per second to 0 meters per second:

$$
\begin{aligned}
v_f^2 &= v_i^2 + 2as \\
(0\ \text{m/s})^2 &= (8\ \text{m/s})^2 + 2\left(-6.86\ \text{m/s}^2\right)s \\
0\ \text{m}^2/\text{s}^2 &= 64\ \text{m}^2/\text{s}^2 - \left(13.72\ \text{m/s}^2\right)s \\
-64\ \text{m}^2/\text{s}^2 &= -\left(13.72\ \text{m/s}^2\right)s \\
4.66\ \text{m} &= s
\end{aligned}
$$

"Constant velocity" means that the acceleration equals 0 meters per second squared. Therefore, the net force in the direction of motion equals 0 newtons. Draw a free-body diagram to see which forces act along the axis of motion (the slope itself):

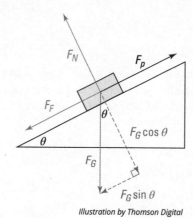

Illustration by Thomson Digital

The force of friction always opposes motion, so it must point in the opposite direction than that in which the crate is moving; if the crate moves *up* the slope, friction must point *down* the slope. The three forces that act along the slope are Jillian's "push" force ($F_p = 50$ newtons), the force of friction (F_F), and the component of the force of gravity along the slope ($F_G \sin \theta = -40$ newtons). The 40 newtons is negative because it points down the slope, in the negative direction. If the sum of the three forces must equal 0 newtons, then

$$F_p + F_F + F_G \sin \theta = 0$$
$$(50\,\text{N}) + F_F + (-40\,\text{N}) = 0$$
$$10\,\text{N} + F_F = 0$$
$$F_F = -10\,\text{N}$$

As expected from the diagram, the friction force is negative: It points "down" the slope.

159.   1.2

The only difference between the two scenarios is the coefficient of kinetic friction, so the free-body diagrams look identical:

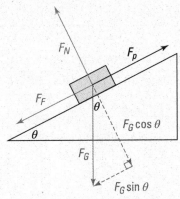

Illustration by Thomson Digital

where F_N is the normal force, F_G is the force of gravity ($F_G = ma_G = mg$ on Earth's surface), F_F is the force of kinetic friction, and F_p is the force with which either boxer pushes his crate. Add up the forces in each direction (x and y) and set their sums equal to the crate's mass times its acceleration along that particular axis ($a_y = 0$ meters per second squared because the crate doesn't accelerate off the mountain's surface or into it). In the y or vertical direction:

$$
\begin{aligned}
F_{y,\text{net}} &= \sum F_y \\
ma_y &= F_N - F_G \cos\theta \\
ma_y &= F_N - mg\cos\theta \\
m(0\,\text{m/s}^2) &= F_N - mg\cos\theta \\
0 &= F_N - mg\cos\theta \\
mg\cos\theta &= F_N
\end{aligned}
$$

And in the x or horizontal direction:

$$
\begin{aligned}
F_{x,\text{net}} &= \sum F_x \\
ma_x &= F_p - F_F - F_G \sin\theta \\
ma_x &= F_p - \mu F_N - mg\sin\theta \\
ma_x &= F_p - \mu(mg\cos\theta) - mg\sin\theta \\
F_p &= ma_x + \mu mg\cos\theta + mg\sin\theta
\end{aligned}
$$

You need to know how many times larger Rocky's force is than Bob's. If F_R is defined as the force with which Rocky pushes his crate, and F_B is defined as the force with which Bob pushes his crate, you need to calculate the ratio $\frac{F_R}{F_B}$. Again, the only difference between the boxers is the value of μ:

$$\frac{F_R}{F_B} = \frac{ma_x + \mu_R mg \cos\theta + mg \sin\theta}{ma_x + \mu_B mg \cos\theta + mg \sin\theta}$$

$$= \frac{m\left(2.6 \text{ m/s}^2\right) + (0.36)\, m\left(9.8 \text{ m/s}^2\right)\cos 21° + m\left(9.8 \text{ m/s}^2\right)\sin 21°}{m\left(2.6 \text{ m/s}^2\right) + (0.18)\, m\left(9.8 \text{ m/s}^2\right)\cos 21° + m\left(9.8 \text{ m/s}^2\right)\sin 21°}$$

$$= \frac{m\left(2.6 \text{ m/s}^2\right) + (0.36)\, m\left(9.8 \text{ m/s}^2\right)(0.934) + m\left(9.8 \text{ m/s}^2\right)(0.358)}{m\left(2.6 \text{ m/s}^2\right) + (0.18)\, m\left(9.8 \text{ m/s}^2\right)(0.934) + m\left(9.8 \text{ m/s}^2\right)(0.358)}$$

$$= \frac{\left(2.6 \text{ m/s}^2\right)m + \left(3.295 \text{ m/s}^2\right)m + \left(3.508 \text{ m/s}^2\right)m}{\left(2.6 \text{ m/s}^2\right)m + \left(1.648 \text{ m/s}^2\right)m + \left(3.508 \text{ m/s}^2\right)m}$$

$$= \frac{\left(9.403 \text{ m/s}^2\right)m}{\left(7.756 \text{ m/s}^2\right)m}$$

$$= \frac{\left(9.403 \text{ m/s}^2\right)}{\left(7.756 \text{ m/s}^2\right)}$$

$$\frac{F_R}{F_B} = 1.2$$

$$F_R = 1.2 F_B$$

Rocky has to push 1.2 times harder than does Bob.

160. 16°

As with almost every force problem, start by drawing a free-body diagram

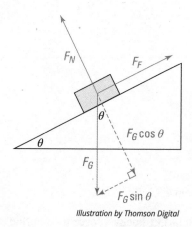

Illustration by Thomson Digital

where F_N is the normal force, F_G is the force of gravity ($F_G = ma_G = mg$ on Earth's surface), and F_F is the force of friction. Friction points up the driveway because the toy car slides down the driveway, and friction always points in the opposite direction from the motion. Now add up the forces in each direction (x and y) and set their sums equal to the mass times the acceleration along that particular axis ($a_y = 0$ meters per second squared because the toy car doesn't accelerate off of the driveway's surface or into it, and $a_x = 0$ meters per second squared because the car is rolling at a constant speed). In the y or vertical direction:

$$F_{y,net} = \sum F_y$$
$$ma_y = F_N - F_G \cos \theta$$
$$ma_y = F_N - mg \cos \theta$$
$$m(0 \text{ m/s}^2) = F_N - mg \cos \theta$$
$$0 = F_N - mg \cos \theta$$
$$mg \cos \theta = F_N$$

And in the x or horizontal direction, where "down the slope" is taken as positive in this solution:

$$F_{x,net} = \sum F_x$$
$$ma_x = F_G \sin \theta - F_F$$
$$ma_x = mg \sin \theta - \mu F_N$$
$$ma_x = mg \sin \theta - \mu(mg \cos \theta)$$
$$ma_x = mg \sin \theta - \mu mg \cos \theta$$
$$a_x = g \sin \theta - \mu g \cos \theta$$
$$0 \text{ m/s}^2 = (9.8 \text{ m/s}^2) \sin \theta - (0.29)(9.8 \text{ m/s}^2) \cos \theta$$
$$0 \text{ m/s}^2 = (9.8 \text{ m/s}^2) \sin \theta - (2.84 \text{ m/s}^2) \cos \theta$$
$$(2.84 \text{ m/s}^2) \cos \theta = (9.8 \text{ m/s}^2) \sin \theta$$
$$(2.84 \text{ m/s}^2) = (9.8 \text{ m/s}^2) \tan \theta$$
$$0.29 = \tan \theta$$
$$\tan^{-1}(0.29) = \theta$$
$$16.2° = \theta$$

161. 18°

Draw the free-body diagram showing the three forces involved in the problem:

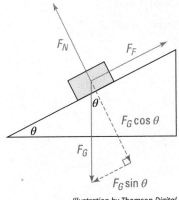

Illustration by Thomson Digital

where F_N is the normal force, F_G is the force of gravity ($F_G = ma_G = mg$ on Earth's surface), and F_F is the force of friction. Friction points up the slope because the snowboarder slides down, and friction always points in the opposite direction from the motion. Now add up the forces in each direction (x and y) and use Newton's second law with $a_y = 0$ meters per second squared because the snowboarder doesn't accelerate off of the mountainside's surface or into it. In the y or vertical direction:

$$F_{y,\text{net}} = \sum F_y$$
$$ma_y = F_N - F_G \cos \theta$$
$$ma_y = F_N - mg \cos \theta$$
$$m\left(0 \, \text{m/s}^2\right) = F_N - mg \cos \theta$$
$$0 = F_N - mg \cos \theta$$
$$mg \cos \theta = F_N$$

And in the x or horizontal direction, where "down the slope" is taken as positive in this solution:

$$F_{x,\text{net}} = \sum F_x$$
$$ma_x = F_G \sin \theta - F_F$$
$$ma_x = mg \sin \theta - \mu F_N$$
$$ma_x = mg \sin \theta - \mu(mg \cos \theta)$$
$$ma_x = mg \sin \theta - \mu mg \cos \theta$$
$$a_x = g \sin \theta - \mu g \cos \theta$$
$$2.1 \, \text{m/s}^2 = \left(9.8 \, \text{m/s}^2\right) \sin \theta - (0.1)\left(9.8 \, \text{m/s}^2\right) \cos \theta$$
$$2.1 = 9.8 \sin \theta - 0.98 \cos \theta$$

At this point the problem becomes one of algebraic manipulation. The following solution makes use of the trigonometric identity $\sin^2 \theta = 1 - \cos^2 \theta$:

$$2.1 = 9.8 \sin \theta - 0.98 \cos \theta$$
$$2.1 + 0.98 \cos \theta = 9.8 \sin \theta$$
$$(2.1 + 0.98 \cos \theta)^2 = (9.8 \sin \theta)^2$$
$$4.41 + 4.12 \cos \theta + 0.96 \cos^2 \theta = 96.04 \sin^2 \theta$$
$$4.41 + 4.12 \cos \theta + 0.96 \cos^2 \theta = 96.04\left(1 - \cos^2 \theta\right)$$
$$4.41 + 4.12 \cos \theta + 0.96 \cos^2 \theta = 96.04 - 96.04 \cos^2 \theta$$
$$97 \cos^2 \theta + 4.12 \cos \theta - 91.63 = 0$$

Now use the quadratic formula to solve for $\cos\theta$:

$$\cos\theta = \frac{-4.12 \pm \sqrt{4.12^2 - 4(97)(-91.63)}}{2(97)}$$

$$= \frac{-4.12 \pm \sqrt{16.97 + 35{,}552}}{194}$$

$$= \frac{-4.12 \pm \sqrt{35{,}569}}{194}$$

$$= \frac{-4.12 \pm 188.6}{194}$$

$$= \frac{-4.12 + 188.6}{194} \text{ or } \frac{-4.12 - 188.6}{194}$$

$$= \frac{184.5}{194} \text{ or } \frac{-192.7}{194}$$

$$\cos\theta = 0.951 \text{ or } -0.993$$

$$\theta = \cos^{-1}(0.951) \text{ or } \cos^{-1}(-0.993)$$

$$= 18° \text{ or } 173°$$

An answer of 173 degrees doesn't make sense in this situation (it's not an acute angle), so the mountain must be angled at an 18-degree inclination.

162.   4.5 m

Draw a free-body diagram to identify the forces acting along the axis of motion (the ramp's surface):

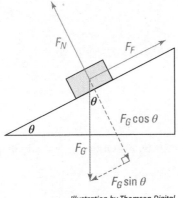

Illustration by Thomson Digital

where F_N is the normal force, F_G is the force of gravity ($F_G = ma_G = mg$ on Earth's surface), and F_F is the force of friction. The two forces acting along the ramp's axis are the force of friction and the horizontal (or x) component of the gravitational force. By Newton's second law, the sum of those forces is equal to the crate's mass times its acceleration along the ramp.

$$F_{x,\text{net}} = \sum F_x$$
$$ma_x = F_G \sin\theta - F_F$$
$$ma_x = mg \sin\theta - F_F$$
$$(5\,\text{kg})a = (5\,\text{kg})(9.8\,\text{m/s}^2)\sin 42° - (30\,\text{N})$$
$$(5\,\text{kg})a = (5\,\text{kg})(9.8\,\text{m/s}^2)(0.669) - 30\,\text{N}$$
$$(5\,\text{kg})a = 32.79\,\text{N} - 30\,\text{N}$$
$$(5\,\text{kg})a = 2.79\,\text{N}$$
$$a = 0.56\,\text{m/s}^2$$

Use the displacement formula to calculate how far away the crate is from its starting position after 4 seconds, keeping in mind that the initial velocity is 0 meters per second:

$$s = v_i t + \frac{1}{2}at^2$$
$$= (0\,\text{m/s})(4\,\text{s}) + \frac{1}{2}(0.56\,\text{m/s}^2)(4\,\text{s})^2$$
$$= 0\,\text{m} + 4.5\,\text{m}$$
$$= 4.5\,\text{m}$$

163. 0.50

First use the kinematic data to solve for the block's acceleration. Use the displacement formula, $s = v_i t + \frac{1}{2}at^2$, where s is the displacement, v_i is the initial velocity, a is the acceleration, and t is the amount of elapsed time:

$$s = v_i t + \frac{1}{2}at^2$$
$$5.8\,\text{m} = (1\,\text{m/s})(1.1\,\text{s}) + \frac{1}{2}a(1.1\,\text{s})^2$$
$$5.8\,\text{m} = 1.1\,\text{m} + (0.605\,\text{s}^2)a$$
$$4.7\,\text{m} = (0.605\,\text{s}^2)a$$
$$7.8\,\text{m/s}^2 = a$$

As with most every force problem, start by drawing a free-body diagram:

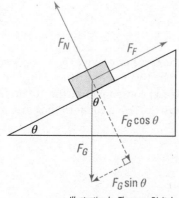

Illustration by Thomson Digital

where F_N is the normal force, F_G is the force of gravity ($F_G = ma_G = mg$ on Earth's surface), and F_F is the force of friction. Friction points up because the block slides down, and friction always points in the opposite direction from the motion. Now add up the forces in each direction (x and y) and set their sums equal to the block's mass times its acceleration along that particular axis ($a_y = 0$ meters per second squared because the block doesn't accelerate off the ramp's surface or into it). In the y or vertical direction:

$$F_{y,\text{net}} = \sum F_y$$
$$ma_y = F_N - F_G \cos \theta$$
$$ma_y = F_N - mg \cos \theta$$
$$m\left(0 \text{ m/s}^2\right) = F_N - mg \cos \theta$$
$$0 = F_N - mg \cos \theta$$
$$mg \cos \theta = F_N$$

And in the x or horizontal direction, where "down the slope" is taken as positive in this solution:

$$F_{x,\text{net}} = \sum F_x$$
$$ma_x = F_G \sin \theta - F_F$$
$$ma_x = mg \sin \theta - \mu F_N$$
$$ma_x = mg \sin \theta - \mu(mg \cos \theta)$$
$$ma_x = mg \sin \theta - \mu mg \cos \theta$$
$$a_x = g \sin \theta - \mu g \cos \theta$$
$$7.8 \text{ m/s}^2 = \left(9.8 \text{ m/s}^2\right) \sin 72° - \mu \left(9.8 \text{ m/s}^2\right) \cos 72°$$
$$7.8 \text{ m/s}^2 = \left(9.8 \text{ m/s}^2\right)(0.951) - \mu \left(9.8 \text{ m/s}^2\right)(0.309)$$
$$7.8 \text{ m/s}^2 = 9.32 \text{ m/s}^2 - \left(3.03 \text{ m/s}^2\right)\mu$$
$$-1.52 \text{ m/s}^2 = -\left(3.03 \text{ m/s}^2\right)\mu$$
$$0.50 = \mu$$

164. 220 m

The skier's journey has two segments, and each requires its own free-body diagram to calculate acceleration. First, for the initial hill,

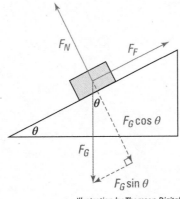

Illustration by Thomson Digital

where F_N is the normal force, F_G is the force of gravity ($F_G = ma_G = mg$ on Earth's surface), and F_F is the force of friction. Friction always opposes motion, so if the skier is moving to the left, the force of friction is exerted to the right throughout each diagram. Add up the forces in each direction (x and y) and set their sums equal to the skier's mass times her acceleration along that particular axis ($a_y = 0$ meters per second squared in each of the three sections because the skier doesn't accelerate off the ground or into it at any point). In the y (or vertical) direction:

$$F_{y,\text{net}} = \sum F_y$$
$$ma_y = F_N - F_G \cos \theta$$
$$ma_y = F_N - mg \cos \theta$$
$$m(0 \text{ m/s}^2) = F_N - mg \cos \theta$$
$$0 = F_N - mg \cos \theta$$
$$mg \cos \theta = F_N$$

And in the x or horizontal direction, where "to the left" is taken as positive throughout this solution:

$$F_{x,\text{net}} = \sum F_x$$
$$ma_x = F_G \sin \theta - F_F$$
$$ma = mg \sin \theta - \mu F_N$$
$$ma = mg \sin \theta - \mu(mg \cos \theta)$$
$$ma = mg \sin \theta - \mu mg \cos \theta$$
$$a = g \sin \theta - \mu g \cos \theta$$
$$= (9.8 \text{ m/s}^2) \sin 70° - (0.08)(9.8 \text{ m/s}^2) \cos 70°$$
$$= (9.8 \text{ m/s}^2)(0.94) - (0.08)(9.8 \text{ m/s}^2)(0.342)$$
$$= 9.21 \text{ m/s}^2 - 0.27 \text{ m/s}^2$$
$$= 8.94 \text{ m/s}^2$$

Use trigonometry to determine the length of the skier's path down the hill:

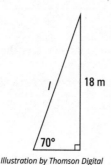

Illustration by Thomson Digital

$$\sin 70° = \frac{18\,\text{m}}{l}$$

$$0.94 = \frac{18\,\text{m}}{l}$$

$$0.94l = 18\,\text{m}$$

$$l = 19.1\,\text{m}$$

To calculate the velocity at the bottom of the hill — and the beginning of the flat section — use the velocity-displacement formula:

$$v_f^2 = v_i^2 + 2as$$

$$
\begin{aligned}
v_f &= \sqrt{v_i^2 + 2as} \\
&= \sqrt{(0\,\text{m/s})^2 + 2(8.94\,\text{m/s}^2)(19.1\,\text{m})} \\
&= \sqrt{0\,\text{m}^2/\text{s}^2 + 341.5\,\text{m}^2/\text{s}^2} \\
&= \sqrt{341.5\,\text{m}^2/\text{s}^2} \\
&= 18.5\,\text{m/s}
\end{aligned}
$$

Second, for the flat section, as the skier continues to the left:

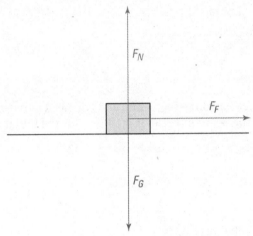

Illustration by Thomson Digital

Follow the previous steps of summing the forces and using Newton's second law. First, in the y direction:

$$
\begin{aligned}
F_{y,\text{net}} &= \sum F_y \\
ma_y &= F_N - F_G \\
ma_y &= F_N - mg \\
m(0\,\text{m/s}^2) &= F_N - mg \\
0 &= F_N - mg \\
mg &= F_N
\end{aligned}
$$

Then in the x direction:

$$F_{x,\text{net}} = \sum F_x$$
$$ma_x = -F_F$$
$$ma = -\mu F_N$$
$$ma = -\mu(mg)$$
$$ma = -\mu mg$$
$$a = -\mu g$$
$$= -(0.08)\left(9.8 \text{ m/s}^2\right)$$
$$= -0.78 \text{ m/s}^2$$

Make one more use of the velocity-displacement formula — this time to figure out the displacement required for the final velocity to equal 0 meters per second (when the skier stops):

$$v_f^2 = v_i^2 + 2as$$
$$(0 \text{ m/s})^2 = (18.5 \text{ m/s})^2 + 2\left(-0.78 \text{ m/s}^2\right)s$$
$$0 \text{ m}^2/\text{s}^2 = 342.25 \text{ m}^2/\text{s}^2 - \left(1.56 \text{ m/s}^2\right)s$$
$$-342.25 \text{ m}^2/\text{s}^2 = -\left(1.56 \text{ m/s}^2\right)s$$
$$219.4 \text{ m} = s \approx 220 \text{ m}$$

165.   **3 m**

The skateboard's journey has three segments, and you need a free-body diagram for each one to determine the board's acceleration. First, for the initial hill,

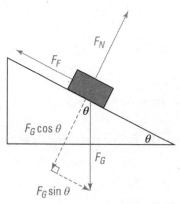

Illustration by Thomson Digital

where F_N is the normal force, F_G is the force of gravity ($F_G = ma_G = mg$ on Earth's surface), and F_F is the force of friction. Friction always opposes motion, so if the skateboard is moving to the right, the force of friction is exerted to the left throughout all the diagrams. Add up the forces in each direction (x and y) and set their sums equal to the board's mass times its acceleration along that particular axis ($a_y = 0$ meters per second squared in each of the three sections because the board doesn't accelerate off the ground or into it at any point). In the y (or vertical) direction:

$$F_{y,\text{net}} = \sum F_y$$
$$ma_y = F_N - F_G \cos \theta$$
$$ma_y = F_N - mg \cos \theta$$
$$m(0 \text{ m/s}^2) = F_N - mg \cos \theta$$
$$0 = F_N - mg \cos \theta$$
$$mg \cos \theta = F_N$$

And in the x or horizontal direction, where "to the right" is taken as positive throughout this solution:

$$F_{x,\text{net}} = \sum F_x$$
$$ma_x = F_G \sin \theta - F_F$$
$$ma = mg \sin \theta - \mu F_N$$
$$ma = mg \sin \theta - \mu(mg \cos \theta)$$
$$ma = mg \sin \theta - \mu mg \cos \theta$$
$$a = g \sin \theta - \mu g \cos \theta$$
$$= (9.8 \text{ m/s}^2) \sin 55° - (0.18)(9.8 \text{ m/s}^2) \cos 55°$$
$$= (9.8 \text{ m/s}^2)(0.819) - (0.18)(9.8 \text{ m/s}^2)(0.574)$$
$$= 8.03 \text{ m/s}^2 - 1.01 \text{ m/s}^2$$
$$= 7.02 \text{ m/s}^2$$

Use trigonometry to determine the length of the skateboard's path down the first hill:

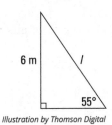

Illustration by Thomson Digital

$$\sin 55° = \frac{6 \text{ m}}{l}$$
$$0.819 = \frac{6 \text{ m}}{l}$$
$$0.819 l = 6 \text{ m}$$
$$l = 7.33 \text{ m}$$

To calculate the board's velocity at the end of the first hill — and the beginning of the flat section — use the velocity-displacement formula:

$$v_f^2 = v_i^2 + 2as$$
$$v_f = \sqrt{v_i^2 + 2as}$$
$$= \sqrt{(0 \text{ m/s})^2 + 2(7.02 \text{ m/s}^2)(7.33 \text{ m})}$$
$$= \sqrt{0 \text{ m}^2/\text{s}^2 + 102.9 \text{ m}^2/\text{s}^2}$$
$$= \sqrt{102.9 \text{ m}^2/\text{s}^2}$$
$$= 10.1 \text{ m/s}$$

Second, for the flat section,

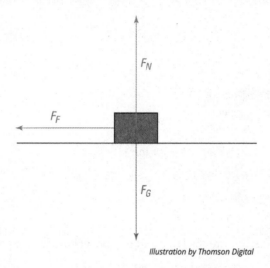

Illustration by Thomson Digital

Follow the previous steps of summing the forces and using Newton's second law. First, in the y direction:

$$F_{y,\text{net}} = \sum F_y$$
$$ma_y = F_N - F_G$$
$$ma_y = F_N - mg$$
$$m(0 \text{ m/s}^2) = F_N - mg$$
$$0 = F_N - mg$$
$$mg = F_N$$

Then in the x direction:

$$F_{x,\text{net}} = \sum F_x$$
$$ma_x = -F_F$$
$$ma = -\mu F_N$$
$$ma = -\mu(mg)$$
$$ma = -\mu mg$$
$$a = -\mu g$$
$$= -(0.18)(9.8 \text{ m/s}^2)$$
$$= -1.76 \text{ m/s}^2$$

Use the velocity-displacement formula again to find the velocity at the end of the flat section — and the initial velocity heading into the final section:

$$v_f^2 = v_i^2 + 2as$$
$$v_f = \sqrt{v_i^2 + 2as}$$
$$= \sqrt{(10.1\,\text{m/s})^2 + 2(-1.76\,\text{m/s}^2)(3.5\,\text{m})}$$
$$= \sqrt{102\,\text{m}^2/\text{s}^2 - 12.3\,\text{m}^2/\text{s}^2}$$
$$= \sqrt{89.7\,\text{m}^2/\text{s}^2}$$
$$= 9.47\,\text{m/s}$$

Finally, for the second hill

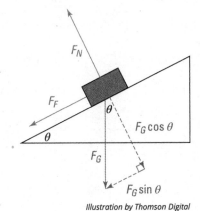

Illustration by Thomson Digital

$$F_{y,\text{net}} = \sum F_y$$
$$ma_y = F_N - F_G \cos \theta$$
$$ma_y = F_N - mg \cos \theta$$
$$m(0\,\text{m/s}^2) = F_N - mg \cos \theta$$
$$0 = F_N - mg \cos \theta$$
$$mg \cos \theta = F_N$$

And in the x or horizontal direction,

$$F_{x,\text{net}} = \sum F_x$$
$$ma_x = -F_G \sin \theta - F_F$$
$$ma = -mg \sin \theta - \mu F_N$$
$$ma = -mg \sin \theta - \mu(mg \cos \theta)$$
$$ma = -mg \sin \theta - \mu mg \cos \theta$$
$$a = -g \sin \theta - \mu g \cos \theta$$
$$= -(9.8\,\text{m/s}^2)\sin 20° - (0.18)(9.8\,\text{m/s}^2)\cos 20°$$
$$= -(9.8\,\text{m/s}^2)(0.342) - (0.18)(9.8\,\text{m/s}^2)(0.94)$$
$$= -3.35\,\text{m/s}^2 - 1.66\,\text{m/s}^2$$
$$= -5.01\,\text{m/s}^2$$

Make one more use of the velocity-displacement formula — this time to figure out the displacement required for the final velocity to equal 0 meters per second (when the skateboard stops moving forward):

$$v_f^2 = v_i^2 + 2as$$
$$(0\,\text{m/s})^2 = (9.47\,\text{m/s})^2 + 2\left(-5.01\,\text{m/s}^2\right)s$$
$$0\,\text{m}^2/\text{s}^2 = 89.68\,\text{m}^2/\text{s}^2 - \left(10.02\,\text{m/s}^2\right)s$$
$$-89.68\,\text{m}^2/\text{s}^2 = -\left(10.02\,\text{m/s}^2\right)s$$
$$8.95\,\text{m} = s$$

Finally, use trigonometry to calculate the final height of the skateboard.

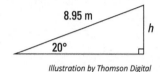

Illustration by Thomson Digital

$$\sin 20° = \frac{h}{8.95\,\text{m}}$$
$$0.342 = \frac{h}{8.95\,\text{m}}$$
$$3.06\,\text{m} = h \approx 3\,\text{m}$$

166.   0 m/s

The highest point in a projectile's arc is the place where the projectile's vertical velocity changes from a positive (upward) value to a negative (downward) one. Therefore, its vertical velocity at the top must be 0 meters per second (the only number between positives and negatives). At the top of its arc, the baseball is no longer moving upward and is about to start moving downward; it is stationary with respect to the vertical axis of motion and has a velocity of 0 meters per second.

167.   0.95 m

First convert the mass into "correct" units:

$$(50\,\text{g})\left(\frac{1\,\text{kg}}{1,000\,\text{g}}\right) = 0.05\,\text{kg}$$

After the Frisbee is launched, two forces act on it, both pointing down:

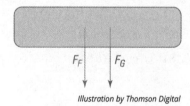

Illustration by Thomson Digital

Use the force form of Newton's second law to calculate the net acceleration on the Frisbee:

$$F_{net} = \sum F$$
$$ma_{net} = -F_F - F_G$$
$$ma_{net} = -F_F - ma_G$$
$$ma_{net} = -F_F - mg$$
$$(0.05 \text{ kg})a_{net} = -(10 \text{ N}) - (0.05 \text{ kg})\left(9.8 \text{ m/s}^2\right)$$
$$(0.05 \text{ kg})a_{net} = -10 \text{ N} - 0.49 \text{ N}$$
$$(0.05 \text{ kg})a_{net} = -10.49 \text{ N}$$
$$a_{net} = -209.8 \text{ m/s}^2$$

Now use the velocity-displacement formula to solve for the Frisbee's maximum height — that is, the location where its vertical velocity equals 0 meters per second:

$$v_f^2 = v_i^2 + 2as$$
$$(0 \text{ m/s})^2 = (20 \text{ m/s})^2 + 2\left(-209.8 \text{ m/s}^2\right)s$$
$$0 \text{ m}^2/\text{s}^2 = 400 \text{ m}^2/\text{s}^2 - \left(419.6 \text{ m/s}^2\right)s$$
$$-400 \text{ m}^2/\text{s}^2 = -\left(419.6 \text{ m/s}^2\right)s$$
$$0.95 \text{ m} = s$$

168. 1.8 km

First convert the force of the lower section's engine into "correct" units:

$$(150 \text{ kN})\left(\frac{1 \times 10^3 \text{ N}}{1 \text{ kN}}\right) = 1.5 \times 10^5 \text{ N}$$

Then analyze the forces acting on the rocket during the stage where both pieces are attached and the engine is firing. Draw a free-body diagram and write out a Newton's second-law equation:

Illustration by Thomson Digital

(F_F is the force of air resistance.)

$$F_{net} = \sum F$$
$$ma = F_{engine} - F_F - F_G$$
$$ma = F_{engine} - \frac{m}{\left(0.06 \text{ s}^2/\text{m}^2\right)l} - ma_G$$
$$ma = F_{engine} - \frac{m}{\left(0.06 \text{ s}^2/\text{m}^2\right)l} - mg$$
$$(13{,}500 \text{ kg})a = \left(1.5 \times 10^5 \text{ N}\right) - \frac{13{,}500 \text{ kg}}{\left(0.06 \text{ s}^2/\text{m}^2\right)(42 \text{ m})} - (13{,}500 \text{ kg})\left(9.8 \text{ m/s}^2\right)$$
$$(13{,}500 \text{ kg})a = 1.5 \times 10^5 \text{ N} - 5{,}357 \text{ N} - 1.323 \times 10^5 \text{ N}$$
$$(13{,}500 \text{ kg})a = 12{,}343 \text{ N}$$
$$a = 0.914 \text{ m/s}^2$$

Use that acceleration in the velocity-time formula to solve for the rocket's velocity after 60 seconds:

$$v_f = v_i + at$$
$$= (0 \text{ m/s}) + \left(0.914 \text{ m/s}^2\right)(60 \text{ s})$$
$$= 0 \text{ m/s} + 54.84 \text{ m/s}$$
$$= 54.84 \text{ m/s}$$

Then use either the displacement formula or the velocity-displacement formula (used here) to solve for the height of the rocket after that first minute:

$$v_f^2 = v_i^2 + 2as$$
$$(54.84 \text{ m/s})^2 = (0 \text{ m/s})^2 + 2\left(0.914 \text{ m/s}^2\right)s$$
$$3{,}007 \text{ m}^2/\text{s}^2 = 0 \text{ m}^2/\text{s}^2 + \left(1.828 \text{ m/s}^2\right)s$$
$$1{,}645 \text{ m} = s$$

So the rocket is 1,645 meters above the ground with a velocity of 54.84 meters per second before the lower stage detaches from the upper one. After they detach, the free-body diagram changes because the engine is no longer running, and, therefore, the second-law equations stop working. When substituting values, note that the mass of the body — as well as its length — is different than it was in the first portion of the solution.

F_F F_G

Illustration by Thomson Digital

$$F_{net} = \sum F$$

$$ma = -F_F - F_G$$

$$ma = -\frac{m}{\left(0.06\ \text{s}^2/\text{m}^2\right)l} - ma_G$$

$$ma = -\frac{m}{\left(0.06\ \text{s}^2/\text{m}^2\right)l} - mg$$

$$a = -\frac{1}{\left(0.06\ \text{s}^2/\text{m}^2\right)l} - g$$

$$a = -\frac{1}{\left(0.06\ \text{s}^2/\text{m}^2\right)(34\ \text{m})} - \left(9.8\ \text{m/s}^2\right)$$

$$a = -0.49\ \text{m/s}^2 - 9.8\ \text{m/s}^2$$

$$a = -10.29\ \text{m/s}^2$$

Because you're interested only in the final height of the lower stage of the rocket, you need to solve for the stage's height when its vertical velocity is 0 meters per second. Use the velocity–displacement formula once more to solve for the height that the rocket travels while under this new acceleration (until its velocity reaches 0):

$$v_f^2 = v_i^2 + 2as$$

$$(0\ \text{m/s})^2 = (54.84\ \text{m/s})^2 + 2\left(-10.29\ \text{m/s}^2\right)s$$

$$0\ \text{m}^2/\text{s}^2 = 3{,}007\ \text{m}^2/\text{s}^2 - \left(20.58\ \text{m/s}^2\right)s$$

$$\left(20.58\ \text{m/s}^2\right)s = 3{,}007\ \text{m}^2/\text{s}^2$$

$$s = 146\ \text{m}$$

Add this displacement to the total you found earlier to find the total displacement above the ground before the lower section starts moving back toward Earth:

$$S_{total} = (1{,}645\ \text{m}) + (146\ \text{m})$$

$$= 1{,}791\ \text{m}$$

Convert to kilometers and round:

$$(1{,}791\ \text{m})\left(\frac{1\ \text{km}}{1{,}000\ \text{m}}\right) = 1.791\ \text{km} \approx 1.8\ \text{km}$$

169. 4.5 km

First convert the force measurements out of kilonewtons:

$$(150\ \text{kN})\left(\frac{1 \times 10^3\ \text{N}}{1\ \text{kN}}\right) = 1.5 \times 10^5\ \text{N}$$

$$(25\ \text{kN})\left(\frac{1 \times 10^3\ \text{N}}{1\ \text{kN}}\right) = 2.5 \times 10^4\ \text{N}$$

Then analyze the forces acting on the rocket during the stage where both pieces are attached and the engine is firing. Draw a free-body diagram and write out a Newton's second-law equation:

(F_F is the force of air resistance.)

$$F_{net} = \sum F$$
$$ma = F_{engine} - F_F - F_G$$
$$ma = F_{engine} - \frac{m}{\left(0.06 \ \text{s}^2/\text{m}^2\right)l} - ma_G$$
$$ma = F_{engine} - \frac{m}{\left(0.06 \ \text{s}^2/\text{m}^2\right)l} - mg$$
$$(13{,}500 \ \text{kg})a = \left(1.5 \times 10^5 \ \text{N}\right) - \frac{13{,}500 \ \text{kg}}{\left(0.06 \ \text{s}^2/\text{m}^2\right)(42 \ \text{m})} - (13{,}500 \ \text{kg})\left(9.8 \ \text{m/s}^2\right)$$
$$(13{,}500 \ \text{kg})a = 1.5 \times 10^5 \ \text{N} - 5{,}357 \ \text{N} - 1.323 \times 10^5 \ \text{N}$$
$$(13{,}500 \ \text{kg})a = 12{,}343 \ \text{N}$$
$$a = 0.914 \ \text{m/s}^2$$

Use that acceleration in the velocity-time formula to solve for the rocket's velocity after 60 seconds:

$$v_f = v_i + at$$
$$= (0 \ \text{m/s}) + \left(0.914 \ \text{m/s}^2\right)(60 \ \text{s})$$
$$= 0 \ \text{m/s} + 54.84 \ \text{m/s}$$
$$= 54.84 \ \text{m/s}$$

Then use either the displacement formula or the velocity-displacement formula (used here) to solve for the height of the rocket after that first minute:

$$v_f^2 = v_i^2 + 2as$$

$$(54.84 \text{ m/s})^2 = (0 \text{ m/s})^2 + 2(0.914 \text{ m/s}^2)s$$

$$3{,}007 \text{ m}^2/\text{s}^2 = 0 \text{ m}^2/\text{s}^2 + (1.828 \text{ m/s}^2)s$$

$$3{,}007 \text{ m}^2/\text{s}^2 = (1.828 \text{ m/s}^2)s$$

$$1{,}645 \text{ m} = s$$

So the rocket is 1,645 meters above the ground with a velocity of 54.84 meters per second before the lower stage detaches from the upper one. After they detach, the second-law equations change given the different mass and body length:

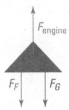

$$F_{\text{net}} = \sum F$$

$$ma = F_{\text{engine}} - F_F - F_G$$

$$ma = F_{\text{engine}} - \frac{m}{(0.06 \text{ s}^2/\text{m}^2)l} - ma_G$$

$$ma = F_{\text{engine}} - \frac{m}{(0.06 \text{ s}^2/\text{m}^2)l} - mg$$

$$(1{,}850 \text{ kg})a = (2.5 \times 10^4 \text{ N}) - \frac{1{,}850 \text{ kg}}{(0.06 \text{ s}^2/\text{m}^2)(8\text{m})} - (1{,}850 \text{ kg})(9.8 \text{ m/s}^2)$$

$$(1{,}850 \text{ kg})a = 2.5 \times 10^4 \text{ N} - 3{,}854 \text{ N} - 1.813 \times 10^4 \text{ N}$$

$$(1{,}850 \text{ kg})a = 3{,}016 \text{ N}$$

$$a = 1.63 \text{ m/s}^2$$

Use the velocity-time formula again to solve for the top stage's velocity after 30 seconds under this rate of acceleration:

$$v_f = v_i + at$$

$$= (54.84 \text{ m/s}) + (1.63 \text{ m/s}^2)(30 \text{ s})$$

$$= 54.84 \text{ m/s} + 48.9 \text{ m/s}$$

$$= 103.74 \text{ m/s}$$

Then use either the displacement formula or the velocity-displacement formula (used here) to solve for the height of the rocket after that first minute:

$$v_f^2 = v_i^2 + 2as$$

$$(103.74 \text{ m/s})^2 = (54.84 \text{ m/s})^2 + 2(1.63 \text{ m/s}^2)s$$

$$10{,}762 \text{ m}^2/\text{s}^2 = 3{,}007 \text{ m}^2/\text{s}^2 + (3.26 \text{ m/s}^2)s$$

$$7{,}755 \text{ m}^2/\text{s}^2 = (3.26 \text{ m/s}^2)s$$

$$2{,}379 \text{ m} = s$$

So the rocket is now $1{,}645 + 2{,}379 = 4{,}024$ meters above the ground with a velocity of 103.74 meters per second before the engine turns off. From there the free-body diagram looks like this:

Use Newton's second law again to solve for the top stage's acceleration during this last section:

$$F_{net} = \sum F$$

$$ma = -F_F - F_G$$

$$ma = -\frac{m}{(0.06 \text{ s}^2/\text{m}^2)l} - ma_G$$

$$ma = -\frac{m}{(0.06 \text{ s}^2/\text{m}^2)l} - mg$$

$$a = -\frac{1}{(0.06 \text{ s}^2/\text{m}^2)l} - g$$

$$a = -\frac{1}{(0.06 \text{ s}^2/\text{m}^2)(8 \text{ m})} - (9.8 \text{ m/s}^2)$$

$$a = -2.08 \text{ m/s}^2 - 9.8 \text{ m/s}^2$$

$$a = -11.88 \text{ m/s}^2$$

Because you're interested only in the final height of the upper stage of the rocket, you need to solve for the stage's height when its vertical velocity is 0 meters per second. Use the velocity-displacement formula once more to solve for the height that the rocket travels while under this new acceleration (until its velocity reaches 0):

$$v_f^2 = v_i^2 + 2as$$

$$(0 \text{ m/s})^2 = (103.74 \text{ m/s})^2 + 2(-11.88 \text{ m/s}^2)s$$

$$0 \text{ m}^2/\text{s}^2 = 10{,}762 \text{ m}^2/\text{s}^2 - (23.76 \text{ m/s}^2)s$$

$$(23.76 \text{ m/s}^2)s = 10{,}762 \text{ m}^2/\text{s}^2$$

$$s = 453 \text{ m}$$

Add this displacement to the total you found earlier to find the total displacement above the ground before the upper section starts moving back toward Earth:

$$s_{\text{total}} = (4{,}024 \text{ m}) + (453 \text{ m})$$
$$= 4{,}477 \text{ m}$$

Convert to kilometers and round:

$$(4{,}477 \text{ m})\left(\frac{1 \text{ km}}{1{,}000 \text{ m}}\right) = 4.477 \text{ km} \approx 4.5 \text{ km}$$

170. 1.4 s

When a projectile launches from and lands at the same vertical position, it reaches its apex (highest point) halfway through the trip — both in terms of horizontal distance traveled and time of flight. Therefore, the rocket reaches its highest point of flight after $\dfrac{2.8 \text{ s}}{2} = 1.4 \text{ s}$.

171. 29 m/s

Weight is vernacular for *force of gravity*, so the force of air resistance F_F is equal to $0.1F_G$, where F_G is the force of gravity. Draw the free-body diagram:

Illustration by Thomson Digital

The friction force F_F (air resistance) points downward because friction always opposes an object's motion, which — on the ball's way to the highest point — is upward. Now use the force form of Newton's second law to calculate the billiard ball's acceleration:

$$F_{\text{net}} = \sum F$$
$$ma_{\text{net}} = -F_F - F_G$$
$$ma_{\text{net}} = -0.1F_G - F_G$$
$$ma_{\text{net}} = -1.1F_G$$
$$ma_{\text{net}} = -1.1ma_G$$
$$ma_{\text{net}} = -1.1mg$$
$$a_{\text{net}} = -1.1g$$
$$a_{\text{net}} = -(1.1)\left(9.8 \text{ m/s}^2\right)$$
$$a_{\text{net}} = -10.78 \text{ m/s}^2$$

The apex (highest point) of a projectile's arc is the place where the projectile's vertical velocity changes from a positive (upward) value to a negative (downward) one. Therefore, its vertical velocity at the top must be 0 meters per second (the only number between positives and negatives). So use the velocity-time formula to solve for the ball's initial velocity at launch, 2.7 seconds prior to it having a "final" velocity of 0 meters per second:

$$v_f = v_i + at$$
$$0\,\text{m/s} = v_i + \left(-10.78\,\text{m/s}^2\right)(2.7\,\text{s})$$
$$0\,\text{m/s} = v_i - 29.1\,\text{m/s}$$
$$29.1\,\text{m/s} = v_i$$

172.   15 s

First convert the force of the engine into "correct" units:

$$(15\,\text{kN})\left(\frac{1\times10^3\,\text{N}}{1\,\text{kN}}\right) = 1.5\times10^4\,\text{N}$$

For the first portion of the upward journey, the engine is on, so two forces are acting on the rocket:

Illustration by Thomson Digital

Use the diagram to write out the force form of Newton's second law:

$$F_{\text{net}} = \sum F$$
$$ma_{\text{net}} = F_{\text{engine}} - F_G$$
$$ma_{\text{net}} = F_{\text{engine}} - ma_G$$
$$ma_{\text{net}} = F_{\text{engine}} - mg$$
$$(1{,}300\,\text{kg})a_{\text{net}} = \left(1.5\times10^4\,\text{N}\right) - (1{,}300\,\text{kg})\left(9.8\,\text{m/s}^2\right)$$
$$(1{,}300\,\text{kg})a_{\text{net}} = 15{,}000\,\text{N} - 12{,}740\,\text{N}$$
$$(1{,}300\,\text{kg})a_{\text{net}} = 2{,}260\,\text{N}$$
$$a_{\text{net}} = 1.74\,\text{m/s}^2$$

Substitute this into the velocity-displacement formula to calculate the rocket's velocity when it reaches a height of 150 meters:

$$v_f^2 = v_i^2 + 2as$$
$$v_f = \sqrt{v_i^2 + 2as}$$
$$= \sqrt{(0 \text{ m/s})^2 + 2(1.74 \text{ m/s}^2)(150 \text{ m})}$$
$$= \sqrt{0 \text{ m}^2/\text{s}^2 + 522 \text{ m}^2/\text{s}^2}$$
$$= \sqrt{522 \text{ m}^2/\text{s}^2}$$
$$= 22.85 \text{ m/s}$$

Use this result along with the acceleration to discover the amount of time this part of the trip lasts by using the velocity-time formula:

$$v_f = v_i + at$$
$$22.85 \text{ m/s} = (0 \text{ m/s}) + (1.74 \text{ m/s}^2)t$$
$$22.85 \text{ m/s} = (1.74 \text{ m/s}^2)t$$
$$13.13 \text{ s} = t$$

When the engine shuts off, however, the only force acting on the rocket is gravity, so the rocket's acceleration from that point is −9.8 meters per second squared. The apex (highest point) of a projectile's arc is the place where the projectile's vertical velocity changes from a positive (upward) value to a negative (downward) one. Therefore, its vertical velocity at the top must be 0 meters per second (the only number between positives and negatives). So use the velocity-time formula once more to find the amount of time it takes for the rocket to slow from 22.85 meters per second to 0:

$$v_f = v_i + at$$
$$0 \text{ m/s} = (22.85 \text{ m/s}) + (-9.8 \text{ m/s}^2)t$$
$$-22.85 \text{ m/s} = -(9.8 \text{ m/s}^2)t$$
$$2.33 \text{ s} = t$$

Add the two times to find the total time it takes for the rocket to reach its apex:

$$t_{\text{total}} = t_1 + t_2$$
$$= (13.13 \text{ s}) + (2.33 \text{ s})$$
$$= 15.46 \text{ s} \approx 15 \text{ s}$$

173.   2.8 N

Immediately after the basketball is released from the shooter's hand, its direction of travel is still 70 degrees with respect to the horizontal. Just as velocity has horizontal and vertical components, so does air resistance (force of friction). To calculate the vertical component of a force — given an angle measured relative to the horizontal direction — use the equation $F_y = F \sin\theta$, where F is the magnitude of the force and θ is the angle relative to the horizontal axis.

$$
\begin{aligned}
F_y &= F \sin\theta \\
&= (3\,\text{N}) \sin 70° \\
&= (3\,\text{N})(0.94) \\
&= 2.8\,\text{N}
\end{aligned}
$$

174.   20 m/s

The highest point in a projectile's arc is the place where the projectile's vertical velocity changes from a positive (upward) value to a negative (downward) one. Therefore, its vertical velocity at the top must be 0 meters per second (the only number between positives and negatives). At that point the only component of the cannonball's velocity is in the horizontal direction, which — because it experiences no forces, and therefore no acceleration, in the horizontal direction — maintains a constant 20 meters per second. Therefore, the cannonball's overall velocity at the top is 20 meters per second.

175.   3.0 s

Start by drawing the pallet's free-body diagram while it's pushed up the ramp:

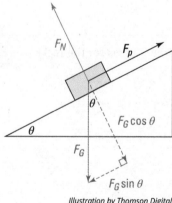

Illustration by Thomson Digital

Write out the force form of Newton's second law in the horizontal — along the ramp's surface — direction (the following solution designates "up the ramp" as the positive direction):

$$F_{x,\text{net}} = \sum F_x$$
$$ma_x = F_p - F_G \sin\theta$$
$$ma = F_p - ma_G \sin\theta$$
$$ma = F_p - mg \sin\theta$$
$$(50\,\text{kg})a = (600\,\text{N}) - (50\,\text{kg})\left(9.8\,\text{m/s}^2\right)\sin 30°$$
$$(50\,\text{kg})a = 600\,\text{N} - (50\,\text{kg})\left(9.8\,\text{m/s}^2\right)(0.5)$$
$$(50\,\text{kg})a = 600\,\text{N} - 245\,\text{N}$$
$$(50\,\text{kg})a = 355\,\text{N}$$
$$a = 7.1\,\text{m/s}^2$$

Then examine the trigonometry of the situation to solve for the ramp's length (along which the above acceleration occurs):

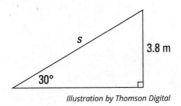

Illustration by Thomson Digital

$$\sin 30° = \frac{3.8\,\text{m}}{s}$$
$$0.5 = \frac{3.8\,\text{m}}{s}$$
$$0.5\,s = 3.8\,\text{m}$$
$$s = 7.6\,\text{m}$$

Use the velocity-displacement formula to find the pallet's velocity at the top of the ramp:

$$v_{fx}^2 = v_{ix}^2 + 2a_x s_x$$
$$v_{fx} = \sqrt{v_{ix}^2 + 2a_x s_x}$$
$$= \sqrt{(0\,\text{m/s})^2 + 2\left(7.1\,\text{m/s}^2\right)(7.6\,\text{m})}$$
$$= \sqrt{0\,\text{m}^2/\text{s}^2 + 107.9\,\text{m}^2/\text{s}^2}$$
$$= \sqrt{107.9\,\text{m}^2/\text{s}^2}$$
$$= 10.4\,\text{m/s}$$

Use the velocity–time formula to calculate the amount of time the slide up the ramp takes:

$$v_{fx} = v_{ix} + a_x t$$

$$10.4 \text{ m/s} = (0 \text{ m/s}) + (7.1 \text{ m/s}^2) t$$

$$10.4 \text{ m/s} = (7.1 \text{ m/s}^2) t$$

$$1.46 \text{ s} = t$$

Now turn your attention to the flight portion of the pallet's journey. Upon "takeoff," the pallet has a horizontal component of velocity equal to

$$\begin{aligned} v_x &= v \cos \theta \\ &= (10.4 \text{ m/s}) \cos 30° \\ &= (10.4 \text{ m/s})(0.866) \\ &= 9.01 \text{ m/s} \end{aligned}$$

The vertical component of velocity is equal to

$$\begin{aligned} v_y &= v \sin \theta \\ &= (10.4 \text{ m/s}) \sin 30° \\ &= (10.4 \text{ m/s})(0.5) \\ &= 5.2 \text{ m/s} \end{aligned}$$

Use the displacement formula to solve for the amount of time the pallet spends in the air; the result is a 3.8-meter fall onto the water's surface. The vertical acceleration is −9.8 meters per second squared because the pallet has left the ramp's surface and is in free fall:

$$s_y = v_{iy} t + \frac{1}{2} a_y t^2$$

$$-3.8 \text{ m} = (5.2 \text{ m/s}) t + \frac{1}{2}(-9.8 \text{ m/s}^2) t^2$$

$$-3.8 \text{ m} = (5.2 \text{ m/s}) t - (4.9 \text{ m/s}^2) t^2$$

$$(4.9 \text{ m/s}^2) t^2 - (5.2 \text{ m/s}) t - 3.8 \text{ m} = 0 \text{ m}$$

$$\begin{aligned} t &= \frac{(5.2 \text{ m/s}) \pm \sqrt{(-5.2 \text{ m/s})^2 - 4(4.9 \text{ m/s}^2)(-3.8 \text{ m})}}{2(4.9 \text{ m/s}^2)} \\ &= \frac{(5.2 \text{ m/s}) \pm \sqrt{27.04 \text{ m}^2/\text{s}^2 + 74.48 \text{ m}^2/\text{s}^2}}{9.8 \text{ m/s}^2} \\ &= \frac{5.2 \text{ m/s} \pm \sqrt{101.52 \text{ m}^2/\text{s}^2}}{9.8 \text{ m/s}^2} \\ &= \frac{5.2 \text{ m/s} \pm 10.1 \text{ m/s}}{9.8 \text{ m/s}^2} \\ &= \frac{5.2 \text{ m/s} + 10.1 \text{ m/s}}{9.8 \text{ m/s}^2} \text{ or } \frac{5.2 \text{m/s} - 10.1 \text{m/s}}{9.8 \text{ m/s}^2} \\ &= \frac{15.3 \text{ m/s}}{9.8} \text{ or } \frac{-4.9 \text{ m/s}}{9.8} \\ &= 1.56 \text{ s or } -0.5 \text{ s} \end{aligned}$$

The negative value doesn't make sense in the context of the problem, so the pertinent result must be 1.56 seconds of flight. Add this to the time the pallet spends sliding and round to the requested place to solve for the final answer:

$$t_{total} = t_{slide} + t_{flight}$$
$$= (1.46 \text{ s}) + (1.56 \text{ s})$$
$$= 3.02 \text{ s} \approx 3.0 \text{ s}$$

176.   183 cm

First convert the mass into "correct" units:

$$(285 \text{ g})\left(\frac{1 \text{ kg}}{1,000 \text{ g}}\right) = 0.285 \text{ kg}$$

Then draw a free–body diagram for the disc and analyze the forces in the vertical directions using Newton's second law.

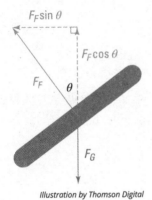

Illustration by Thomson Digital

Geometric principles reveal the following relationships between the angles:

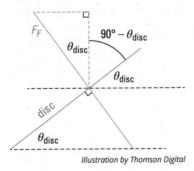

Illustration by Thomson Digital

So in the vertical direction,

$$F_{y,net} = \sum F_y$$
$$ma_y = F_F \cos\theta - F_G$$
$$ma_y = F_F \cos\theta - ma_G$$
$$ma_y = F_F \cos\theta - mg$$
$$(0.285\text{ kg})a_y = (1.2\text{ N})\cos 35° - (0.285\text{ kg})\left(9.8\text{ m/s}^2\right)$$
$$(0.285\text{ kg})a_y = (1.2\text{ N})(0.819) - (0.285\text{ kg})\left(9.8\text{ m/s}^2\right)$$
$$(0.285\text{ kg})a_y = 0.983\text{ N} - 2.793\text{ N}$$
$$(0.285\text{ kg})a_y = -1.81\text{ N}$$
$$a_y = -6.35\text{ m/s}^2$$

The vertical velocity of a projectile at its highest point is 0 meters per second. So $v_{fy} = 0$ m/s. The initial velocity in the vertical direction is equal to $v_i \sin\theta$, where v_i is the initial velocity and θ is the angle that the velocity vector makes to the horizontal — in this case, 20 degrees. Therefore,

$$v_{iy} = v_i \sin\theta$$
$$= (5\text{ m/s}) \sin 20°$$
$$= (5\text{ m/s})(0.342)$$
$$= 1.71\text{ m/s}$$

Use the velocity-displacement formula to solve for the distance the disc rises from its initial launch height:

$$v_{fy}^2 = v_{iy}^2 + 2a_y s_y$$
$$(0\text{ m/s})^2 = (1.71\text{ m/s})^2 + 2\left(-6.35\text{ m/s}^2\right)s_y$$
$$0\text{ m}^2/\text{s}^2 = 2.92\text{ m}^2/\text{s}^2 - \left(12.7\text{ m/s}^2\right)s_y$$
$$-2.92\text{ m}^2/\text{s} = -\left(12.7\text{ m/s}^2\right)s_y$$
$$0.23\text{ m} = s_y$$

Finally, add this to the disc's starting height and convert into the requested units of centimeters:

$$(1.6\text{ m}) + (0.23\text{ m}) = 1.83\text{ m}$$
$$(1.83\text{ m})\left(\frac{100\text{ cm}}{1\text{ m}}\right) = 183\text{ cm}$$

177.   **4.9 m**

First convert the mass into "correct" units:

$$(285\text{ g})\left(\frac{1\text{ kg}}{1,000\text{ g}}\right) = 0.285\text{ kg}$$

Then draw a free-body diagram for the disc and analyze the forces in the vertical directions using Newton's second law.

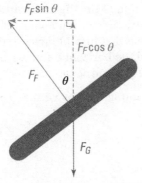

Illustration by Thomson Digital

Geometric principles reveal the following relationships between the angles:

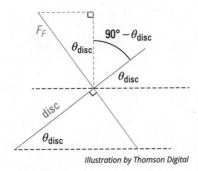

Illustration by Thomson Digital

So in the vertical direction,

$$
\begin{aligned}
F_{y,\text{net}} &= \sum F_y \\
ma_y &= F_F \cos\theta - F_G \\
ma_y &= F_F \cos\theta - ma_G \\
ma_y &= F_F \cos\theta - mg \\
(0.285\ \text{kg})a_y &= (1.2\ \text{N})\cos 12° - (0.285\ \text{kg})\left(9.8\ \text{m/s}^2\right) \\
(0.285\ \text{kg})a_y &= (1.2\ \text{N})(0.978) - (0.285\ \text{kg})\left(9.8\ \text{m/s}^2\right) \\
(0.285\ \text{kg})a_y &= 1.174\ \text{N} - 2.793\ \text{N} \\
(0.285\ \text{kg})a_y &= -1.619\ \text{N} \\
a_y &= -5.68\ \text{m/s}^2
\end{aligned}
$$

And in the horizontal direction,

$$
\begin{aligned}
F_{x,\text{net}} &= \sum F_x \\
ma_x &= -F_F \sin\theta \\
(0.285\ \text{kg})a_x &= -(1.2\ \text{N})\sin 12° \\
(0.285\ \text{kg})a_x &= -(1.2\ \text{N})(0.208) \\
(0.285\ \text{kg})a_x &= -0.25\ \text{N} \\
a_x &= -0.88\ \text{m/s}^2
\end{aligned}
$$

These are the accelerations for the first portion of the flight — when the wind blows. When the wind stops blowing, the force of friction is eliminated, and gravity is the only force acting on the disc. Therefore, from that instant onward, the horizontal acceleration is 0 meters per second squared (no horizontal forces), and the vertical acceleration is the standard −9.8 meters per second squared.

Start by calculating the components of the initial velocity vector:

$$v_x = v \cos \theta$$
$$= (5 \text{ m/s}) \cos 33°$$
$$= (5 \text{ m/s})(0.839)$$
$$= 4.2 \text{ m/s}$$

$$v_y = v \sin \theta$$
$$= (5 \text{ m/s}) \sin 33°$$
$$= (5 \text{ m/s})(0.545)$$
$$= 2.7 \text{ m/s}$$

Then sum up the kinematic data for the first portion of the flight:

$$\mathbf{s} = (s_x, s_y)$$
$$\mathbf{v} = (4.2 \text{ m/s}, 2.7 \text{ m/s})$$
$$\mathbf{a} = (-0.88 \text{ m/s}^2, -5.68 \text{ m/s}^2)$$

Use the displacement formula $\mathbf{s} = \mathbf{v}_i t + \frac{1}{2}\mathbf{a}t^2$ to find the location of the disc after 1.3 seconds. The horizontal-component form of the equation yields

$$s_x = v_{ix}t + \frac{1}{2}a_x t^2$$
$$= (4.2 \text{ m/s})(1.3 \text{ s}) + \frac{1}{2}(-0.88 \text{ m/s}^2)(1.3 \text{ s})^2$$
$$= 5.46 \text{ m} - 0.74 \text{ m}$$
$$= 4.72 \text{ m}$$

And the vertical-component form gives you

$$s_y = v_{iy}t + \frac{1}{2}a_y t^2$$
$$= (2.7 \text{ m/s})(1.3 \text{ s}) + \frac{1}{2}(-5.68 \text{ m/s}^2)(1.3 \text{ s})^2$$
$$= 3.51 \text{ m} - 4.8 \text{ m}$$
$$= -1.29 \text{ m}$$

This doesn't seem to make sense, but remember that this is the *change* in the vertical position compared with the launch position, which was 1.6 meters above the ground. Therefore, the disc's height after 1.3 seconds is actually $(1.6 \text{ m}) + (-1.29 \text{ m}) = 0.31 \text{ m}$ above the ground.

Now use the velocity-time formula $\mathbf{v}_f = \mathbf{v}_i + \mathbf{a}t$ to find the horizontal and vertical velocities of the disc after 1.3 seconds. First, horizontally:

$$v_{fx} = v_{ix} + a_x t$$
$$= (4.2 \text{ m/s}) + \left(-0.88 \text{ m/s}^2\right)(1.3 \text{ s})$$
$$= 4.2 \text{ m/s} - 1.14 \text{ m/s}$$
$$= 3.06 \text{ m/s}$$

Then, vertically:

$$v_{fy} = v_{iy} + a_y t$$
$$= (2.7 \text{ m/s}) + \left(-5.68 \text{ m/s}^2\right)(1.3 \text{ s})$$
$$= 2.7 \text{ m/s} - 7.38 \text{ m/s}$$
$$= -4.68 \text{ m/s}$$

So, when the wind stops blowing, the disc is 0.31 meters above the ground with a vertical velocity of −4.68 meters per second. Use the displacement formula to calculate the time it takes for the disc to fall 0.31 meters with an acceleration due only to the force of gravity:

$$s_y = v_{iy}t + \frac{1}{2}a_y t^2$$

$$-0.31 \text{ m} = (-4.68 \text{ m/s})t + \frac{1}{2}\left(-9.8 \text{ m/s}^2\right)t^2$$

$$-0.31 \text{ m} = -(4.68 \text{ m/s})t - \left(4.9 \text{ m/s}^2\right)t^2$$

$$\left(4.9 \text{ m/s}^2\right)t^2 + (4.68 \text{ m/s})t - 0.31 \text{ m} = 0 \text{ m}$$

$$t = \frac{-(4.68 \text{ m/s}) \pm \sqrt{(4.68 \text{ m/s})^2 - 4\left(4.9 \text{ m/s}^2\right)(-0.31 \text{ m})}}{2\left(4.9 \text{ m/s}^2\right)}$$

$$= \frac{-(4.68 \text{ m/s}) \pm \sqrt{21.9 \text{ m}^2/\text{s}^2 + 6.08 \text{ m}^2/\text{s}^2}}{\left(9.8 \text{ m/s}^2\right)}$$

$$= \frac{-(4.68 \text{ m/s}) \pm \sqrt{27.98 \text{ m}^2/\text{s}^2}}{\left(9.8 \text{ m/s}^2\right)}$$

$$= \frac{-(4.68 \text{ m/s}) \pm 5.29 \text{ m/s}}{\left(9.8 \text{ m/s}^2\right)}$$

$$= \frac{-(4.68 \text{ m/s}) + 5.29 \text{ m/s}}{\left(9.8 \text{ m/s}^2\right)} \text{ or } \frac{-(4.68 \text{ m/s}) - 5.29 \text{ m/s}}{\left(9.8 \text{ m/s}^2\right)}$$

$$= \frac{0.61 \text{ m/s}}{\left(9.8 \text{ m/s}^2\right)} \text{ or } \frac{-9.97 \text{ m/s}}{\left(9.8 \text{ m/s}^2\right)}$$

$$= 0.062 \text{ s or } -1.01 \text{ s}$$

The negative value doesn't make sense in the context of the problem, so the disc must spend another 0.062 seconds in flight. Use the displacement formula a final time — in the horizontal direction, where the acceleration is 0 — to solve for the additional horizontal distance traveled in that time:

$$s_x = v_{ix}t + \frac{1}{2}a_x t^2$$
$$= (3.06 \text{ m/s})(0.062 \text{ s}) + \frac{1}{2}\left(0 \text{ m/s}^2\right)(0.062 \text{ s})^2$$
$$= 0.19 \text{ m} + 0 \text{ m}$$
$$= 0.19 \text{ m}$$

The horizontal displacement for the first 1.3 seconds is 4.72 meters; adding the last 0.19 meters brings the final distance across the field to $(4.72 \text{ m}) + (0.19 \text{ m}) = 4.91 \text{ m}$.

178. 1.2 s

First convert the velocity vector into its components:

$$v_x = v\cos\theta$$
$$= (12 \text{ m/s})\cos 23°$$
$$= (12 \text{ m/s})(0.921)$$
$$= 11.05 \text{ m/s}$$

$$v_y = v\sin\theta$$
$$= (12 \text{ m/s})\sin 23°$$
$$= (12 \text{ m/s})(0.391)$$
$$= 4.69 \text{ m/s}$$

Then summarize the data related to the displacement, velocity, and acceleration vectors:

$$\mathbf{s} = \left(s_x, -1.25 \text{ m}\right)$$
$$\mathbf{v} = (11.05 \text{ m/s}, 4.69 \text{ m/s})$$
$$\mathbf{a} = \left(a_x, -9.8 \text{ m/s}^2\right)$$

The horizontal acceleration is "unknown" because it involves dealing with the force of air resistance. However, you don't need its value to solve this question, because time in flight is dependent on the vertical information. Use the displacement formula in the vertical direction to solve for the time the baseball spends in the air:

$$s_y = v_{iy}t + \frac{1}{2}a_y t^2$$

$$-1.25\,\text{m} = (4.69\,\text{m/s})t + \frac{1}{2}\left(-9.8\,\text{m/s}^2\right)t^2$$

$$-1.25\,\text{m} = (4.69\,\text{m/s})t - \left(4.9\,\text{m/s}^2\right)t^2$$

$$\left(4.9\,\text{m/s}^2\right)t^2 - (4.69\,\text{m/s})t - 1.25\,\text{m} = 0\,\text{m}$$

$$t = \frac{(4.69\,\text{m/s}) \pm \sqrt{(-4.69\,\text{m/s})^2 - 4\left(4.9\,\text{m/s}^2\right)(-1.25\,\text{m})}}{2\left(4.9\,\text{m/s}^2\right)}$$

$$= \frac{(4.69\,\text{m/s}) \pm \sqrt{22\,\text{m}^2/\text{s}^2 + 24.5\,\text{m}^2/\text{s}^2}}{\left(9.8\,\text{m/s}^2\right)}$$

$$= \frac{(4.69\,\text{m/s}) \pm \sqrt{46.5\,\text{m}^2/\text{s}^2}}{\left(9.8\,\text{m/s}^2\right)}$$

$$= \frac{(4.69\,\text{m/s}) \pm 6.82\,\text{m/s}}{\left(9.8\,\text{m/s}^2\right)}$$

$$= \frac{(4.69\,\text{m/s}) + 6.82\,\text{m/s}}{\left(9.8\,\text{m/s}^2\right)} \text{ or } \frac{(4.69\,\text{m/s}) - 6.82\,\text{m/s}}{\left(9.8\,\text{m/s}^2\right)}$$

$$= \frac{11.51\,\text{m/s}}{\left(9.8\,\text{m/s}^2\right)} \text{ or } \frac{-2.13\,\text{m/s}}{\left(9.8\,\text{m/s}^2\right)}$$

$$= 1.2\,\text{s} \text{ or } -0.2\,\text{s}$$

The negative value doesn't make sense in the context of the problem, so the answer must be 1.2 seconds of flight.

179. 3.6 m/s

First draw a free-body diagram of the block as it travels up the ramp:

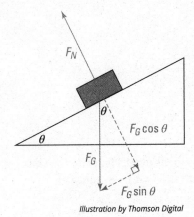

Illustration by Thomson Digital

Use the force form of Newton's second law to calculate the block's acceleration in the direction parallel to the ramp's surface (the following solution uses "up the ramp" as the positive direction):

$$F_{x,net} = \sum F_x$$
$$ma_x = -F_G \sin\theta$$
$$ma_x = -ma_G \sin\theta$$
$$ma_x = -mg \sin\theta$$
$$a_x = -g \sin\theta$$
$$= -(9.8 \text{ m/s}^2)\sin 18°$$
$$= -(9.8 \text{ m/s}^2)(0.309)$$
$$= -3.03 \text{ m/s}^2$$

Use trigonometry along with the known dimensions of the ramp to solve for the distance that the block slides up the ramp before launch:

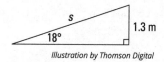

Illustration by Thomson Digital

$$\sin 18° = \frac{1.3 \text{ m}}{s}$$
$$0.309 = \frac{1.3 \text{ m}}{s}$$
$$0.309s = 1.3 \text{ m}$$
$$s = 4.21 \text{ m}$$

Now use the velocity–displacement formula to solve for the block's velocity just as it leaves the ramp's surface:

$$v_{fx}^2 = v_{ix}^2 + 2a_x s_x$$
$$v_{fx} = \sqrt{v_{ix}^2 + 2a_x s_x}$$
$$= \sqrt{(6.3 \text{ m/s})^2 + 2(-3.03 \text{ m/s}^2)(4.21 \text{ m})}$$
$$= \sqrt{39.69 \text{ m}^2/\text{s}^2 - 25.51 \text{ m}^2/\text{s}^2}$$
$$= \sqrt{14.18 \text{ m}^2/\text{s}^2}$$
$$= 3.77 \text{ m/s}$$

Then calculate the horizontal component of this velocity (the block continues at the same 18-degree angle the instant after leaving the ramp):

$$v_x = v \cos\theta$$
$$= (3.77 \text{ m/s})\cos 18°$$
$$= (3.77 \text{ m/s})(0.951)$$
$$= 3.59 \text{ m/s} \approx 3.6 \text{ m/s}$$

180. 7.8 m

First draw a free-body diagram of the block as it travels up the ramp:

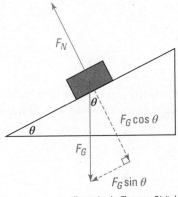

Illustration by Thomson Digital

Use the force form of Newton's second law to calculate the block's acceleration in the direction parallel to the ramp's surface (the following solution uses "up the ramp" as the positive direction):

$$F_{x,\text{net}} = \sum F_x$$
$$ma_x = -F_G \sin\theta$$
$$ma_x = -ma_G \sin\theta$$
$$ma_x = -mg \sin\theta$$
$$a_x = -g \sin\theta$$
$$= -\left(9.8 \text{ m/s}^2\right)\sin 25°$$
$$= -\left(9.8 \text{ m/s}^2\right)(0.423)$$
$$= -4.15 \text{ m/s}^2$$

Now use the velocity-displacement formula to solve for the block's velocity just as it leaves the ramp's surface:

$$v_{fx}^2 = v_{ix}^2 + 2a_x s_x$$
$$v_{fx} = \sqrt{v_{ix}^2 + 2a_x s_x}$$
$$= \sqrt{(10 \text{ m/s})^2 + 2\left(-4.15 \text{ m/s}^2\right)(3.5 \text{ m})}$$
$$= \sqrt{100 \text{ m}^2/\text{s}^2 - 29 \text{ m}^2/\text{s}^2}$$
$$= \sqrt{71 \text{ m}^2/\text{s}^2}$$
$$= 8.43 \text{ m/s}$$

Then calculate the horizontal and vertical components of this velocity (the block continues at the same 25-degree angle the instant after leaving the ramp):

$$v_x = v \cos \theta$$
$$= (8.43 \text{ m/s}) \cos 25°$$
$$= (8.43 \text{ m/s})(0.906)$$
$$= 7.64 \text{ m/s}$$

$$v_y = v \sin \theta$$
$$= (8.43 \text{ m/s}) \sin 25°$$
$$= (8.43 \text{ m/s})(0.423)$$
$$= 3.57 \text{ m/s}$$

To solve for the time the block spends aloft, you need to first calculate its starting height. Use trigonometry along with the known dimensions of the ramp to solve for h:

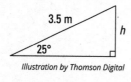

Illustration by Thomson Digital

$$\sin 25° = \frac{h}{3.5 \text{ m}}$$

$$0.423 = \frac{h}{3.5 \text{ m}}$$

$$1.48 \text{ m} = h$$

So the block falls 1.48 meters during its flight. Substitute that data into the displacement formula to solve for the time before the block lands on the ground (you'll need the quadratic formula along the way):

$$s_y = v_{iy}t + \frac{1}{2}a_y t^2$$

$$-1.48 \text{ m} = (3.57 \text{ m/s})t + \frac{1}{2}\left(-9.8 \text{ m/s}^2\right)t^2$$

$$-1.48 \text{ m} = (3.57 \text{ m/s})t - \left(4.9 \text{ m/s}^2\right)t^2$$

$$\left(4.9 \text{ m/s}^2\right)t^2 - (3.57 \text{ m/s})t - 1.48 \text{ m} = 0 \text{ m}$$

$$t = \frac{3.57 \text{ m/s} \pm \sqrt{(-3.57 \text{ m/s})^2 - 4\left(4.9 \text{ m/s}^2\right)(-1.48 \text{ m})}}{2\left(4.9 \text{ m/s}^2\right)}$$

$$= \frac{3.57 \text{ m/s} \pm \sqrt{12.74 \text{ m}^2/\text{s}^2 + 29.01 \text{ m}^2/\text{s}^2}}{9.8 \text{ m/s}^2}$$

$$= \frac{3.57 \text{ m/s} \pm \sqrt{41.75 \text{ m}^2/\text{s}^2}}{9.8 \text{ m/s}^2}$$

$$= \frac{3.57 \text{ m/s} \pm 6.46 \text{ m/s}}{9.8 \text{ m/s}^2}$$

$$= \frac{3.57 \text{ m/s} + 6.46 \text{ m/s}}{9.8 \text{ m/s}^2} \text{ or } \frac{3.57 \text{ m/s} - 6.46 \text{ m/s}}{9.8 \text{ m/s}^2}$$

$$= \frac{10.03 \text{ m/s}}{9.8 \text{ m/s}^2} \text{ or } \frac{-2.89 \text{ m/s}}{9.8 \text{ m/s}^2}$$

$$= 1.02 \text{ s or } -0.29 \text{ s}$$

The negative value doesn't make sense in the context of the problem, so the answer must be 1.02 seconds of flight.

Finally, use the displacement formula again — this time in the horizontal direction — to solve for the block's distance from the ramp after leaving its surface. After the block leaves the ramp, the net horizontal force on the block is 0 newtons; therefore, the net acceleration in the horizontal direction is 0 meters per second squared:

$$s_x = v_{ix}t + \frac{1}{2}a_x t^2$$
$$= (7.64 \text{ m/s})(1.02 \text{ s}) + \frac{1}{2}\left(0 \text{ m/s}^2\right)(1.02 \text{ s})^2$$
$$= 7.79 \text{ m} + 0 \text{ m}$$
$$= 7.79 \text{ m} \approx 7.8 \text{ m}$$

181. 0.010 m/s

Speed is distance traveled divided by the time it takes to travel the distance. The distance traveled by the tip of the second hand is the circumference C of a circle with a radius of r, or $C = 2\pi r$.

In this problem, the radius is the length of the second hand, so $r = 0.10$. The time it takes for the tip of the second hand to travel once around the circumference of the circle is $T = 60$ seconds.

Divide the distance by the time to find the speed v:

$$v = \frac{C}{T}$$
$$= \frac{2\pi r}{T}$$
$$= \frac{2\pi \times 0.10 \text{ m}}{60 \text{ s}}$$
$$= 0.010 \text{ m/s}$$

182. 0.21 rad/s

In 15 seconds the child completes one half of a revolution, which is expressed as $\Delta\theta = \pi$ radians. The time it takes is $\Delta t = 15$ seconds, so the angular speed is

$$\omega = \frac{\Delta\theta}{\Delta t}$$
$$= \frac{\pi \text{ rad}}{15 \text{ s}}$$
$$= 0.21 \text{ rad/s}$$

183. 1.4×10^2 revolutions

For constant angular motion, the angle θ through which an object turns is $\theta = \omega t$.

In this case, $\omega = 15$ radians per second. Therefore, the angle through which the blades turn in $t = 60$ seconds is

$$\theta = \omega t$$
$$= 15 \text{ rad/s} \times 60 \text{ s}$$
$$= 900 \text{ rad}$$

There are 2π radians in a revolution, so the conversion factor from radians to revolutions is $1 = \dfrac{1 \text{ rev}}{2\pi \text{ rad}}$.

So, the number of revolutions made by the blade is

$$\theta = 900 \text{ rad} \times \frac{1 \text{ rev}}{2\pi \text{ rad}}$$
$$= 1.4 \times 10^2 \text{ rev}$$

184. 0.785 rad

A full circle has an angle of 2π radians. To find the angle α that is one eighth of this, you must divide 2π by 8:

$$\alpha = \frac{2\pi \text{ rad}}{8}$$
$$= 0.785 \text{ rad}$$

185. $\pi/3$ rad

The angle between horizontal and vertical is 90 degrees, or a quarter of a circle. Because there are 2π radians in a circle, there are $2\pi/4 = \pi/2$ radians in 90 degrees. The angle $\Delta\theta$ through which the drawbridge moved is then two-thirds of $\pi/2$, or

$$\Delta\theta = \frac{2}{3} \times \frac{\pi}{2} \text{ rad}$$
$$= \frac{\pi}{3} \text{ rad}$$

186.   3.9 rad

You're told that the first two slices make a quarter of a whole pizza, which means that they make a quarter of a circle. There are 2π radians in a circle, so this angle is

$$\theta_1 = \frac{2\pi}{4} \text{ rad}$$

$$= \frac{\pi}{2} \text{ rad}$$

The last three slices each make an eighth of a circle, so they each make an angle

$$\theta_2 = \frac{2\pi}{8} \text{ rad}$$

$$= \frac{\pi}{4} \text{ rad}$$

The total angle made when the slices are laid side by side is

$$\theta = \theta_1 + 3\theta_2$$

$$= \frac{\pi}{2} \text{ rad} + 3\frac{\pi}{4} \text{ rad}$$

$$= \frac{5\pi}{4} \text{ rad}$$

Using $\pi = 3.14$, this gives

$$\theta = \frac{5\pi}{4} \text{ rad}$$

$$= \frac{5}{4} \times 3.14 \text{ rad}$$

$$= 3.9 \text{ rad}$$

187.   0.87 rad

The total angle θ through which the roller coaster travels is the difference between the final and initial angles: $\theta = \theta_f - \theta_i$.

At the initial angle, the roller coaster is heading down, traveling at an angle of 15 degrees below the horizontal: $\theta_i = -15°$. At the final angle, the roller coaster is heading up, so the angle is 35 degrees above the horizontal: $\theta_f = +35°$. So, the total angle is

$$\theta = \theta_f - \theta_i$$

$$= 35° - \left(-15°\right)$$

$$= 50°$$

To convert this to radians, use the conversion factor $1 = \dfrac{2\pi \text{ rad}}{360°}$. The final answer is therefore

$$\theta = 50° \times \dfrac{2\pi \text{ rad}}{360°}$$
$$= 0.87 \text{ rad}$$

188. 4.0×10^2 km

The angular speed of the moon is $\omega = \dfrac{\Delta\theta}{\Delta t}$. For this problem, $\Delta\theta = 2\pi$ radians (one complete revolution) and $\Delta t = 27$ days, so the angular speed is

$$\omega = \dfrac{\Delta\theta}{\Delta t}$$
$$= \dfrac{2\pi}{27 \text{ d}}$$

To find the angle through which the astronaut rotates in one day, solve the equation above for $\Delta\theta_{ast}$ (the subscript *ast* means that this is the angle through which the astronaut rotates in one day, not the angle through which the moon rotates in 27 days). The result is

$$\omega = \dfrac{\Delta\theta_{ast}}{\Delta t_{ast}}$$
$$\Delta\theta_{ast} = \omega\Delta t_{ast}$$

where $\Delta t_{ast} = 1$ day is the time during which the astronaut rotates. Insert the known quantities to find

$$\Delta\theta_{ast} = \omega\Delta t_{ast}$$
$$= \dfrac{2\pi \text{ rad}}{27 \text{ d}} \times 1 \text{ d}$$
$$= \dfrac{2\pi}{27} \text{ rad}$$

To find the distance that the astronaut travels, use the equation relating arc length to arc radius and the angle subtended by the arc: $s = r\Delta\theta_{ast}$. Use $r = 1{,}734$ kilometers to find

$$s = r\Delta\theta_{ast}$$
$$= 1{,}734 \text{ km} \times \dfrac{2\pi}{27} \text{ rad}$$
$$= 4.0 \times 10^2 \text{ km}$$

189. 2.4×10^2 rad/hr

The angular speed is related to the tangential velocity and the radius of curvature by $v = r\omega$. Solve this for ω and insert the known quantities of $r = 1/4$ mile and $v = 60$ miles per hour. The result is

$$v = r\omega$$
$$\omega = \frac{v}{r}$$
$$= \frac{60 \text{ mi/hr}}{1/4 \text{ mi}}$$
$$= 2.4 \times 10^2 \text{ rad/hr}$$

190. 0.022 s

Recall that, in a complete circle, there are 2π radians. If there are 34 separators evenly spaced around the wheel, then the angle between each separator is $\Delta\theta = \frac{2\pi}{34}$ radians.

The angular speed of the wheel is $\omega = 8.3$ radians/second. You can solve the equation relating angular speed to time and angle for the time, and then insert the known quantities to find the time. This gives

$$\omega = \frac{\Delta\theta}{\Delta t}$$
$$\Delta t = \frac{\Delta\theta}{\omega}$$
$$= \frac{2\pi}{34} \text{ rad} \times \frac{1}{8.3 \text{ rad/s}}$$
$$= 0.022 \text{ s}$$

191. 4.1×10^3 m

First convert the given angular speed to radians per second. To do this, use the conversion factors $1 = \frac{2\pi \text{ rad}}{1 \text{ rev}}$ and $1 = \frac{1 \text{ min}}{60 \text{ s}}$. This gives an angular speed of

$$\omega = 3,000 \frac{\text{rev}}{\text{min}} \times \frac{2\pi \text{ rad}}{1 \text{ rev}} \times \frac{1 \text{ min}}{60 \text{ s}}$$
$$= 314 \text{ rad/s}$$

In $\Delta t = 10$ seconds, the angle through which the propeller rotates is

$$\omega = \frac{\Delta\theta}{\Delta t}$$
$$\Delta\theta = \omega\Delta t$$
$$= 314 \text{ rad/s} \times 10 \text{ s}$$
$$= 3,140 \text{ rad}$$

Now, use the equation relating distance to the radius of curvature and the angle to find the distance travelled by the propeller tip. The radius of the circle described by the tip of the propeller is $r = 1.3$ meters. The result is

$$s = r\Delta\theta$$
$$= 1.3\,\text{m} \times 3{,}140\,\text{rad}$$
$$= 4.1 \times 10^3\,\text{m}$$

192.   4.8 rad/s

Angular acceleration is defined by $\alpha = \frac{\Delta\omega}{\Delta t}$ where $\Delta\omega$ is the final angular speed minus the initial angular speed and Δt is the time over which the angular speed changes. You know the initial angular speed of the fan blades, so you can write $\Delta\omega = \omega_f - \omega_i$ where $\omega_i = 3.0$ radians per second. Solve the equation for acceleration for the final angular speed and plug in the known quantities to get the answer. The result is

$$\alpha = \frac{\Delta\omega}{\Delta t}$$
$$= \frac{\omega_f - \omega_i}{\Delta t}$$
$$\alpha\Delta t = \omega_f - \omega_i$$
$$\omega_f = \alpha\Delta t + \omega_i$$
$$= 1.2\,\text{rad/s}^2 \times 1.5\,\text{s} + 3.0\,\text{rad/s}$$
$$= 4.8\,\text{rad/s}$$

193.   8.0 rad/s² to left

The angular acceleration is related to the linear acceleration by $\alpha = \frac{a}{r}$. In this case, $a = 2.8$ meters per second per second and $r = 0.35$ meters. Plug these quantities into the equation:

$$\alpha = \frac{a}{r}$$
$$= \frac{2.8\,\text{m/s}^2}{0.35\,\text{m}}$$
$$= 8.0\,\text{rad/s}^2$$

By using the right-hand rule, you find that the direction of the angular acceleration is to the left of the car when facing the direction in which the car moves.

194. 1.2 rad

The acceleration of the merry-go-round is $\alpha = 0.30$ radians/second, and it accelerates for $\Delta t = 2.8$ seconds. Its initial angular speed is $\omega_i = 0$ radians per second. Plug these values into the equation for angular displacement to get

$$\theta = \omega_i t + \frac{1}{2}\alpha t^2$$
$$= \frac{1}{2} \times 0.30 \text{ rad/s}^2 \times (2.8 \text{ s})^2$$
$$= 1.2 \text{ rad}$$

195. 130 rad

The initial angular speed of your tires is $\omega_i = \frac{v_i}{r}$ where $v_i = 22$ meters per second and $r = 0.37$ meters meters. The angular acceleration of your tires is $\alpha = \frac{a}{r}$ where $a = -2.7$ meters per second per second is the tangential acceleration of your tires. The minus sign indicates that this acceleration is in the direction opposite the direction of the initial velocity.

You accelerate for $t = 2.6$ seconds. Plug these quantities into the equation for angular displacement to find

$$\theta = \omega_i t + \frac{1}{2}\alpha t^2$$
$$= \frac{v_i}{r}t + \frac{1}{2}\alpha t^2$$
$$= \frac{22 \text{ m/s}}{0.37 \text{ m}} \times 2.6 \text{ s} + \frac{1}{2} \times (-7.3 \text{ rad/s}^2) \times (2.6 \text{ s})^2$$
$$= 130 \text{ rad}$$

196. 0.19 rad/s

Your angular acceleration is $\alpha = \frac{a}{r}$ where $a = 3.4$ meters per second per second is your tangential acceleration and $r = 25$ meters is the radius of the curved part of the track. Angular velocity is related to angular acceleration by $\Delta\omega = \alpha\Delta t$ where $\Delta\omega = \omega_f - \omega_i$ is your change in angular speed and $\Delta t = 1.4$ seconds is the time over which you accelerate. For this problem, the initial speed is $\omega_i = 0$ radians per second. Combining these two equations and plugging in the known quantities gives

$$\Delta\omega = \alpha\Delta t$$
$$\omega_f - \omega_i = \frac{a}{r}\Delta t$$
$$\omega_f = \frac{a}{r}\Delta t$$
$$= \frac{3.4 \text{ m/s}^2}{25 \text{ m}} \times 1.4 \text{ s}$$
$$= 0.19 \text{ rad/s}$$

197. 0.015 m/s²

First, convert the given quantities to meters and seconds. The time it takes to complete the turn is

$$\Delta t = 12 \, \text{min} \times \frac{60 \, \text{s}}{1 \, \text{min}}$$
$$= 720 \, \text{s}$$

The radius of the turn is

$$r = 0.50 \, \text{mi} \times \frac{1609 \, \text{m}}{1 \, \text{mi}}$$
$$= 804.5 \, \text{m}$$

The angular speed of the boat is $\omega = \frac{\Delta\theta}{\Delta t}$ where $\Delta\theta = \pi$ radians because there are π radians in 180 degrees. Express the centripetal acceleration in terms of the angular speed using $v = \omega r$ and $a = \frac{v^2}{r}$.

Combining these two equations and the equation above for angular speed gives

$$a = \frac{v^2}{r}$$
$$= \frac{(\omega r)^2}{r}$$
$$= \omega^2 r$$
$$= \left(\frac{\Delta\theta}{\Delta t}\right)^2 r$$

Insert the known quantities:

$$a = \left(\frac{\Delta\theta}{\Delta t}\right)^2 r$$
$$= \left(\frac{\pi \, \text{rad}}{720 \, \text{s}}\right)^2 \times 804.5 \, \text{m}$$
$$= 0.015 \, \text{m/s}^2$$

198. 20 m/s²

The tangential speed of the tip of the lasso is $v = \omega r$ where $\omega = 3.8$ radians per second is the angular speed and $r = 1.4$ meters is the radial distance from the center of the circle (that is, your hand) to the tip of the lasso. The centripetal acceleration is

$$a = \frac{v^2}{r}$$
$$= \frac{(\omega r)^2}{r}$$
$$= \omega^2 r$$

Plug in the known quantities to find

$$a = \omega^2 r$$
$$= (3.8 \text{ rad/s})^2 \times 1.4 \text{ m}$$
$$= 20 \text{ m/s}^2$$

199. **0.32 m**

The maximum centripetal acceleration is $a = 3.8$ meters per second per second, and the maximum speed at which the slot cars can go without flying off the track is $v = 1.1$ meters per second. Solve the equation for centripetal acceleration for the radius and insert these quantities. The result is

$$a = \frac{v^2}{r}$$
$$r = \frac{v^2}{a}$$
$$= \frac{(1.1 \text{ m/s})^2}{3.8 \text{ m/s}^2}$$
$$= 0.32 \text{ m}$$

200. **26 m/s**

The equation for centripetal acceleration is $a = \frac{v^2}{r}$. Solve this for the tangential speed v to find

$$a = \frac{v^2}{r}$$
$$v = \pm\sqrt{ar}$$

Insert the given values for maximum centripetal acceleration ($a = 9.8$ meters per second squared) and the curve radius ($r = 70$ meters) to find

$$v = \pm\sqrt{ar}$$
$$= \pm\sqrt{9.8 \text{ m/s}^2 \times 70 \text{ m}}$$
$$= \pm 26 \text{ m/s}$$

Choose the positive value because you're looking for speed, which is always positive. So, your maximum speed is 26 meters per second.

201. 0.13 N

You complete 3 revolutions in $\Delta t = 9.0$ seconds. Each revolution is 2π radians, so your angular speed is

$$\omega = \frac{\Delta\theta}{\Delta t}$$
$$= \frac{3 \times 2\pi \text{ rad}}{9.0 \text{ s}}$$
$$= 2.09 \text{ rad/s}$$

The equation for centripetal acceleration is $a = \frac{v^2}{r}$. The tangential speed v is related to angular speed by $v = \omega r$, so the equation for centripetal acceleration becomes

$$a = \frac{v^2}{r}$$
$$= \frac{(\omega r)^2}{r}$$
$$= \omega^2 r$$

Use this equation for acceleration in Newton's second law to find the centripetal force on your big toe:

$$F = ma$$
$$= m\omega^2 r$$

Plug in $r = 0.85$ meters, $m = 0.035$ kilograms, and the angular speed from above to find the answer:

$$F = m\omega^2 r$$
$$= 0.035 \text{ kg} \times (2.09 \text{ rad/s})^2 \times 0.85 \text{ m}$$
$$= 0.13 \text{ N}$$

202. 3.6×10^{22} N

In one orbit, Earth travels a distance of $C = 2\pi r$ where $r = 1.5 \times 10^{11}$ meters. The time it takes to complete one orbit is one year, which in seconds is

$$t = 1 \text{ y}$$
$$= 1 \text{ y} \times \frac{365 \text{ d}}{\text{y}} \times \frac{24 \text{ h}}{\text{d}} \times \frac{60 \text{ min}}{\text{h}} \times \frac{60 \text{ s}}{\text{min}}$$
$$= 3.15 \times 10^7 \text{ s}$$

The orbital speed of Earth is $v = \frac{C}{t}$.

The centripetal acceleration is $a = \frac{v^2}{r}$.

Plug these equations into Newton's second law to find the force:

$$F = ma$$
$$= m\frac{v^2}{r}$$
$$= m\frac{(C/t)^2}{r}$$
$$= m\frac{(2\pi r)^2}{t^2 r}$$
$$= m\frac{4\pi^2 r}{t^2}$$
$$= \left(6.0 \times 10^{24} \text{ kg}\right) \frac{4\pi^2 \times 1.5 \times 10^{11} \text{ m}}{\left(3.15 \times 10^7 \text{ s}\right)^2}$$
$$= 3.6 \times 10^{22} \text{ N}$$

203. 80 m

From Newton's second law, force is related to acceleration by $F = ma$ where m is the mass of the object being accelerated. The acceleration in this case is the centripetal acceleration, which is related to tangential speed by $a = \frac{v^2}{r}$ where r is the radius of the curve through which the object moves. Combining these two equations to eliminate the acceleration gives

$$F = ma$$
$$= m\frac{v^2}{r}$$

Solving this equation for the radius of the turn gives

$$F = m\frac{v^2}{r}$$
$$r = m\frac{v^2}{F}$$

In this case, $F = 10,000$ newtons, $v = 20$ meters per second, and $m = 2,000$ kilograms. Plugging these values into the equation above gives

$$r = m\frac{v^2}{F}$$
$$= 2,000 \text{ kg} \times \frac{(20 \text{ m/s})^2}{10,000 \text{ N}}$$
$$= 80 \text{ m}$$

204. 7.5×10^3 N

From Newton's second law, force is related to acceleration by $F = ma$ where m is the mass of the object being accelerated. The acceleration in this case is the centripetal acceleration, which is related to tangential speed by $a = \frac{v^2}{r}$ where r is the radius of the curve through which the object moves. Combining these two equations to eliminate the acceleration gives

$$F = ma$$
$$= m\frac{v^2}{r}$$

In this case, you're given the angular speed of the centrifuge, so use the equation $v = \omega r$ to express force in terms of angular speed. The result is

$$F = m\frac{v^2}{r}$$
$$= m\frac{(\omega r)^2}{r}$$
$$= m\omega^2 r$$

In this case, $m = 0.050$ kilograms, $\omega = 1{,}000$ radians per second, and $r = 0.15$ meters. Plugging these values into the equation above gives

$$F = m\omega^2 r$$
$$= 0.050 \text{ kg} \times (1{,}000 \text{ rad/s})^2 \times 0.15 \text{ m}$$
$$= 7.5 \times 10^3 \text{ N}$$

205.   0 m/s

Because the force is applied perpendicular to the velocity, there is no force in the direction of the velocity, so the magnitude of the velocity (in other words, the speed) won't change. The jet ski will therefore execute circular motion at a constant tangential speed. The tangential speed of the jet ski doesn't change during the turn, so the difference between the initial and final tangential speed is zero.

206. $\sqrt{2}$

When friction is the only force contributing to the centripetal force, the mass cancels out:

$$F_F = F_C$$
$$\mu F_N = \frac{mv^2}{r}$$
$$\mu(mg) = \frac{mv^2}{r}$$
$$\mu g = \frac{v^2}{r}$$
$$\mu g r = v^2$$

$v^2 = \mu g r$, where v is the maximum velocity attainable without slipping, μ is the coefficient of friction between the tires and the road, and r is the radius of the road's curvature. According to the formula, μ is proportional to the square of v, or, conversely, v is proportional to $\sqrt{\mu}$. So if μ increases by a factor of 2, then v must also increase — but by a factor of $\sqrt{2}$.

207.   0 m/s

Without friction, a car can't maintain a curved path because no force is pointing toward the "center of the circle." Therefore, nothing provides the necessary centripetal force to keep the car going in a circular path.

208. 59.7 m

First, convert the given velocity to the desired units of meters per second:

$$\left(50\ \frac{km}{h}\right)\left(\frac{1{,}000\ m}{1\ km}\right)\left(\frac{1\ h}{60\ min}\right)\left(\frac{1\ min}{60\ s}\right)=13.89\ \frac{m}{s}.$$

Because the road is level (that is, completely horizontal), the only force that can possibly account for the centripetal force (F_C) necessary to keep the car moving in a circular path is friction (F_F). Therefore:

$$F_F = F_C$$
$$\mu F_N = \frac{mv^2}{r}$$

The car is not accelerating in the vertical direction, so the net force in the vertical direction equals 0 newtons. Summing up the two forces in that direction, the normal force (F_N) and the gravitational force (F_G), results in:

$$F_N - F_G = 0$$
$$F_N - mg = 0$$
$$F_N = mg$$

Substituting this result into the formula relating the centripetal force to the frictional force yields:

$$F_F = F_C$$
$$\mu F_N = \frac{mv^2}{r}$$
$$\mu mg = \frac{mv^2}{r}$$
$$\mu g = \frac{v^2}{r}$$
$$(0.33)\left(9.8\ m/s^2\right) = \frac{(13.89\ m/s)^2}{r}$$
$$3.23\ m/s^2 = \frac{192.9\ m^2/s^2}{r}$$
$$r = \frac{192.9\ m^2/s^2}{3.23\ m/s^2}$$
$$r = 59.7\ m$$

209. 410 m

First, convert "incorrect" units. This question includes both the rover's velocity, which needs to be in meters per second, and the radius of Mars, which needs to be in meters:

$$v_{rover} : \left(120\ \frac{km}{h}\right)\left(\frac{1{,}000\ m}{1\ km}\right)\left(\frac{1\ h}{60\ min}\right)\left(\frac{1\ min}{60\ s}\right)=33.3\ \frac{m}{s}$$

$$r_{Mars} : \left(3.37\times10^3\ km\right)\left(\frac{1\times10^3\ m}{1\ km}\right)=3.37\times10^6\ m$$

Because the superhighway is flat (that is, completely horizontal), the only force that can provide the centripetal force (F_C) necessary to keep the rover moving in a circular path is friction (F_F). Therefore:

$$F_F = F_C$$

$$\mu F_N = \frac{m_{rover} v_{rover}^2}{r_{road}}$$

The rover is not accelerating in the vertical direction, so the net force in the vertical direction equals 0 newtons. Summing up the two forces in that direction, the normal force (F_N) and the gravitational force (F_G), results in:

$$F_N - F_G = 0$$

$$F_N - \frac{G m_{rover} m_{Mars}}{r_{Msrs}^2} = 0$$

$$F_N = \frac{G m_{rover} m_{Mars}}{r_{Msrs}^2}$$

Substituting this result into the formula relating the centripetal force to the frictional force yields:

$$F_F = F_C$$

$$\mu F_N = \frac{m_{rover} v_{rover}^2}{r_{road}}$$

$$\mu \frac{G m_{rover} m_{Mars}}{r_{Mars}^2} = \frac{m_{rover} v_{rover}^2}{r_{road}}$$

$$\mu \frac{G m_{Mars}}{r_{Mars}^2} = \frac{v_{rover}^2}{r_{road}}$$

$$(0.72) \frac{\left(6.67 \times 10^{-11}\ \text{N} \cdot \text{m}^2/\text{kg}^2\right)\left(6.42 \times 10^{23}\ \text{kg}\right)}{\left(3.37 \times 10^6\ \text{m}\right)^2} = \frac{(33.3\ \text{m/s})^2}{r}$$

$$2.71 = \frac{1{,}111\ \text{m}}{r}$$

$$r = \frac{1{,}111\ \text{m}}{2.71}$$

$$r = 409\ \text{m}$$

$$\approx 410\ \text{m}$$

210. 0°

The normal force — the force exerted by a surface on an object touching it — is equal to the force of gravity only when the surface is exactly horizontal. The angle of inclination is thus 0 degrees.

211. 7 m/s

Start by drawing the forces acting on the car. No friction is present, so the only two forces are the force of gravity, pulling straight down, and the normal force, or the force with which the ground is pushing the car, directed perpendicularly away from the ground. Make sure to break your normal force into its components so that you can sum forces in both the horizontal direction and the vertical direction.

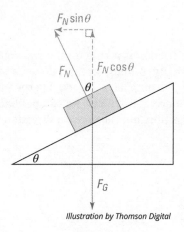

Because the car is not accelerating upward or downward, the sum of the vertical forces must equal 0:

$$F_N \cos \theta - F_G = 0$$

Knowing that the force of gravity near Earth's surface is equal to mg, you can rewrite this equation and rearrange it to solve for the unknown normal force:

$$F_N \cos \theta - mg = 0$$
$$F_N \cos \theta = mg$$
$$F_N = \frac{mg}{\cos \theta}$$

The sum of the horizontal forces provides the centripetal force keeping the car in its circular motion: $F_N \sin \theta = F_C$.

Now use the result you obtained earlier for the normal force and the centripetal force equation ($F_C = \frac{mv^2}{r}$, where m is the mass of the object, v is the object's velocity, and r is the radius of the curve the object is traversing) to solve the equation.

$$F_N \sin \theta = F_C$$
$$\left(\frac{mg}{\cos \theta} \right) \sin \theta = \frac{mv^2}{r}$$
$$mg \tan \theta = \frac{mv^2}{r}$$
$$g \tan \theta = \frac{v^2}{r}$$
$$\left(9.8 \text{ m/s}^2 \right) \tan 10° = \frac{v^2}{25 \text{ m}}$$
$$1.73 \text{ m/s}^2 = \frac{v^2}{25 \text{ m}}$$
$$43.2 \text{ m/s}^2 = v^2$$
$$6.6 \text{ m/s} = v$$

Then round to the requested amount: $6.6 \text{ m/s} \approx 7 \text{ m/s}$.

212. 75°

Start by drawing a force diagram. No friction is acting on the snowmobile, so the only two forces present are the force of gravity, pulling straight down, and the normal force, or the force with which the ground is pushing back, directed perpendicularly away from the ground. Make sure to break your normal force into its components so that you can sum forces in both the horizontal direction and the vertical direction.

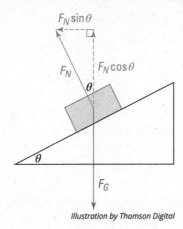

Illustration by Thomson Digital

Because the snowmobile is not accelerating upward or downward, the sum of the vertical forces must equal 0: $F_N \cos \theta - F_G = 0$.

Knowing that the force of gravity near Earth's surface is equal to mg, you can rewrite this equation and rearrange it to solve for the unknown normal force:

$$F_N \cos \theta - mg = 0$$
$$F_N \cos \theta = mg$$
$$F_N = \frac{mg}{\cos \theta}$$

The sum of the horizontal forces provides the centripetal force keeping the snowmobile in its circular motion: $F_N \sin \theta = F_C$.

Now use the result you obtained earlier for the normal force and the centripetal force equation ($F_C = \frac{mv^2}{r}$, where m is the mass of the object, v is the object's velocity, and r is the radius of the curve the object is traversing) to solve the equation.

$$F_N \sin \theta = F_C$$
$$\left(\frac{mg}{\cos \theta}\right) \sin \theta = \frac{mv^2}{r}$$
$$mg \tan \theta = \frac{mv^2}{r}$$
$$g \tan \theta = \frac{v^2}{r}$$
$$\left(9.8 \text{ m/s}^2\right) \tan \theta = \frac{(15 \text{ m/s})^2}{(6 \text{ m})}$$
$$9.8 \tan \theta = 37.5$$
$$\tan \theta = 3.83$$
$$\theta = \tan^{-1}(3.83) = 75.4°$$

213. 12.6 m/s^2

Start by drawing the forces acting on the vehicle. Three forces are involved in this situation: the force of gravity, pulling straight down; the normal force, or the force with which the ground is pushing against the vehicle, directed perpendicularly away from the ground; and the force of friction, directed in the opposite direction of the vehicle's motion. Speeding along, the vehicle prefers to keep going "straight," which means heading "up" the bank, so you need to draw your friction force heading "down" the bank. Make sure to break your normal and friction forces into their components so that you can sum forces in the horizontal direction and in the vertical direction.

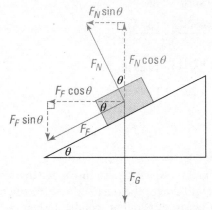

Illustration by Thomson Digital

Because the vehicle is not accelerating upward or downward, the sum of the vertical forces must equal 0: $F_N \cos \theta - F_F \sin \theta - F_G = 0$.

Knowing that $F_F = \mu F_N$ (μ is the coefficient of friction) and that the force of gravity is the product of mass and the acceleration due to gravity (a_G), you can rewrite this equation and rearrange it to solve for the unknown normal force:

$$F_N \cos \theta - \mu F_N \sin \theta - ma_G = 0$$
$$F_N \cos \theta - \mu F_N \sin \theta = ma_G$$
$$F_N (\cos \theta - \mu \sin \theta) = ma_G$$
$$F_N = \frac{ma_G}{(\cos \theta - \mu \sin \theta)}$$

The sum of the horizontal forces provides the centripetal force keeping the vehicle in its circular motion:

$$F_N \sin \theta + F_F \cos \theta = F_C$$
$$F_N \sin \theta + \mu F_N \cos \theta = F_C$$
$$F_N (\sin \theta + \mu \cos \theta) = F_C$$

Before jumping into the calculations, make sure you convert the velocity to "correct" units: $\left(130 \frac{\text{km}}{\text{h}} \right) \left(\frac{1,000 \text{ m}}{1 \text{ km}} \right) \left(\frac{1 \text{ h}}{60 \text{ min}} \right) \left(\frac{1 \text{ min}}{60 \text{ s}} \right) = 36.1 \text{ m/s}.$

Now use the result you obtained earlier for the normal force and the centripetal force equation ($F_C = \dfrac{mv^2}{r}$, where m is the mass of the object, v is the object's velocity, and r is the radius of the curve the object is traversing) to solve the equation.

$$F_N(\sin\theta + \mu\cos\theta) = F_C$$

$$\frac{ma_G}{(\cos\theta - \mu\sin\theta)}(\sin\theta + \mu\cos\theta) = \frac{mv^2}{r}$$

$$ma_G\left(\frac{\sin\theta + \mu\cos\theta}{\cos\theta - \mu\sin\theta}\right) = \frac{mv^2}{r}$$

$$a_G\left(\frac{\sin\theta + \mu\cos\theta}{\cos\theta - \mu\sin\theta}\right) = \frac{v^2}{r}$$

$$a_G\left[\frac{\sin 42° + 0.18\cos 42°}{\cos 42° - 0.18\sin 42°}\right] = \frac{(36.1\,\text{m/s})^2}{(80\,\text{m})}$$

$$a_G\left(\frac{0.803}{0.623}\right) = 16.29\,\text{m/s}^2$$

$$1.29a_G = 16.29\,\text{m/s}^2$$

$$a_G = 12.6\,\text{m/s}^2$$

214.   29 km/h

Start by drawing the forces acting on the car. Three forces are involved in this situation: the force of gravity, pulling straight down; the normal force, or the force with which the ground is pushing against the car, directed perpendicularly away from the ground; and the force of friction, directed in the opposite direction of the car's motion. If the car isn't moving fast enough, it will slide *down* the bank, so you need to draw your friction force heading up the bank. Make sure to break your normal and friction forces into their components so that you can sum forces in the horizontal direction and in the vertical direction.

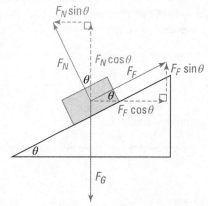

Illustration by Thomson Digital

Because the car is not accelerating upward or downward, the sum of the vertical forces must equal 0: $F_N\cos\theta + F_F\sin\theta - F_G = 0$.

Knowing that $F_F = \mu F_N$ (μ is the coefficient of friction) and that the force of gravity near Earth's surface is equal to mg, you can rewrite this equation and rearrange it to solve for the unknown normal force:

$$F_N \cos\theta + \mu F_N \sin\theta - mg = 0$$
$$F_N \cos\theta + \mu F_N \sin\theta = mg$$
$$F_N (\cos\theta + \mu \sin\theta) = mg$$
$$F_N = \frac{mg}{(\cos\theta + \mu \sin\theta)}$$

The sum of the horizontal forces provides the centripetal force keeping the car in its circular motion. Friction is pointing away from the "center of the circle" and is therefore antiparallel to the centripetal force:

$$F_N \sin\theta - F_F \cos\theta = F_C$$
$$F_N \sin\theta - \mu F_N \cos\theta = F_C$$
$$F_N (\sin\theta - \mu \cos\theta) = F_C$$

Now use the result you obtained earlier for the normal force and the centripetal force equation ($F_C = \frac{mv^2}{r}$, where m is the mass of the object, v is the object's velocity, and r is the radius of the curve the object is traversing) to solve the equation.

$$F_N (\sin\theta + \mu \cos\theta) = F_C$$
$$\frac{mg}{(\cos\theta + \mu \sin\theta)}(\sin\theta - \mu \cos\theta) = \frac{mv^2}{r}$$
$$mg\left(\frac{\sin\theta - \mu \cos\theta}{\cos\theta + \mu \sin\theta}\right) = \frac{mv^2}{r}$$
$$g\left(\frac{\sin\theta - \mu \cos\theta}{\cos\theta + \mu \sin\theta}\right) = \frac{v^2}{r}$$
$$(9.8 \text{ m/s}^2)\left[\frac{\sin 25° - 0.08 \cos 25°}{\cos 25° + 0.08 \sin 25°}\right] = \frac{v^2}{(18 \text{ m})}$$
$$(9.8 \text{ m/s}^2)\left(\frac{0.35}{0.94}\right) = \frac{v^2}{18 \text{ m}}$$
$$3.65 \text{ m/s}^2 = \frac{v^2}{18 \text{ m}}$$
$$65.7 \text{ m/s}^2 = v^2$$
$$8.1 \text{ m/s} = v$$

Finally, because the requested units of the answer are not meters per second, you must convert:

$$\left(8.1 \frac{\text{m}}{\text{s}}\right)\left(\frac{1 \text{ km}}{1{,}000 \text{ m}}\right)\left(\frac{60 \text{ s}}{1 \text{ min}}\right)\left(\frac{60 \text{ min}}{1 \text{ h}}\right) = 29.2 \frac{\text{km}}{\text{h}}$$

215.   2 N

The force of gravity exerted between objects is proportional to each object's mass. If B's mass is halved — with A's mass remaining unchanged — then the gravitational force between A and B is also halved: $\frac{1}{2}(4\text{ N}) = 2\text{ N}$.

216. 6.0×10^{-8} N

Before you can substitute all the given values into the law of universal gravitation, you need to convert the distance between the ball bearings into meters to match the units in the gravitational constant, G: $(10\text{ cm})\left(\dfrac{1\text{ m}}{100\text{ cm}}\right) = 0.1\text{ m}$.

$$
\begin{aligned}
F_g &= \frac{Gm_1m_2}{r^2} \\
&= \frac{\left(6.67 \times 10^{-11}\text{ N} \cdot \text{m}^2/\text{kg}^2\right)(3.00\text{ kg})(3.00\text{ kg})}{(0.1\text{ m})^2} \\
&= \frac{6.00 \times 10^{-10}\text{ N}}{1.0 \times 10^{-2}} \\
&= 6.0 \times 10^{-8}\text{ N}
\end{aligned}
$$

217. 9.1×10^7 m/s^2

With its engines off, the only force that the starship feels is the gravitational force attracting it to Planet X. Therefore, the net force on the starship

$(F_{\text{starship}} = m_{\text{starship}} a_{\text{starship}})$ must be equal to the force of gravity between the ship and the planet $(F_g = \dfrac{Gm_{\text{starship}}m_{\text{planet}}}{r^2})$. r represents the distance between the centers of the two objects: The distance from the center of Planet X to its surface is 65,000 kilometers, and the distance from the surface to the starship is another 2,500 kilometers, making the total distance between the planet and the starship 67,500 kilometers — or, more importantly, given the units situation, 67,500,000 meters (6.75×10^7 m). Substituting all the data into the equations leaves you with:

$$
\begin{aligned}
F_{\text{starship}} &= F_g \\
m_{\text{starship}} a_{\text{starship}} &= \frac{Gm_{\text{starship}}m_{\text{planet}}}{r^2} \\
(9,000\text{ kg})\, a_{\text{starship}} &= \frac{\left(6.67 \times 10^{-11}\text{ N} \cdot \text{m}^2/\text{kg}^2\right)(9,000\text{ kg})\left(6.2 \times 10^{33}\text{ kg}\right)}{\left(6.75 \times 10^7\text{ m}\right)^2} \\
(9,000\text{ kg})\, a_{\text{starship}} &= 8.17 \times 10^{11}\text{ N} \\
a_{\text{starship}} &= 9.1 \times 10^7\text{ m/s}^2
\end{aligned}
$$

(You can save yourself a little handwriting by noticing that because m_starship appears on both sides of the equation in the second line, you can divide it from both sides to leave you with $a_\text{starship} = \dfrac{Gm_\text{planet}}{r^2}$, an equation with the unknown already separated and only three substitutions to enter.)

218. $\quad$ **7.8×10^5 km**

Before you attempt any calculations, convert any values with "wrong" units. In astronomical situations, this most always involves distances because the physics-formula-friendly meters are very small relative to intra- and interstellar positioning:

$$r_\text{Cerberus} = \left(1.8 \times 10^4 \text{ km}\right)\left(\frac{1 \times 10^3 \text{ m}}{1 \text{ km}}\right) = 1.8 \times 10^7 \text{ m}$$

$$r_\text{Hades} = \left(2.5 \times 10^5 \text{ km}\right)\left(\frac{1 \times 10^3 \text{ m}}{1 \text{ km}}\right) = 2.5 \times 10^8 \text{ m}$$

$$s_\text{Cerberus-to-Hades} = \left(5.78 \times 10^6 \text{ km}\right)\left(\frac{1 \times 10^3 \text{ m}}{1 \text{ km}}\right) = 5.78 \times 10^9 \text{ m}$$

Note that s is the standard variable used for displacement.

Orion is being pulled in one direction by Cerberus and in the other direction by Hades. If *Orion*'s net acceleration is 0, then the net force (the sum of all the forces) on *Orion* must also be 0 — 0 newtons. If Cerberus is located to *Orion*'s left and Hades is located to *Orion*'s right, then the gravitational force on *Orion* from Cerberus is a negative value and the gravitational force from Hades is a positive one (it doesn't matter which asteroid you choose to be on which side of the starship — it matters only that one force be negative and the other positive because they are "pulling" Orion in *opposite* directions).

Let d represent the distance from *Orion* to the surface of Cerberus. Therefore, the distance from *Orion* to the surface of Hades is $5.78 \times 10^9 - d$ (the distance from Cerberus to *Orion* plus the distance from *Orion* to Hades must equal the distance from Cerberus to Hades).

Remembering that the law of universal gravitation requires distances pertaining to the *centers* of objects. You can equate the two forces on *Orion* as follows:

$$-F_\text{fromCerberus} + F_\text{fromHades} = 0$$

$$-\frac{Gm_\text{O}m_\text{C}}{r_\text{OC}^2} + \frac{Gm_\text{O}m_\text{H}}{r_\text{OH}^2} = 0$$

$$Gm_\text{O}\left(-\frac{m_\text{C}}{r_\text{OC}^2} + \frac{m_\text{H}}{r_\text{OH}^2}\right) = 0$$

The subscripts O, H, and C stand for *Orion*, *Hades*, and *Cerberus*, respectively.

Because Gm_O can never equal 0, the only factor in the equation that *can* equal 0 is in the parentheses, so:

$$-\frac{m_C}{r_{OC}^2}+\frac{m_H}{r_{OH}^2}=0$$

$$\frac{m_H}{r_{OH}^2}=\frac{m_C}{r_{OC}^2}$$

$$\frac{8.2\times10^7\text{ kg}}{\left[2.5\times10^8\text{ m}+\left(5.78\times10^9\text{ m}-d\right)\right]^2}=\frac{1.88\times10^6\text{ kg}}{\left(1.8\times10^7\text{ m}+d\right)^2}$$

$$\frac{8.2\times10^7\text{ kg}}{\left(6.03\times10^9\text{ m}-d\right)^2}=\frac{1.88\times10^6\text{ kg}}{\left(1.8\times10^7\text{ m}+d\right)^2}$$

$$\left[\frac{\left(1.8\times10^7\text{ m}+d\right)^2}{8.2\times10^7\text{ kg}}\right]\left[\frac{8.2\times10^7\text{ kg}}{\left(6.03\times10^9\text{ m}-d\right)^2}\right]=\left[\frac{1.88\times10^6\text{ kg}}{\left(1.8\times10^7\text{ m}+d\right)^2}\right]\left[\frac{\left(1.8\times10^7\text{ m}+d\right)^2}{8.2\times10^7\text{ kg}}\right]$$

$$\frac{\left(1.8\times10^7\text{ m}+d\right)^2}{\left(6.03\times10^9\text{ m}-d\right)^2}=\frac{1.88\times10^6}{8.2\times10^7}$$

$$\left(\frac{1.8\times10^7\text{ m}+d}{6.03\times10^9\text{ m}-d}\right)^2=0.0229$$

$$\frac{1.8\times10^7\text{ m}+d}{6.03\times10^9\text{ m}-d}=\sqrt{.0229}=0.151$$

$$1.8\times10^7\text{ m}+d=0.151\left(6.03\times10^9\text{ m}-d\right)$$

$$1.8\times10^7\text{ m}+d=9.11\times10^8\text{ m}-0.151d$$

$$1.151d=8.93\times10^8\text{ m}$$

$$d=7.76\times10^8\text{ m}$$

Whew! Now convert that distance to kilometers, and you're done:

$$\left(7.8\times10^8\text{m}\right)\left(\frac{1\text{ km}}{1\times10^3\text{ m}}\right)=7.8\times10^5\text{ km}$$

219. $\frac{1}{4}g$

The acceleration due to gravity at the surface of a planet (or other large body) is proportional to the mass of the planet and inversely proportional to the square of the planet's radius. The mass remains the same, but the radius doubles — meaning that the square of the radius increases by a factor of 4. Therefore, the acceleration due to gravity decreases by a factor of 4 (that is, it's divided by 4).

220. 100 kg

Although the weight of (the force of gravity on) the package differs on the two planets' surfaces, the mass does not differ. The mass of an object stays constant regardless of the forces acting upon it.

221.   3

The phrase "how many times greater" indicates that you need to use a ratio to solve this question. You need to know the ratio of Jupiter's gravitational acceleration compared with Earth's: $\dfrac{a_{g,\text{Jupiter}}}{a_{g,\text{Earth}}}$.

Use the relationship that $a_g = \dfrac{Gm}{r^2}$, where m is the mass of the particular planet and r is the radius of the planet:

$$
\begin{aligned}
\frac{a_{g,\text{Jupiter}}}{a_{g,\text{Earth}}} &= \frac{\dfrac{Gm_{\text{Jupiter}}}{r^2_{\text{Jupiter}}}}{\dfrac{Gm_{\text{Earth}}}{r^2_{\text{Earth}}}} \\[2mm]
&= \frac{Gm_{\text{Jupiter}}}{r^2_{\text{Jupiter}}} \div \frac{Gm_{\text{Earth}}}{r^2_{\text{Earth}}} \\[2mm]
&= \frac{Gm_{\text{Jupiter}}}{r^2_{\text{Jupiter}}} \cdot \frac{r^2_{\text{Earth}}}{Gm_{\text{Earth}}} \\[2mm]
&= \frac{m_{\text{Jupiter}}\, r^2_{\text{Earth}}}{m_{\text{Earth}}\, r^2_{\text{Jupiter}}} \\[2mm]
&= \frac{\left(1.9 \times 10^{27}\ \text{kg}\right)\left(6.4 \times 10^{6}\ \text{m}\right)^2}{\left(5.98 \times 10^{24}\ \text{kg}\right)\left(6.99 \times 10^{7}\ \text{m}\right)^2} \\[2mm]
&= 2.66
\end{aligned}
$$

Therefore, the acceleration of gravity at Jupiter's surface is 2.66 (or 3, rounded) times as great as that at Earth's surface.

222. $5.5 \times 10^{-3}\,\text{m/s}^2$

You can figure out the acceleration of the object on the outer ring from the tangential velocity v and the radius of the ring: $a_{\text{net}} = \dfrac{v^2}{r_{\text{station}}}$.

Note that the tangential velocity of a point on the outer ring is currently the only value you don't have at your immediate disposal, so use the relationship between distance traveled (in this case, the circumference of the circle made by the object in one rotation of the station) and the time it takes to travel that distance (30 minutes):

$$
\begin{aligned}
d &= vt \\
2\pi r &= vt \\
2\pi (450\ \text{m}) &= v(1{,}800\ \text{s}) \\
1.57\ \text{m/s} &= v
\end{aligned}
$$

Recall that the circumference of a circle C is equal to $2\pi r$ and convert the time to the desired units $\left(30\ \text{min} \cdot \dfrac{60\ \text{s}}{1\ \text{min}} = 1{,}800\ \text{s}\right)$.

Now you can substitute everything into the formula you derived earlier to calculate the acceleration:

$$a_{net} = \frac{v^2}{r_{station}}$$

$$= \frac{(1.57 \text{ m/s})^2}{(450 \text{ m})}$$

$$= 5.5 \times 10^{-3} \text{ m/s}^2$$

223. 2.8 s

To solve this problem, you need to calculate the acceleration near/at the surface of Mars. Because $a_g = \frac{Gm}{r^2}$, you need to know the mass of Mars (given) and the radius (which you have to deduce using the given volume). The volume of a sphere is $V_{sphere} = \frac{4}{3}\pi r^3$ (where r is radius), so start by solving for Mars's planetary radius:

$$V_{sphere} = \frac{4}{3}\pi r^3$$

$$1.63 \times 10^{11} \text{ km}^3 = \frac{4}{3}\pi r^3$$

$$1.63 \times 10^{11} \text{ km}^3 = 4.189 r^3$$

$$3.89 \times 10^{10} \text{ km}^3 = r^3$$

$$\sqrt[3]{3.89 \times 10^{10} \text{ km}^3} = r$$

$$3,389 \text{ km} = r$$

This result is in kilometers, so you need to convert it to meters before proceeding:

$$(3,389 \text{ km})\left(\frac{1,000 \text{ m}}{1 \text{ km}}\right) = 3,389,000 \text{ m} = 3.389 \times 10^6 \text{ m}.$$

Therefore,

$$a_{g,Mars} = \frac{Gm_{Mars}}{r_{Mars}^2}$$

$$= \frac{\left(6.67 \times 10^{-11} \text{ N} \cdot \text{m}^2/\text{kg}^2\right)\left(6.42 \times 10^{23} \text{ kg}\right)}{\left(3.389 \times 10^6 \text{ m}\right)^2}$$

$$= 3.73 \text{ m/s}^2$$

You can now calculate how long it takes for an object to fall 100 meters on Mars's surface. The formula for displacement of an object in the vertical direction is $s_y = v_y t + \frac{1}{2}a_y t^2$, where s is displacement, v is initial velocity, a is acceleration, and t is time. So,

$$s_y = v_y t + \frac{1}{2}a_y t^2$$

$$-100 \text{ m} = (0 \text{ m/s})t + \frac{1}{2}\left(-3.73 \text{ m/s}^2\right)t^2$$

$$-100 \text{ m} = \left(-1.865 \text{ m/s}^2\right)t^2$$

$$53.6 \text{ s}^2 = t^2$$

$$7.32 \text{ s} = t$$

Compare this to Earth, where only the value of the acceleration changes from the previous calculation:

$$s_y = v_y t + \frac{1}{2} a_y t^2$$
$$-100 \text{ m} = (0 \text{ m/s})t + \frac{1}{2}\left(-9.8 \text{ m/s}^2\right)t^2$$
$$-100 \text{ m} = \left(-4.9 \text{ m/s}^2\right)t^2$$
$$20.4 \text{ s}^2 = t^2$$
$$4.52 \text{ s} = t$$

The difference is $7.32 \text{ s} - 4.52 \text{ s} = 2.8 \text{ s}$.

224. $\frac{v}{\sqrt{3}}$

By the equation relating orbital velocity to the distance between an orbiting body and its center of revolution, velocity is inversely proportional to the square root of distance. If the distance is *multiplied* by a factor of 3, the velocity is *divided* by a factor of $\sqrt{3}$. Therefore, Jupiter's velocity is equal to Mars's velocity divided by $\sqrt{3}$:

$$v_{\text{Jupiter}} = v_{\text{Mars}} \div \sqrt{3}$$
$$= \frac{v}{\sqrt{3}}$$

225. 7.67×10^3 m/s

Use the orbital velocity equation: $v = \sqrt{\dfrac{Gm}{r}}$.

G is the gravitational constant, m is the mass of the central body — in this case, Earth — and r is the distance from the satellite to Earth's center (in meters). First, convert 400 kilometers to the correct units: $(400 \text{ km})\left(\dfrac{1,000 \text{ m}}{1 \text{ km}}\right) = 400,000 \text{ m} = 4 \times 10^5 \text{ m}$

If the radius of Earth is 6.38×10^6 m, then the distance r from the center of Earth to the satellite is:

$$r = r_{\text{Earth}} + h$$
$$= 6.38 \times 10^6 \text{ m} + 4 \times 10^5 \text{ m}$$
$$= 6.78 \times 10^6 \text{ m}$$

Here, h is the altitude of the satellite above the Earth's surface.

Therefore, substituting that value along with the mass of Earth into the orbital velocity equation yields:

$$v = \sqrt{\frac{Gm}{r}}$$
$$= \sqrt{\frac{\left(6.67 \times 10^{-11} \text{ N} \cdot \text{m}^2/\text{kg}^2\right)\left(5.98 \times 10^{24} \text{ kg}\right)}{\left(6.78 \times 10^6 \text{ m}\right)}}$$
$$= \sqrt{5.883 \times 10^7 \text{ m}^2/\text{s}^2}$$
$$= 7.67 \times 10^3 \text{ m/s}$$

226.

2.8×10^5 km

Use the orbital velocity equation, $v = \sqrt{\dfrac{Gm}{r}}$, where m is the mass of the orbited body (here, Earth) and r is the distance the orbiting object (the satellite) is separated from the center of the body it's orbiting.

$$v = \sqrt{\frac{Gm}{r}}$$

$$v^2 = \frac{Gm}{r}$$

$$(1,200 \text{ m/s})^2 = \frac{\left(6.67 \times 10^{-11} \text{ N} \cdot \text{m}^2/\text{kg}^2\right)\left(5.98 \times 10^{24} \text{ kg}\right)}{r}$$

$$1.44 \times 10^6 \text{ m}^2/\text{s}^2 = \frac{3.989 \times 10^{14} \text{ N} \cdot \text{m}^2/\text{kg}}{r}$$

$$1.44 \times 10^6 \text{ m}^2/\text{s}^2 = \frac{3.989 \times 10^{14} \text{ kg} \cdot \text{m/s}^2 \cdot \text{m}^2/\text{kg}}{r}$$

$$1.44 \times 10^6 \text{ m}^2/\text{s}^2 = \frac{3.989 \times 10^{14} \text{ m}^3/\text{s}^2}{r}$$

$$r = \frac{3.989 \times 10^{14} \text{ m}^3/\text{s}^2}{1.44 \times 10^6 \text{ m}^2/\text{s}^2}$$

$$r = 2.8 \times 10^8 \text{ m}$$

Now convert to kilometers:

$$\left(2.8 \times 10^8 \text{ m}\right)\left(\frac{1 \text{ km}}{1 \times 10^3 \text{ m}}\right) = 2.8 \times 10^5 \text{ km}$$

227.

1.68×10^3 km/h

Geosynchronous orbit about a planet requires an orbiting body to have an orbital time equal to the planet's daily period. You're given the period of the satellite's orbit in Venusian days (d_{Venus}) in terms of Earth days (d_{Earth}), but you need a few conversions to calculate the value in the proper units — seconds:

$$\left(1 \, d_V\right)\left(\frac{243 \, d_E}{1 \, d_V}\right)\left(\frac{24 \text{h}}{1 \, d_E}\right)\left(\frac{60 \text{ min}}{1 \text{ h}}\right)\left(\frac{60 \text{ s}}{1 \text{ min}}\right) = 2.10 \times 10^7 \text{ s}$$

Combining this info with your knowledge of Venus's mass allows you to calculate the distance between this satellite and Venus's center.

Use $T^2 = \dfrac{4\pi^2}{Gm} r^3$, commonly referred to as Kepler's third law, where T is the orbital period, r is the distance between the centers of the orbiter and the orbited, and m is the mass of the orbited body — in this case, Venus.

$$T^2 = \frac{4\pi^2}{Gm} r^3$$

$$\left(2.10 \times 10^7 \text{ s}\right)^2 = \frac{4\pi^2}{\left(6.67 \times 10^{-11} \text{ N} \cdot \text{m}^2/\text{kg}^2\right)\left(4.88 \times 10^{24} \text{ kg}\right)} r^3$$

$$4.41 \times 10^{14} \text{ s}^2 = \left(1.21 \times 10^{-13} \text{ s}^2/\text{m}^3\right) r^3$$

$$3.64 \times 10^{27} \text{ m}^3 = r^3$$

$$1.5 \times 10^9 \text{ m} = r$$

Finally, use this distance in the orbital velocity equation, $v = \sqrt{\dfrac{Gm}{r}}$, to solve for the satellite's velocity:

$$v = \sqrt{\frac{Gm}{r}}$$

$$= \sqrt{\frac{\left(6.67 \times 10^{-11}\ \mathrm{N \cdot m^2/kg^2}\right)\left(4.88 \times 10^{24}\ \mathrm{kg}\right)}{\left(1.5 \times 10^9\ \mathrm{m}\right)}}$$

$$= \sqrt{2.17 \times 10^5\ \mathrm{m^2/s^2}}$$

$$= 466\ \mathrm{m/s}$$

Convert to units of kilometers per hour:

$$\left(466\ \frac{\mathrm{m}}{\mathrm{s}}\right)\left(\frac{1\ \mathrm{km}}{1{,}000\ \mathrm{m}}\right)\left(\frac{60\ \mathrm{s}}{1\ \mathrm{min}}\right)\left(\frac{60\ \mathrm{min}}{1\ \mathrm{h}}\right) = 1.68 \times 10^3\ \frac{\mathrm{km}}{\mathrm{h}}$$

228. 1.8

Kepler's third law states that, given two orbiting bodies A and B, their periods (*T*) of revolution and distances (*r*) from the object they're revolving about are related by this equation: $\dfrac{T_A^2}{T_B^2} = \dfrac{r_A^3}{r_B^3}$.

As long as the units match in a ratio, you don't have to convert them to "correct" physics units, so you don't need to convert the astronomical unit values to kilometer values. Given that the period of Earth's revolution is 1 Earth year, the equation is easily solved for Mars's revolution in the same units, which is what you want:

$$\frac{T_{\mathrm{Earth}}^2}{T_{\mathrm{Mars}}^2} = \frac{r_{\mathrm{Earth}}^3}{r_{\mathrm{Mars}}^3}$$

$$\frac{(1\ \mathrm{Earth\ year})^2}{T_{\mathrm{Mars}}^2} = \frac{(1\ \mathrm{a.u.})^3}{(1.5\ \mathrm{a.u.})^3}$$

$$\frac{(1\ \mathrm{Earth\ year})^2}{T_{\mathrm{Mars}}^2} = \frac{1}{3.375}$$

$$3.375\,(\mathrm{Earth\ year})^2 = T_{\mathrm{Mars}}^2$$

$$\sqrt{3.375\,(\mathrm{Earth\ year})^2} = T_{\mathrm{Mars}}$$

$$1.8\ \mathrm{Earth\ years} = T_{\mathrm{Mars}}$$

229. 3.90×10^5 km

Use the equation relating orbital period to orbital position, $T^2 = \dfrac{4\pi^2}{Gm} r^3$, where *T* is the orbital period, *r* is the distance between the centers of the orbiter and the orbited, and *m* is the mass of the orbited body — in this case, Earth. Therefore, the given value of the moon's mass is extraneous.

Before substituting values into the equation, make sure to convert the moon's orbital period to the desired units — seconds:

$$(28\ \mathrm{d})\left(\frac{24\ \mathrm{h}}{1\ \mathrm{d}}\right)\left(\frac{60\ \mathrm{min}}{1\ \mathrm{h}}\right)\left(\frac{60\ \mathrm{s}}{1\ \mathrm{min}}\right) = 2.42 \times 10^6\ \mathrm{s}$$

$$T^2 = \frac{4\pi^2}{Gm}r^3$$

$$\left(2.42\times10^6 \text{ s}\right)^2 = \frac{4\pi^2}{\left(6.67\times10^{-11} \text{ N}\cdot\text{m}^2/\text{kg}^2\right)\left(5.98\times10^{24} \text{ kg}\right)}r^3$$

$$\left(2.42\times10^6 \text{ s}\right)^2 = \frac{39.48}{\left(6.67\times10^{-11} \text{ kg}\cdot\text{m/s}^2\cdot\text{m}^2/\text{kg}^2\right)\left(5.98\times10^{24} \text{ kg}\right)}r^3$$

$$5.85\times10^{12} \text{ s}^2 = \left(9.898\times10^{-14} \text{ s}^2/\text{m}^3\right)r^3$$

$$5.91\times10^{25} \text{ m}^3 = r^3$$

$$\sqrt[3]{5.91\times10^{25} \text{ m}^3} = r$$

$$3.90\times10^8 \text{ m} = r$$

Finally, convert to the requested unit — kilometers:

$$\left(3.90\times10^8 \text{ m}\right)\left(\frac{1 \text{ km}}{1\times10^3 \text{ m}}\right) = 3.90\times10^5 \text{ km}$$

230. 1.55 h

Use the equation relating orbital period to orbital position, $T^2 = \frac{4\pi^2}{Gm}r^3$, where T is the orbital period, r is the distance between the centers of the orbiter and the orbited, and m is the mass of the orbited body — in this case, Earth, which has a mass of 5.98×10^{24} kilograms. Add the 420 kilometers to Earth's radius to calculate the total distance between the ISS and Earth's center, and then convert to meters:

$$r = r_{\text{Earth}} + h$$
$$= 6.38\times10^3 \text{ km} + 420 \text{ km}$$
$$= \left(6.80\times10^3 \text{ km}\right)\left(\frac{1\times10^3 \text{ m}}{1 \text{ km}}\right) = 6.80\times10^6 \text{ m}$$

Now substitute that into the orbital equation, and you find the ISS's orbital period (in units of seconds):

$$T^2 = \frac{4\pi^2}{Gm}r^3$$

$$T = \sqrt{\frac{4\pi^2 r^3}{Gm}}$$

$$= 2\pi\sqrt{\frac{\left(6.80\times10^6 \text{ m}\right)^3}{\left(6.67\times10^{-11} \text{ N}\cdot\text{m}^2/\text{kg}^2\right)\left(5.98\times10^{24} \text{ kg}\right)}}$$

$$= 2\pi\sqrt{\frac{3.14\times10^{20} \text{ m}^3}{\left(6.67\times10^{-11} \text{ kg}\cdot\text{m/s}^2\cdot\text{m}^2/\text{kg}^2\right)\left(5.98\times10^{24} \text{ kg}\right)}}$$

$$= 2\pi\sqrt{7.883\times10^5 \text{ s}^2}$$

$$= 5{,}579 \text{ s}$$

Finally, convert to hours: $(5{,}579 \text{ s})\left(\frac{1 \text{ min}}{60 \text{ s}}\right)\left(\frac{1 \text{ h}}{60 \text{ min}}\right) = 1.55$ h.

231.

$6.02 \times 10^4 \, \mathrm{d_E}$

Use the equation relating orbital period to orbital position, $T^2 = \dfrac{4\pi^2}{Gm} r^3$, where T is the orbital period, r is the distance between the centers of the orbiter and the orbited, and m is the mass of the orbited body — in this case, the sun. You can ignore the radii of Neptune and the sun because they are at least four orders of magnitude (10^4) smaller than the distance between them; therefore, they have no effect on the final value to three significant digits.

The distance from the sun to Neptune is the only one that really matters for Neptune's orbit, so: $r_{\text{Neptune}} = \left(4.5 \times 10^9 \, \text{km}\right)\left(\dfrac{1 \times 10^3 \, \text{m}}{1 \, \text{km}}\right) = 4.5 \times 10^{12} \, \text{m}$.

Therefore, substituting all the known values for Neptune relative to its revolution about the sun yields:

$$T^2 = \frac{4\pi^2}{Gm} r^3$$

$$= \frac{4\pi^2 \left(4.5 \times 10^{12} \, \text{m}\right)^3}{\left(6.67 \times 10^{-11} \, \text{N} \cdot \text{m}^2/\text{kg}^2\right)\left(1.991 \times 10^{30} \, \text{kg}\right)}$$

$$= 2.709 \times 10^{19} \, \text{s}^2$$

$$T = \sqrt{2.709 \times 10^{19} \, \text{s}^2}$$

$$= 5.205 \times 10^9 \, \text{s}$$

Finally, convert that amount of time to Earth days ($\mathrm{d_E}$):

$$\left(5.205 \times 10^9 \, \text{s}\right)\left(\frac{1 \, \text{min}}{60 \, \text{s}}\right)\left(\frac{1 \, \text{h}}{60 \, \text{min}}\right)\left(\frac{1 \, \mathrm{d_E}}{24 \, \text{h}}\right) = 6.02 \times 10^4 \, \mathrm{d_E}.$$

232.

$\dfrac{1}{\sqrt{6}} V$

The velocity required to barely maintain contact with the top of a vertical loop is proportional to the square root of the gravitational acceleration ($v = \sqrt{r a_g}$, where r is the radius of the loop and a_g is the acceleration due to gravity — g, in Earth's case).

If the gravitational acceleration is reduced by a factor of 6, the velocity is also reduced, but by a factor of $\sqrt{6}$. Therefore, the velocity required on the moon is $V \div \sqrt{6} = \dfrac{1}{\sqrt{6}} V$.

233.

0.86 m/s

Before you use the vertical-loop velocity equation ($v = \sqrt{rg}$, where r is the radius of the loop and g is the gravitational acceleration on Earth's surface), you need to find the radius from the diameter in the correct units. Radius is half the diameter, or 7.5 centimeters. Converting to meters yields $(7.5 \, \text{cm})\left(\dfrac{1 \, \text{m}}{100 \, \text{cm}}\right) = 0.075 \, \text{m}.$

Therefore, the minimum velocity the mouse needs to make it past the top of the loop without falling is:

$$v = \sqrt{rg}$$
$$= \sqrt{(0.075\,\text{m})(9.8\,\text{m/s}^2)}$$
$$= \sqrt{0.735\,\text{m}^2/\text{s}^2}$$
$$= 0.86\,\text{m/s}$$

234. **0.25 km**

To use the vertical-loop velocity formula, $v = \sqrt{rg}$, you need to first convert the given velocity to units of meters per second (to match the meters per second squared of g):

$$\left(126\,\frac{\text{km}}{\text{h}}\right)\left(\frac{1{,}000\,\text{m}}{1\,\text{km}}\right)\left(\frac{1\,\text{h}}{60\,\text{min}}\right)\left(\frac{1\,\text{min}}{60\,\text{s}}\right) = 35\,\frac{\text{m}}{\text{s}}$$

Therefore:

$$v = \sqrt{rg}$$
$$35\,\text{m/s} = \sqrt{(9.8\,\text{m/s}^2)\,r}$$
$$35^2\,\text{m}^2/\text{s}^2 = (9.8\,\text{m/s}^2)\,r$$
$$1{,}225\,\text{m} = 9.8r$$
$$125\,\text{m} = r$$

Finally, multiply by 2 to get the diameter and then convert to kilometers and round to the nearest hundredth: $(125\,\text{m}) \times 2 \times \left(\dfrac{1\,\text{km}}{1{,}000\,\text{m}}\right) = 0.25\,\text{km}$.

235. **68.6 N**

The only difference between the top and bottom of the vertical loop in terms of the forces lies in the directions of the tension forces (F_T). At the top of the loop, the tension points downward, toward the center of rotation (where Fred's arm connects to his shoulder). At the bottom of the loop, tension points upward. The force of gravity (F_G) always points down, and centripetal force (F_C) is defined as pointing toward the center of the circle being traveled.

So at the top of the circle, both F_T and F_G point in the same direction, toward the center of the loop — parallel to F_C. Therefore, $F_{T,\text{top}} + F_G = F_C$. And at the bottom of the circle, F_T points toward the center of the loop — parallel to F_C — and F_G points away from the center — antiparallel to F_C. So $F_{T,\text{bottom}} - F_G = F_C$. Knowing that $F_G = ma_G = mg$ (on Earth's surface) and $F_C = \dfrac{mv^2}{r}$, you can easily solve for the two tension forces.

First, convert any units that aren't "correct" — in this question, the "80 centimeters" is the only issue: $(80 \text{ cm})\left(\dfrac{1 \text{ m}}{100 \text{ cm}}\right) = 0.08 \text{ m}$.

Then solve for $F_{T,\text{top}}$ and $F_{T,\text{bottom}}$:

$$F_{T,\text{top}} + F_G = F_C$$
$$F_{T,\text{top}} = F_C - F_G$$
$$= \frac{mv^2}{r} - mg$$
$$= \frac{(3.5 \text{ kg})(3 \text{ m/s})^2}{(0.8 \text{ m})} - (3.5 \text{ kg})\left(9.8 \text{ m/s}^2\right)$$
$$= 39.4 \text{ N} - 34.3 \text{ N}$$
$$= 5.1 \text{ N}$$

$$F_{T,\text{top}} - F_G = F_C$$
$$F_{T,\text{top}} = F_C + F_G$$
$$= \frac{mv^2}{r} + mg$$
$$= \frac{(3.5 \text{ kg})(3 \text{ m/s})^2}{(0.8 \text{ m})} + (3.5 \text{ kg})\left(9.8 \text{ m/s}^2\right)$$
$$= 39.4 \text{ N} + 34.3 \text{ N}$$
$$= 73.7 \text{ N}$$

Finally, calculate the positive difference: $\left| F_{T,\text{top}} - F_{T,\text{bottom}} \right| = |5.1 - 73.7| = |-68.6| = 68.6 \text{ N}$.

236. $1.7 \times 10^2 \text{ kg} / \text{m}^3$

Density is mass per unit volume, or $\rho = \dfrac{m}{V}$. In this case, the mass is

$$m = 300 \text{ g} \times \frac{1 \text{ kg}}{1{,}000 \text{ g}}$$
$$= 0.300 \text{ kg}$$

The volume is

$$V = 30 \text{ cm} \times 10 \text{ cm} \times 6.0 \text{ cm} \times \left(\frac{1 \text{ m}}{100 \text{ cm}}\right)^3$$
$$= 0.0018 \text{ m}^3$$

Insert these values into the equation for density to find the density of the cake:

$$\rho = \frac{m}{V}$$
$$= \frac{0.300 \text{ kg}}{0.0018 \text{ m}^3}$$
$$= 1.7 \times 10^2 \text{ kg/m}^3$$

237. $1.6 \times 10^3 \text{ kg/m}^3$

Density is mass per unit volume, or $\rho = \frac{m}{V}$. The total mass of the box plus cough drops is

$$m = 1.0 \text{ g} + 30 \times 2.2 \text{ g}$$
$$= 67 \text{ g} \times \frac{1 \text{ kg}}{1,000 \text{ g}}$$
$$= 0.067 \text{ kg}$$

The volume of the box is

$$V = 1.0 \text{ cm} \times 5.0 \text{ cm} \times 8.5 \text{ cm} \times \left(\frac{1 \text{ m}}{100 \text{ cm}} \right)^3$$
$$= 4.25 \times 10^{-5} \text{ m}^3$$

Insert these values into the equation for density to find the density of the box:

$$\rho = \frac{m}{V}$$
$$= \frac{0.067 \text{ kg}}{4.25 \times 10^{-5} \text{ m}^3}$$
$$= 1.6 \times 10^3 \text{ kg/m}^3$$

238. 0.721

Specific gravity is the density of a material divided by the density of water at $4°C$, which is 1,000 kilograms per cubic meter. The equation for specific gravity is

specific gravity$_x = \frac{\rho_x}{1,000 \text{ kg/m}^3}$. The subscript x refers to the material in question.

You know that $\rho_{\text{gasoline}} = 721 \text{ kg/m}^3$. Insert this value into the equation for specific gravity to find

$$\text{specific gravity}_{\text{gasoline}} = \frac{\rho_{\text{gasoline}}}{1,000 \text{ kg/m}^3}$$
$$= \frac{721 \text{ kg/m}^3}{1,000 \text{ kg/m}^3}$$
$$= 0.721$$

239.

$1.7 \times 10^2 \text{ kg/m}^3$

Specific gravity is the density of a material divided by the density of water at $4°C$, which is 1,000 kilograms per cubic meter. The equation for specific gravity is

specific gravity $_x = \dfrac{\rho_x}{1{,}000 \text{ kg/m}^3}$, where the subscript x refers to the material in question. You know that when you stuff the pillows into the box, their specific gravity triples. Mathematically, this gives the following equation: specific gravity $_{\text{pillowsinbox}} = 3 \times$ specific gravity $_{\text{pillows}}$.

Using the equation for specific gravity, this equation becomes

$$\text{specific gravity}_{\text{pillowsinbox}} = 3 \times \text{specific gravity}_{\text{pillows}}$$

$$\frac{\rho_{\text{pillowsinbox}}}{1{,}000 \text{ kg/m}^3} = 3 \times \frac{\rho_{\text{pillows}}}{1{,}000 \text{ kg/m}^3}$$

$$\rho_{\text{pillowsinbox}} = 3 \times \rho_{\text{pillows}}$$
$$= 3 \times 55 \text{ kg/m}^3$$
$$= 1.7 \times 10^2 \text{ kg/m}^3$$

240.

0.36 N

Pressure is the force per unit area: $P = \dfrac{F}{A}$. Solving this equation for the force gives

$$P = \frac{F}{A}$$
$$F = PA$$

Insert the given values of $P = 1.2 \text{ Pa}$ and $A = 0.3 \text{ m}^2$ to find the magnitude of the force:

$$F = PA$$
$$= 1.2 \text{ Pa} \times 0.30 \text{ m}^2$$
$$= 0.36 \text{ N}$$

241.

$7.5 \times 10^4 \text{ Pa}$

Pressure is the force per unit area: $P = \dfrac{F}{A}$. You know that the total downward force is 300 newtons. The total area over which this force is applied is 4 times the cross-sectional area of one leg, which is

$$A = 4 \times 10 \text{ cm}^2 \times \left(\frac{1 \text{ m}}{100 \text{ cm}} \right)^2$$
$$= 4.0 \times 10^{-3} \text{ m}^2$$

Inserting these values into the equation for pressure gives

$$P = \frac{F}{A}$$
$$= \frac{300 \text{ N}}{4.0 \times 10^{-3} \text{ m}^2}$$
$$= 7.5 \times 10^4 \text{ Pa}$$

242. $3.3 \times 10^4 \, Pa$

The change in pressure is given by $\Delta P = \rho g h$. In this case, $\rho = 1{,}000$ kilograms per cubic meter and $h = 3.4$ meters. Insert these values into the previous equation:

$$\Delta P = \rho g h$$
$$= 1{,}000 \times 9.8 \text{ m/s}^2 \times 3.4 \text{ m}$$
$$= 3.3 \times 10^4 \text{ Pa}$$

243. 1.8 m

The change in pressure is given by $\Delta P = \rho g h$. In this case, $\Delta P = 18{,}000$ pascals. Solve this equation for the height h:

$$\Delta P = \rho g h$$
$$h = \frac{\Delta P}{\rho g}$$

Insert the known values into the previous equation:

$$h = \frac{\Delta P}{\rho g}$$
$$= \frac{18{,}000 \text{ Pa}}{1{,}000 \text{ kg/m}^3 \times 9.8 \text{ m/s}^2}$$
$$= 1.8 \text{ m}$$

244. $1.1 \times 10^5 \, Pa$

The change in pressure is given by $\Delta P = \rho g h$. By using $\Delta P = P_{\text{bottom}} - P_{\text{top}}$, the first equation becomes $P_{\text{bottom}} - P_{\text{top}} = \rho g h$. You know that $P_{\text{bottom}} = 130{,}000$ pascals, so you can solve for P_{top}:

$$P_{\text{bottom}} - P_{\text{top}} = \rho g h$$
$$P_{\text{top}} = P_{\text{bottom}} - \rho g h$$
$$= 130{,}000 \text{ Pa} - 1{,}000 \text{ kg/m}^3 \times 9.8 \text{ m/s}^2 \times 2.5 \text{ m}$$
$$= 1.1 \times 10^5 \text{ Pa}$$

245. $1.8 \times 10^2 \, N$

Pascal's principle gives $\frac{F_1}{A_1} = \frac{F_2}{A_2}$. In this case, you know that $F_1 = 38$ newtons, $A_1 = 21$ square centimeters, and $A_2 = 100$ square centimeters. Solve the previous equation for F_2 and insert the given values to find the force on the second piston:

$$\frac{F_1}{A_1} = \frac{F_2}{A_2}$$
$$F_2 = A_2 \frac{F_1}{A_1}$$
$$= 100 \text{ cm}^2 \times \frac{38 \text{ N}}{21 \text{ cm}^2}$$
$$= 1.8 \times 10^2 \text{ N}$$

246. $F_2' = \frac{2}{3} F_2$

Apply Pascal's principle to the new system (in other words, apply it after F_1 is doubled and A_2 is reduced by a factor of 3). Using primes to indicate the new quantities, you have $\dfrac{F_1'}{A_1'} = \dfrac{F_2'}{A_2'}$.

You know that $F_1' = 2F_1$, $A_2' = A_2/3$, and $A_1' = A_1$, where the unprimed quantities are the original quantities. Use these equations in the previous equation and solve for F_2':

$$\frac{F_1'}{A_1'} = \frac{F_2'}{A_2'}$$

$$\frac{2F_1}{A_1} = \frac{F_2'}{A_2/3}$$

$$\frac{2F_2}{\cancel{A_2}} = \frac{F_2'}{\cancel{A_2}/3}$$

$$F_2' = \frac{2}{3} F_2$$

247. **20,000 Pa**

According to Pascal's principle, the change in pressure in an enclosed system is the same throughout the system. Therefore, the pressure on any piston in the system is the same: 20,000 pascals.

248. **7.75 kg**

Archimedes's principle tells you that the weight of the water displaced is equal to the buoyancy force: $W_{\text{waterdisplaced}} = F_{\text{buoyancy}}$. To keep the wood afloat, the buoyancy force must have the same magnitude as the force of gravity on the block, so $F_{\text{buoyancy}} = m_{\text{wood}}g$.

The volume of water displaced is

$$\begin{aligned}
V_{\text{waterdisplaced}} &= 0.053\ \text{m} \times 0.34\ \text{m} \times 0.43\ \text{m} \\
&= 0.00775\ \text{m}^3
\end{aligned}$$

So the mass of water displaced is

$$\begin{aligned}
m_{\text{waterdisplaced}} &= \rho_{\text{water}} V_{\text{waterdisplaced}} \\
&= 1{,}000\ \text{kg/m}^3 \times 0.00775\ \text{m}^3 \\
&= 7.75\ \text{kg}
\end{aligned}$$

Thus, the weight of the water displaced is $W_{\text{waterdisplaced}} = m_{\text{waterdisplaced}}g$. The weight of the water displaced must equal the buoyancy force, so

$$W_{\text{waterdisplaced}} = F_{\text{buoyancy}}$$
$$m_{\text{waterdisplaced}}g = m_{\text{wood}}g$$
$$m_{\text{waterdisplaced}} = m_{\text{wood}}$$
$$7.75\,\text{kg} =$$

Thus, the mass of the piece of wood is 7.75 kilograms.

249.   35 N

The buoyancy force is the mass of the water displaced multiplied by the acceleration due to gravity: $F_{\text{buoyancy}} = m_{\text{waterdisplaced}}g$.

The volume of water displaced is half the volume of the basketball:

$$V_{\text{waterdisplaced}} = \frac{1}{2}V_{\text{basketball}}$$
$$= \frac{1}{2}\frac{4}{3}\pi r^3$$
$$= \frac{2}{3}\pi r^3$$

Here, $r = 12$ cm. In meters, the radius is

$$r = 12\,\cancel{\text{cm}} \times \frac{1\,\text{m}}{100\,\cancel{\text{cm}}}$$
$$= 0.12\,\text{m}$$

Using the equation for density, the mass of water displaced is $m_{\text{waterdisplaced}} = \rho_{\text{water}}V_{\text{waterdisplaced}}$. The buoyancy force is

$$F_{\text{buoyancy}} = m_{\text{waterdisplaced}}g$$
$$= \rho_{\text{water}}V_{\text{waterdisplaced}}g$$
$$= \rho_{\text{water}} \times \frac{2}{3}\pi r^3 \times g$$
$$= 1{,}000\,\text{kg/m}^3 \times \frac{2}{3}\pi(0.12\,\text{m})^3 \times 9.8\,\text{m/s}^2$$
$$= 35\,\text{N}$$

250.   0.14 m³

The weight of the additional water displaced is equal to the combined weight of the two extra people who got into the boat:

$$m_{\text{water}}g = 2 \times 690\,\text{N}$$
$$= 1{,}380\,\text{N}$$

The mass of the water displaced is then

$$m_{water}g = 1,380 \text{ N}$$
$$m_{water} = \frac{1,380 \text{ N}}{g}$$
$$= \frac{1,380 \text{ N}}{9.8 \text{ m/s}^2}$$
$$= 141 \text{ kg}$$

Solve the equation for density for the volume of water displaced and use this result for the mass of water displaced to find the answer:

$$\rho_{water} = \frac{m_{water}}{V_{water}}$$
$$V_{water} = \frac{m_{water}}{\rho_{water}}$$
$$= \frac{141 \text{ kg}}{1,000 \text{ kg/m}^3}$$
$$= 0.14 \text{ m}^3$$

251. (E) Liquid A is more viscous than liquid B.

Liquid A is more viscous than liquid B because viscous fluids experience more friction between liquid layers when they flow. This fact reduces the flow rate, so liquid A spills out more slowly than liquid B.

252. (A) The flow is irrotational.

Imagine that you drew a dot on the left side of the marker. Upstream of the curve, this dot would face to the left. Downstream of the curve, the same is true. The flow is thus irrotational. You can also understand this by realizing that the marker does not spin about an axis that goes through the marker; it rotates about an axis that is external to the marker.

253. 4.0 m/s

Use the equation of continuity: $A_1v_1 = A_2v_2$. You know that $A_2 = \frac{1}{5}A_1$ and that $v_1 = 0.80$ meters per second. Solve the continuity equation for A_2:

$$A_1v_1 = A_2v_2$$
$$= \frac{1}{5}A_1v_2$$
$$v_2 = 5v_1$$
$$= 5 \times 0.80 \text{ m/s}$$
$$= 4.0 \text{ m/s}$$

254. 0.25 m

Use the equation of continuity: $A_1v_1 = A_2v_2$. In terms of the stream width w and depth d, the cross-sectional area of the stream is $A = wd$. Use this in the continuity equation:

$$A_1v_1 = A_2v_2$$
$$w_1dv_1 = w_2dv_2$$
$$w_1v_1 = w_2v_2$$

You know that $v_2 = 3v_1$ so you can solve the continuity equation for the width w_2:

$$w_1v_1 = w_2v_2$$
$$w_1\cancel{v_1} = w_2 \times 3\cancel{v_1}$$
$$w_2 = \frac{w_1}{3}$$
$$w_2 = \frac{0.76\,\text{m}}{3}$$
$$= 0.25\,\text{m}$$

255. 5.0

You know that $Q_1 = 2.5Q_2$, where Q_1 and Q_2 are the volume flow rate through pipes 1 and 2, respectively. In terms of cross-sectional area and flow speed, the volume flow rate is $Q = Av$. When combined with the previous equation, you get

$$Q_1 = 2.5Q_2$$
$$A_1v_1 = 2.5A_2v_2$$

You also know that $A_1 = \frac{1}{2}A_2$. Combine these equations to find

$$A_1v_1 = 2.5A_2v_2$$
$$\frac{1}{2}\cancel{A_2}v_1 = 2.5\cancel{A_2}v_2$$
$$\frac{v_1}{v_2} = 2 \times 2.5$$
$$= 5.0$$

256. 1.2×10^{-3} m³/s

Flow rate is the volume of fluid moving past a point per unit time. The volume of fluid that enters the pool is

$$V = 3.0\,\text{m} \times 20\,\text{m} \times 5.0\,\text{m}$$
$$= 300\,\text{m}^3$$

The time it takes for this volume of water to enter the pool is

$$t = 3\ \cancel{d} \times \frac{24\ \cancel{h}}{1\ \cancel{d}} \times \frac{3600\ \text{s}}{1\ \cancel{h}}$$
$$= 2.59 \times 10^5\ \text{s}$$

The volume flow rate is therefore

$$Q = \frac{V}{t}$$
$$= \frac{300\ \text{m}^3}{2.59 \times 10^5\ \text{s}}$$
$$= 1.2 \times 10^{-3}\ \text{m}^3/\text{s}$$

257.   5.2 m/s

Use Bernoulli's equation: $P_1 + \frac{1}{2}\rho v_1^2 + \rho g y_1 = P_2 + \frac{1}{2}\rho v_2^2 + \rho g y_2$.

P, v, ρ, and y are the pressure, speed, density, and height, respectively, of a fluid. The subscripts 1 and 2 refer to two different points. In this case, let point 1 be on the surface of the lake and point 2 be at the outlet of the hole in the dam. The pressure at each point is just atmospheric pressure, so $P_1 = P_2$.

The hole is 1.4 meters below the lake, so $y_1 - y_2 = 1.4$ m. Because the hole is "small," you can assume that the level of the lake doesn't change much as water leaks out of the hole, so $v_1 = 0$ meters per second. Using these equations in Bernoulli's equation, you can solve for the speed of the fluid at point 2:

$$P_1 + \frac{1}{2}\rho v_1^2 + \rho g y_1 = P_2 + \frac{1}{2}\rho v_2^2 + \rho g y_2$$
$$\rho g (y_1 - y_2) = \frac{1}{2}\rho v_2^2$$
$$v_2 = \pm\sqrt{2g(y_1 - y_2)}$$
$$= \pm\sqrt{2 \times 9.8\ \text{m/s}^2 \times 1.4\ \text{m}}$$
$$= \pm 5.2\ \text{m/s}$$

Because you're interested in the speed of the water, which is a positive quantity, use the plus sign in the equation. Thus, the speed of the water coming out of the hole is 5.2 meters per second.

258.   1.9 m/s

Use Bernoulli's equation: $P_1 + \frac{1}{2}\rho v_1^2 + \rho g y_1 = P_2 + \frac{1}{2}\rho v_2^2 + \rho g y_2$.

P, v, ρ, and y are the pressure, speed, density, and height, respectively, of a fluid. The subscripts 1 and 2 refer to two different points. In this case, let point 1 be on the ground and point 2 be at 1.3 meters above the ground. At both points, the pressure is atmospheric pressure, so $P_1 = P_2 = 101,000$ pascals. The difference in heights between points 1 and 2 is $y_2 - y_1 = 1.3$ m.

Using these equations, you can solve Bernoulli's equation for the speed v_2:

$$P_1 + \frac{1}{2}\rho v_1^2 + \rho g y_1 = P_2 + \frac{1}{2}\rho v_2^2 + \rho g y_2$$

$$\frac{1}{2}\rho v_1^2 + \rho g y_1 - \rho g y_2 = \frac{1}{2}\rho v_2^2$$

$$v_2^2 = v_1^2 + 2gy_1 - 2gy_2$$

$$v_2 = \pm\sqrt{v_1^2 + 2g\left(y_1 - y_2\right)}$$

$$= \pm\sqrt{(5.4\ \text{m/s})^2 + 2\times 9.8\ \text{m/s}^2 \times (-1.3\ \text{m})}$$

$$= 1.9\ \text{m/s}$$

259. $4.4 \times 10^2\ \text{kg/m}^3$

Use Bernoulli's equation: $P_1 + \frac{1}{2}\rho v_1^2 + \rho g y_1 = P_2 + \frac{1}{2}\rho v_2^2 + \rho g y_2$.

P, v, P, and y are the pressure, speed, density, and height, respectively, of a fluid. The subscripts 1 and 2 refer to two different points. In this case, let point 1 be the lower horizontal part of the pipe and point 2 be the upper horizontal part of the pipe. Solve Bernoulli's equation for the density ρ:

$$P_1 + \frac{1}{2}\rho v_1^2 + \rho g y_1 = P_2 + \frac{1}{2}\rho v_2^2 + \rho g y_2$$

$$\rho\left(\frac{1}{2}v_1^2 + gy_1 - \frac{1}{2}v_2^2 - gy_2\right) = P_2 - P_1$$

$$\rho = \frac{2\left(P_2 - P_1\right)}{v_1^2 - v_2^2 + 2g\left(y_1 - y_2\right)}$$

Use the following values:

$$v_1 = 3.9\ \text{m/s}$$
$$v_2 = 1.2\ \text{m/s}$$
$$P_1 = 110{,}000\ \text{Pa}$$
$$P_2 = 101{,}000\ \text{Pa}$$
$$y_1 - y_2 = -2.8\ \text{m}$$

The result is

$$\rho = \frac{2\left(P_2 - P_1\right)}{v_1^2 - v_2^2 + 2g\left(y_1 - y_2\right)}$$

$$= \frac{2\left(101{,}000\ \text{Pa} - 110{,}000\ \text{Pa}\right)}{(3.9\ \text{m/s})^2 - (1.2\ \text{m/s})^2 - 2\times 9.8\ \text{m/s}^2 \times 2.8\ \text{m}}$$

$$= 4.4 \times 10^2\ \text{kg/m}^3$$

260. 0.78 m

Use Bernoulli's equation: $P_1 + \frac{1}{2}\rho v_1^2 + \rho g y_1 = P_2 + \frac{1}{2}\rho v_2^2 + \rho g y_2$.

P, v, P, and y are the pressure, speed, density, and height, respectively, of a fluid. The subscripts 1 and 2 refer to two different points. In this case, assign point 1 to be the location of the pump and point 2 to be the high point at which the water speed is given. Call the position of the pump $y_1 = 0$ meters. You are given the following quantities:

$$P_1 = 115,000 \, \text{Pa}$$
$$P_2 = 110,000 \, \text{Pa}$$
$$v_1 = 2.5 \, \text{m/s}$$
$$v_2 = 1.0 \, \text{m/s}$$

Solve Bernoulli's equation for the height y_2 and insert these values:

$$P_1 + \frac{1}{2}\rho v_1^2 + \rho g y_1 = P_2 + \frac{1}{2}\rho v_2^2 + \rho g y_2$$

$$P_1 + \frac{1}{2}\rho v_1^2 - P_2 - \frac{1}{2}\rho v_2^2 = \rho g y_2$$

$$y_2 = \frac{P_1 - P_2 + \frac{1}{2}\rho\left(v_1^2 - v_2^2\right)}{\rho g}$$

$$= \frac{115,000 \, \text{Pa} - 110,000 \, \text{Pa} + \frac{1}{2} \times 1,000 \, \text{kg/m}^3 \times \left[(2.5 \, \text{m/s})^2 - (1.0 \, \text{m/s})^2\right]}{1,000 \, \text{kg/m}^3 \times 9.8 \, \text{m/s}^2}$$

$$= 0.78 \, \text{m}$$

261. 0.94

Apply Bernoulli's equation with $y_1 = y_2$:

$$P_1 + \frac{1}{2}\rho v_1^2 + \rho g y_1 = P_2 + \frac{1}{2}\rho v_2^2 + \rho g y_2$$

$$P_1 + \frac{1}{2}\rho v_1^2 = P_2 + \frac{1}{2}\rho v_2^2$$

You want to know when the pressure at point 2 goes to 0, so setting $P_2 = 0$ pascals reduces Bernoulli's equation to $P_1 + \frac{1}{2}\rho v_1^2 = \frac{1}{2}\rho v_2^2$. Solving this for the speed v_2 gives $v_2 = \pm\sqrt{\dfrac{2P_1}{\rho} + v_1^2}$.

Inserting the given values gives

$$v_2 = \pm\sqrt{\frac{2P_1}{\rho} + v_1^2}$$

$$= \pm\sqrt{\frac{2 \times 12,000 \, \text{Pa}}{1,060 \, \text{kg/m}^3} + (0.30 \, \text{m/s})^2}$$

$$= \pm 4.77 \, \text{m/s}$$

Use the positive solution because you're interested in a speed. The fraction of the aorta's cross-sectional area that is *open* to allow blood circulation is

$$A_1 v_1 = A_2 v_2$$

$$\frac{A_2}{A_1} = \frac{v_1}{v_2}$$

$$= \frac{0.30 \, \text{m/s}}{4.77 \, \text{m/s}}$$

$$= 0.0629$$

The fraction f of the aorta that *blocks* circulation is therefore

$$f = 1 - \frac{A_2}{A_1}$$
$$= 1 - 0.629$$
$$= 0.94$$

262. −24%

Apply the continuity equation:

$$A_1 v_1 = A_2 v_2$$
$$\frac{v_2}{v_1} = \frac{A_1}{A_2}$$

Let subscript 1 refer to the wide part of the channel and subscript 2 refer to the narrow part. The ratio of the areas of the two channels is

$$\frac{A_1}{A_2} = \frac{1.0\,\text{m} \times 0.20\,\text{m}}{0.33\,\text{m} \times 0.80\,\text{m}}$$
$$= 0.758$$

The percent change Δ is

$$\Delta = \frac{v_2 - v_1}{v_1} \times 100\%$$
$$= \left(\frac{v_2}{v_1} - 1 \right) \times 100\%$$
$$= \left(\frac{A_1}{A_2} - 1 \right) \times 100\%$$
$$= (0.758 - 1) \times 100\%$$
$$= -24\%$$

263. 5.0 cm

Apply the continuity equation: $A_{1+2} v_{1+2} = A_3 v_3$. Let the subscript 1+2 refer to pipes 1 and 2, and let the subscript 3 refer to pipe 3. Because you want the speeds to be the same in pipe 3 as in pipes 1 and 2, you have $v_{1+2} = v_3$, so $A_{1+2} = A_3$.

The total cross-sectional area of pipes 1 and 2 is $A_{1+2} = A_1 + A_2 = \pi r_1^2 + \pi r_2^2$.

The cross-sectional area of pipe 3 is $A_3 = \pi r_3^2$.

Combining these equations and solving for the radius of pipe 3 gives

$$A_{1+2} = A_3$$
$$\pi r_1^2 + \pi r_2^2 = \pi r_3^2$$
$$r_3 = \pm \sqrt{r_1^2 + r_2^2}$$
$$= \pm \sqrt{(3.0\,\text{cm})^2 + (4.0\,\text{cm})^2}$$
$$= \pm 5.0\,\text{cm}$$

Because you're looking for a distance, use the positive solution, which makes the radius of the third pipe 5.0 centimeters.

264. 3.8×10^{-3} m

The volume flow rate at the pipe outlet is $Q_2 = A_2 v_2$. You can use Bernoulli's equation to find the speed v_2 of the water at the pipe outlet. Assume that the speed v_1 at the inlet is essentially 0. Bernoulli's equation then takes this form:

$$P_1 + \frac{1}{2}\rho v_1^2 + \rho g y_1 = P_2 + \frac{1}{2}\rho v_2^2 + \rho g y_2$$

$$P_1 + \rho g y_1 = P_2 + \frac{1}{2}\rho v_2^2 + \rho g y_2$$

Solving for v_2 gives

$$P_1 + \rho g y_1 = P_2 + \frac{1}{2}\rho v_2^2 + \rho g y_2$$

$$P_1 - P_2 + \rho g y_1 - \rho g y_2 = \frac{1}{2}\rho v_2^2$$

$$v_2^2 = 2\left[\frac{P_1 - P_2}{\rho} + g(y_1 - y_2)\right]$$

$$v_2 = \pm\sqrt{2\left[\frac{P_1 - P_2}{\rho} + g(y_1 - y_2)\right]}$$

The pressure at the outlet is $P_2 = 101{,}000$ pascals, so the pressure difference is

$$P_1 - P_2 = 101{,}000\,\text{Pa} + \rho g d - 101{,}000\,\text{Pa}$$

$$= \rho g d$$

Here, $d = 1.2$ meters. The height difference between the inlet and outlet is $y_1 - y_2 = 25$ meters. Using these values, the speed at the outlet is

$$v_2 = \pm\sqrt{2\left[\frac{P_1 - P_2}{\rho} + g(y_1 - y_2)\right]}$$

$$= \pm\sqrt{2\left[\frac{\rho g d}{\rho} + g(y_1 - y_2)\right]}$$

$$= \pm\sqrt{2g(d + y_1 - y_2)}$$

$$= \pm\sqrt{2 \times 9.8\,\text{m/s}^2 \times (1.2\,\text{m} + 25\,\text{m})}$$

$$= \pm 22.7\,\text{m/s}$$

Use this value in the equation for volume flow rate and solve for the cross-sectional area of the pipe:

$$Q_2 = A_2 v_2$$

$$A_2 = \frac{Q_2}{v_2}$$

$$= \frac{0.001\,\text{m}^3/\text{s}}{22.7\,\text{m/s}}$$

$$= 4.41 \times 10^{-5}\,\text{m}^2$$

The radius of the pipe must be

$$A_2 = \pi r^2$$
$$r = \pm\sqrt{\frac{A_2}{\pi}}$$
$$= \pm\sqrt{\frac{4.41 \times 10^{-5} \text{ m}^2}{\pi}}$$
$$= 3.8 \times 10^{-3} \text{ m}$$

Because you're looking for distance, use the positive answer.

265. 5.7×10^{-4} m^3/s

The volume flow rate is $Q = Av$. Using $A = \pi r^2$ gives $Q = \pi r^2 v$. You know that $v = 0.20$ meters per second and

$$r = 3.0 \text{ cm} \times \frac{1 \text{ m}}{100 \text{ cm}}$$
$$= 0.030 \text{ m}$$

Insert these values into the equation for volume flow rate:

$$Q = \pi r^2 v$$
$$= \pi \times (0.030 \text{ m})^2 \times 0.20 \text{ m/s}$$
$$= 5.7 \times 10^{-4} \text{ m}^3/\text{s}$$

266. 0 J

There is no net force on the book if the book is not accelerating. A constant speed (velocity) means that the acceleration equals 0, and, because $F_{net} = ma_{net}$, if $a_{net} = 0$, then $F_{net} = 0$ as well. Because work is the product of force and distance, if the force is 0, then the work done must also be 0.

267. 32.9 J

First convert the distance between the shelves into the proper units of meters:

$$(40 \text{ cm})\left(\frac{1 \text{ m}}{100 \text{ cm}}\right) = 0.4 \text{ m}$$

$W = Fd$, where W is the work done on an object, F is the force exerted on the object, and d is the distance the object moves. If the shelves are each 0.4 meters apart, the distance between three shelves is $(3)(0.4\text{ m}) = 1.2\text{ m}$. To figure out the force that Roger exerts to lift the book, draw a free-body diagram to find F_L, where F_L is the "lift" force that Roger exerts.

Illustration by Thomson Digital

$$F_{y,\text{net}} = \sum F_y$$
$$ma_{y,\text{net}} = F_L - F_G$$
$$ma = F_L - ma_G$$
$$ma = F_L - mg$$
$$(2.8\text{ kg})\left(0\text{ m/s}^2\right) = F_L - (2.8\text{ kg})\left(9.8\text{ m/s}^2\right)$$
$$0\text{ N} = F_L - 27.44\text{ N}$$
$$27.44\text{ N} = F_L$$

Then substitute this into the formula for work, along with the previously calculated distance.

$$W = Fd$$
$$= (27.44\text{ N})(1.2\text{ m})$$
$$= 32.9\text{ J}$$

268. **285 kJ**

First convert the velocities to the proper units of meters:

$$\left(25\,\frac{\text{km}}{\text{h}}\right)\left(\frac{1{,}000\text{ m}}{1\text{ km}}\right)\left(\frac{1\text{ h}}{60\text{ min}}\right)\left(\frac{1\text{ min}}{60\text{ s}}\right) = 6.94\,\frac{\text{m}}{\text{s}}$$

$$\left(40\,\frac{\text{km}}{\text{h}}\right)\left(\frac{1{,}000\text{ m}}{1\text{ km}}\right)\left(\frac{1\text{ h}}{60\text{ min}}\right)\left(\frac{1\text{ min}}{60\text{ s}}\right) = 11.1\,\frac{\text{m}}{\text{s}}$$

Use the velocity-time formula to solve for the car's net acceleration:

$$v_f = v_i + at$$
$$11.1\text{ m/s} = (6.94\text{ m/s}) + a(5\text{ s})$$
$$4.16\text{ m/s} = (5\text{ s})a$$
$$0.832\text{ m/s}^2 = a$$

And use the velocity-displacement formula to solve for the distance the car moves in the 5-second time interval:

$$v_f^2 = v_i^2 + 2as$$

$$(11.1\,\text{m/s})^2 = (6.94\,\text{m/s})^2 + 2\left(0.832\,\text{m/s}^2\right)s$$

$$123.2\,\text{m}^2/\text{s}^2 = 48.16\,\text{m}^2/\text{s}^2 + \left(1.664\,\text{m/s}^2\right)s$$

$$75.04\,\text{m}^2/\text{s}^2 = \left(1.664\,\text{m/s}^2\right)s$$

$$45.1\,\text{m} = s$$

To calculate the work, you need both a distance and a force. Use a free-body diagram to help solve for the force.

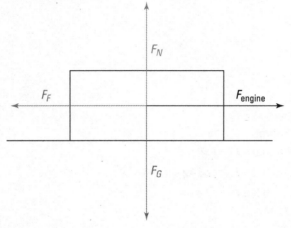

Illustration by Th+omson Digital

Use Newton's second law in the horizontal and vertical directions. First, in the y direction,

$$F_{y,\text{net}} = \sum F_{y,\text{net}}$$

$$ma_y = F_N - F_G$$

$$ma_y = F_N - ma_G$$

$$ma_y = F_N - mg$$

$$m\left(0\,\text{m/s}^2\right) = F_N - mg$$

$$0\,\text{N} = F_N - mg$$

$$mg = F_N$$

And then in the x direction,

$$F_{x,\text{net}} = \sum F_x$$

$$ma_x = F_{\text{engine}} - F_F$$

$$ma_x = F_{\text{engine}} - \mu F_N$$

$$ma_x = F_{\text{engine}} - \mu mg$$

$$(800\,\text{kg})\left(0.832\,\text{m/s}^2\right) = F_{\text{engine}} - (0.72)(800\,\text{kg})\left(9.8\,\text{m/s}^2\right)$$

$$665.6\,\text{N} = F_{\text{engine}} - 5{,}644.8\,\text{N}$$

$$6{,}310.4\,\text{N} = F_{\text{engine}}$$

$W = Fd$ where W is the work done on an object, F is the force exerted on the object, and d is the distance that the object moves. So:

$$W = Fd$$
$$= (6{,}310.4 \text{ N})(45.1 \text{ m})$$
$$= 2.846 \times 10^5 \text{ J}$$

Finally, convert your answer into the requested units, kilojoules:

$$\left(2.846 \times 10^5 \text{ J}\right)\left(\frac{1 \text{ kJ}}{1 \times 10^3 \text{ J}}\right) = 2.846 \times 10^2 \text{ kJ}$$
$$= 284.6 \text{ kJ}$$

269. $\left[90°, 180°\right]$

In the definition of work, $W = Fd \cos \theta$, where W is work, F is force, d is distance, and θ is the angle between the force and distance vectors. F and d are always taken as positive-value measurements. Only the cosine term can introduce a negative term, and the cosine of angles between 90 and 180 degrees (and including 180 degrees) is a negative number.

270. 5,250 J

Use the work formula, $W = Fd \cos \theta$, where W is the amount of work done on an object, F is the magnitude of the force exerted on the object, d is the distance the object moves, and θ is the angle between the force and distance vectors. Because the 500-newton force is parallel to the hill's surface, it must be in the same direction as the distance the sled travels while it's being pushed. In that case, $\theta = 0$ degrees. To solve for the distance the sled travels, you need to know the sled's acceleration given the initial and final velocities, and you need to know the time. Start with the velocity-time formula:

$$v_f = v_i + at$$
$$4.2 \text{ m/s} = (0 \text{ m/s}) + a(5 \text{ s})$$
$$4.2 \text{ m/s} = (5 \text{ s})a$$
$$0.84 \text{ m/s}^2 = a$$

Follow up by using the displacement formula or the velocity-displacement formula (used here) to solve for the displacement/distance traveled:

$$v_f^2 = v_i^2 + 2as$$
$$(4.2 \text{ m/s})^2 = (0 \text{ m/s})^2 + 2\left(0.84 \text{ m/s}^2\right)s$$
$$17.64 \text{ m}^2/\text{s}^2 = 0 \text{ m}^2/\text{s}^2 + \left(1.68 \text{ m/s}^2\right)s$$
$$10.5 \text{ m} = s$$

Finally, substitute this into the work formula along with the information given in the problem setup.

$$W = Fd \cos \theta$$
$$= (500\,\text{N})(10.5\,\text{m}) \cos 0°$$
$$= (500\,\text{N})(10.5\,\text{m})(1)$$
$$= 5{,}250\,\text{J}$$

---

271.   −26 kJ

First calculate the sled's acceleration, which changes in velocity from 4.2 meters per second to 0 meters per second at the end of the hill. To do so, you need to find out how far the sled travels. Use trigonometry to calculate the entire length of the hill's surface:

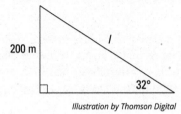

Illustration by Thomson Digital

$$\sin 32° = \frac{200\,\text{m}}{l}$$
$$0.53 = \frac{200\,\text{m}}{l}$$
$$0.53l = 200\,\text{m}$$
$$l = 377.4\,\text{m}$$

You can determine how far the sled travels while the father pushes it by first finding the acceleration with the velocity–time formula:

$$v_f = v_i + at$$
$$4.2\,\text{m/s} = (0\,\text{m/s}) + a(5\,\text{s})$$
$$4.2\,\text{m/s} = (5\,\text{s})a$$
$$0.84\,\text{m/s}^2 = a$$

Then use the displacement formula or the velocity–displacement formula (used here) to solve for the displacement:

$$v_f^2 = v_i^2 + 2as$$
$$(4.2\,\text{m/s})^2 = (0\,\text{m/s})^2 + 2\left(0.84\,\text{m/s}^2\right)s$$
$$17.64\,\text{m}^2/\text{s}^2 = 0\,\text{m}^2/\text{s}^2 + \left(1.68\,\text{m/s}^2\right)s$$
$$10.5\,\text{m} = s$$

That leaves $377.4 \text{ m} - 10.5 \text{ m} = 366.9 \text{ m}$ for the no-push portion of the sled's descent. Then, to find the amount of force that friction exerts after the pushing stops, draw a free-body diagram:

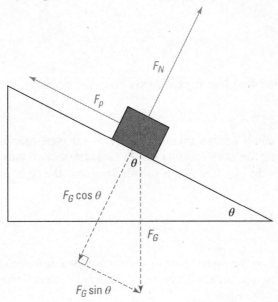

Illustration by Thomson Digital

Use Newton's second law in the vertical direction ($F_G = ma_G = mg$ on the surface of Earth):

$$F_{y,\text{net}} = \sum F_y$$
$$ma_y = F_N - F_G \cos\theta$$
$$ma_y = F_N - mg \cos\theta$$
$$(65 \text{ kg})\left(0 \text{ m/s}^2\right) = F_N - (65 \text{ kg})\left(9.8 \text{ m/s}^2\right)\cos 32°$$
$$(65 \text{ kg})\left(0 \text{ m/s}^2\right) = F_N - (65 \text{ kg})\left(9.8 \text{ m/s}^2\right)(0.848)$$
$$0 \text{ N} = F_N - 540.2 \text{ N}$$
$$540.2 \text{ N} = F_N$$

Use the definition of friction, $F_F = \mu F_N$, to solve for the force of friction.

$$F_F = \mu F_N$$
$$= (0.13)(540.2 \text{ N})$$
$$= 70.23 \text{ N}$$

Then use that value in the work formula. Because the force of friction runs opposite to the motion of an object, the angle between the force and distance vectors is 180 degrees.

$$W = Fd \cos\theta$$
$$W_F = F_F d \cos\theta$$
$$= (70.23 \text{ N})(366.9 \text{ m})\cos 180°$$
$$= (70.23 \text{ N})(366.9 \text{ m})(-1)$$
$$= -2.58 \times 10^4 \text{ J}$$

Finally, convert into the desired units:

$$\left(-2.58 \times 10^4 \text{ J}\right)\left(\frac{1 \text{ kJ}}{1 \times 10^3 \text{ J}}\right) = -25.8 \text{ kJ}$$

272. **69° south of west**

First convert the distance given in centimeters into the proper physics units of meters: $(89 \text{ cm})\left(\dfrac{1 \text{ m}}{100 \text{ cm}}\right) = 0.89 \text{ m}$.

With that distance and the given amount of work on the goal, calculate how much force moved the goal using the work formula $W = Fd$ (W is the work done on an object, F is the force exerted on the object, and d is the distance that the object moves):

$$W = Fd$$
$$(23 \text{ J}) = F(0.89 \text{ m})$$
$$25.8 \text{ N} = F$$

Forces are vectors, and the easiest way to add vectors is to break them into component form. For the 25-newton force, with its angle relative to the positive x-axis (in other words, due east) of 180 degrees:

$$\mathbf{F}_1 = \left(F_{x1}, F_{y1}\right)$$
$$= \left(F_1 \cos \theta_1, F_1 \sin \theta_1\right)$$
$$= \left[(25 \text{ N}) \cos 180°, (25 \text{ N}) \sin 180°\right]$$
$$= [(25 \text{ N})(-1), (25 \text{ N})(0)]$$
$$= (-25 \text{ N}, 0 \text{ N})$$

And for the 2-newton force:

$$\mathbf{F}_2 = \left(F_{x2}, F_{y2}\right)$$
$$= \left(F_2 \cos \theta_2, F_2 \sin \theta_2\right)$$
$$= [(2 \text{ N}) \cos \theta, (2 \text{ N}) \sin \theta]$$

Add the components:

$$F_x = F_{x1} + F_{x2}$$
$$= (-25 \text{ N}) + (2 \text{ N}) \cos \theta$$
$$= (2 \cos \theta - 25) \text{ N}$$
$$F_y = F_{y1} + F_{y2}$$
$$= 0 \text{ N} + (2 \text{ N}) \sin \theta$$
$$= (2 \sin \theta) \text{ N}$$

Finally, use the Pythagorean theorem to relate the magnitude of the resultant vector to the individual components using the trigonometric identity $\sin^2\theta + \cos^2\theta = 1$:

$$F^2 = F_x^2 + F_y^2$$
$$(25.8\,\text{N})^2 = ((2\cos\theta - 25)\,\text{N})^2 + ((2\sin\theta)\,\text{N})^2$$
$$665.6\,\text{N}^2 = 4\cos^2\theta\,\text{N}^2 - 100\cos\theta\,\text{N}^2 + 625\,\text{N}^2 + 4\sin^2\theta\,\text{N}^2$$
$$40.6\,\text{N}^2 = 4\sin^2\theta\,\text{N}^2 + 4\cos^2\theta\,\text{N}^2 - 100\cos\theta\,\text{N}^2$$
$$40.6\,\text{N}^2 = 4\left(\sin^2\theta + \cos^2\theta\right)\text{N}^2 - 100\cos\theta\,\text{N}^2$$
$$40.6\,\text{N}^2 = 4(1)\,\text{N}^2 - 100\cos\theta\,\text{N}^2$$
$$40.6\,\text{N}^2 = 4\,\text{N}^2 - 100\cos\theta\,\text{N}^2$$
$$36.6\,\text{N}^2 = -100\cos\theta\,\text{N}^2$$
$$-0.366 = \cos\theta$$
$$\cos^{-1}(-0.366) = \theta$$
$$111.4° = \theta$$

Unfortunately, a positive 111 degrees is above — or north of — the x-axis, and the problem stipulates that the goal moved toward the southern half of the ice. You need to make use of the fact that the cosine of any positive angle is equal to the cosine of the negative of that angle. So, the cosine of 111 degrees is equal to the cosine of –111 degrees, which is *south* of the x-axis. This is the correct direction; converting this to a direction on a compass rose, this is 69 degrees below west — or 69 degrees south of west.

273.   –40 J

When a force and a distance don't point in the same direction, use the "official" work formula $W = Fd\cos\theta$, where W is the work done on an object, F is the force exerted on the object, d is the distance that the object moves, and θ is the angle between the direction of the force and the direction of the distance. West and east are separated by 180 degrees, so the work in this situation is

$$W = Fd\cos\theta$$
$$= (12.8\,\text{N})(3.1\,\text{m})\cos 180°$$
$$= (12.8\,\text{N})(3.1\,\text{m})(-1)$$
$$= -39.7\,\text{J}$$

274.   –370 J

Although gravitational force is present and is essential to solve the problem, you don't need a force diagram because gravity is the only force that concerns you (you don't need to find a net force on the crate). If $W = Fd\cos\theta$, then $W_G = F_Gd\cos\theta$, where W_G is the work done by gravity, F_G is the force of gravity, d is the distance that the object moves, and θ is the angle between the direction of the force (straight down) and the direction of the distance (straight up). The angle between straight up and straight down is 180 degrees. Furthermore, because $F = ma$, $F_G = ma_G$. And $a_G = g$ near the surface of

Earth. Put all that together with the combined mass of the crate and lemons to solve for the amount of work gravity does:

$$W_G = F_G d \cos \theta$$
$$= mgd \cos \theta$$
$$= [(5 \text{ kg}) + 32(0.8 \text{ kg})] \left(9.8 \text{ m/s}^2\right) (1.25 \text{ m}) \cos 180°$$
$$= [5 \text{ kg} + 25.6 \text{ kg}] \left(9.8 \text{ m/s}^2\right) (1.25 \text{ m})(-1)$$
$$= (30.6 \text{ kg}) \left(9.8 \text{ m/s}^2\right) (1.25 \text{ m})(-1)$$
$$= -374.85 \text{ J} \approx -370 \text{ J}$$

275.   65 J

Gravity acts on both blocks, so it does work on both blocks. Therefore, you need a force, distance, and angle between the two for both the block of unknown mass and the block that weighs 1 kilogram. Start by drawing the free-body diagrams:

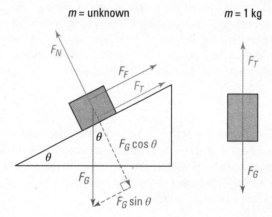

Illustration by Thomson Digital

Use the force form of Newton's second law to solve for the force that gravity exerts on each mass. Start with the 1-kilogram mass. ("Down the ramp" is used as the positive direction in this solution, so, relative to the 1-kilogram mass, that means "up" is the positive direction of motion.)

$$F_{y,\text{net}} = \sum F_y$$
$$ma_y = F_T - F_G$$
$$ma_y = F_T - ma_G$$
$$ma_y = F_T - mg$$
$$(1 \text{ kg})\left(1.3 \text{ m/s}^2\right) = F_T - (1 \text{ kg})\left(9.8 \text{ m/s}^2\right)$$
$$1.3 \text{ N} = F_T - 9.8 \text{ N}$$
$$11.1 \text{ N} = F_T$$

Now proceed to analyze the y (or vertical) components for the block of unknown mass:

$$F_{y,\text{net}} = \sum F_y$$
$$ma_y = F_N - F_G \cos \theta$$
$$ma_y = F_N - ma_G \cos \theta$$
$$ma_y = F_N - mg \cos \theta$$
$$m\left(0 \text{ m/s}^2\right) = F_N - m\left(9.8 \text{ m/s}^2\right)\cos 38°$$
$$m\left(0 \text{ m/s}^2\right) = F_N - m\left(9.8 \text{ m/s}^2\right)(0.788)$$
$$0 \text{ N} = F_N - \left(7.72 \text{ m/s}^2\right)m$$
$$\left(7.72 \text{ m/s}^2\right)m = F_N$$

For the x (or horizontal) components, you can substitute both of your previous results to figure out the mass of the block on the inclined plane:

$$F_{x,\text{net}} = \sum F_x$$
$$ma_x = F_G \sin \theta - F_T - F_F$$
$$ma_x = mg \sin \theta - F_T - \mu F_N$$
$$m\left(1.3 \text{ m/s}^2\right) = m\left(9.8 \text{ m/s}^2\right)\sin 38° - (11.1\text{ N}) - (0.2)\left(\left(7.72 \text{ m/s}^2\right)m\right)$$
$$m\left(1.3 \text{ m/s}^2\right) = m\left(9.8 \text{ m/s}^2\right)(0.616) - 11.1\text{ N} - \left(1.54 \text{ m/s}^2\right)m$$
$$\left(1.3 \text{ m/s}^2\right)m = \left(6.04 \text{ m/s}^2\right)m - 11.1\text{ N} - \left(1.54 \text{ m/s}^2\right)m$$
$$\left(1.3 \text{ m/s}^2\right)m = \left(4.5 \text{ m/s}^2\right)m - 11.1\text{ N}$$
$$-\left(3.2 \text{ m/s}^2\right)m = -11.1\text{ N}$$
$$m = 3.47 \text{ kg}$$

So, the force of gravity on the 3.47-kilogram block is

$$F_G = mg$$
$$= (3.47 \text{ kg})\left(9.8 \text{ m/s}^2\right)$$
$$= 34.01 \text{ N}$$

And the force of gravity on the 1-kilogram block is

$$F_G = mg$$
$$= (1 \text{ kg})\left(9.8 \text{ m/s}^2\right)$$
$$= 9.8 \text{ N}$$

To find the distance the system moves in 3 seconds, use the displacement formula:

$$s = v_i t + \frac{1}{2}at^2$$
$$= (0 \text{ m/s})(3 \text{ s}) + \frac{1}{2}\left(1.3 \text{ m/s}^2\right)(3 \text{ s})^2$$
$$= 0 \text{ m} + 5.85 \text{ m}$$
$$= 5.85 \text{ m}$$

The angle between F_G on the 1-kilogram block (straight down) and its motion (straight up) is 180 degrees. The angle between F_G on the 3.47-kilogram block (straight down) and its motion (which can be discovered using the following geometric relationship) is $90 - \theta = 90 - 38 = 52°$.

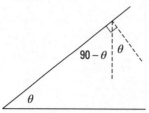

Illustration by Thomson Digital

So, the entire work done by gravity is

$$W_G = W_{G1} + W_{G2}$$
$$= F_{G1}d \cos \theta_1 + F_{G1}d \cos \theta_1$$
$$= (34.01 \text{ N})(5.85 \text{ m}) \cos 52° + (9.8 \text{ N})(5.85 \text{ m}) \cos 180°$$
$$= (34.01 \text{ N})(5.85 \text{ m})(0.616) + (9.8 \text{ N})(5.85 \text{ m})(-1)$$
$$= 122.56 \text{ J} - 57.33 \text{ J}$$
$$= 65.23 \text{ J}$$

276. 4K

An object's kinetic energy is directly proportional to its mass. If the mass is multiplied by a factor of 4, then so is the kinetic energy.

277. 290 km/h

First convert the mass into "correct" physics units: $(28 \text{ g})\left(\dfrac{1 \text{ kg}}{1,000 \text{ g}}\right) = 0.028 \text{ kg}$.

Then use the formula for kinetic energy, $K = \frac{1}{2}mv^2$, where m is the mass of the object and v is its velocity.

$$K = \frac{1}{2}mv^2$$
$$90 \text{ J} = \frac{1}{2}(0.028 \text{ kg})v^2$$
$$90 \text{ J} = (0.014 \text{ kg})v^2$$
$$6,428.6 \text{ m}^2/\text{s}^2 = v^2$$
$$\sqrt{6,428.6 \text{ m}^2/\text{s}^2} = v$$
$$80.18 \text{ m/s} = v$$

Finally, convert into the correct units:

$$\left(80.18\frac{\text{m}}{\text{s}}\right)\left(\frac{1 \text{ km}}{1,000 \text{ m}}\right)\left(\frac{60 \text{ s}}{1 \text{ min}}\right)\left(\frac{60 \text{ min}}{1 \text{ h}}\right) = 288.6 \frac{\text{km}}{\text{h}}$$

278.   0.7 J

If the hanging block's mass is 1 kilogram, then the sitting block's mass is 2 kilograms: $2(1\,\text{kg}) = 2\,\text{kg}$.

To calculate the hanging block's velocity after 1.2 seconds — which you need to calculate its kinetic energy — you first need to know the acceleration of the system. Draw a free-body diagram for each block and analyze the forces using Newton's second law.

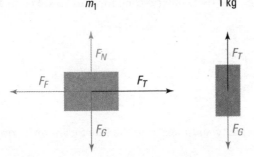

The following computations define the positive direction as "clockwise" for the system: the sitting block moving to the right, and the hanging block falling.

For the sitting block, the forces in the vertical direction produce the following result:

$$F_{y,\text{net}} = \sum F_y$$
$$m_1 a_y = F_N - F_G$$
$$m_1 a_y = F_N - m_1 a_G$$
$$m_1 a_y = F_N - m_1 g$$
$$(2\,\text{kg})\left(0\,\text{m/s}^2\right) = F_N - (2\,\text{kg})\left(9.8\,\text{m/s}^2\right)$$
$$0\,\text{N} = F_N - 19.6\,\text{N}$$
$$19.6\,\text{N} = F_N$$

As for the horizontal forces on the sitting block:

$$F_{x,\text{net}} = \sum F_x$$
$$m_1 a_x = F_T - F_F$$
$$m_1 a_x = F_T - \mu F_N$$
$$(2\,\text{kg})a = F_T - (0.35)(19.6\,\text{N})$$
$$(2\,\text{kg})a = F_T - 6.86\,\text{N}$$
$$(2\,\text{kg})a + 6.86\,\text{N} = F_T$$

The hanging block's diagram contains no horizontal forces, so you only need to add the vertical ones:

$$F_{y,\text{net}} = \sum F_y$$
$$m_2 a_y = F_G - F_T$$
$$m_2 a_y = m_2 a_G - F_T$$
$$m_2 a_y = m_2 g - F_T$$
$$(1\,\text{kg})a = (1\,\text{kg})\left(9.8\,\text{m/s}^2\right) - F_T$$
$$(1\,\text{kg})a = 9.8\,\text{N} - F_T$$
$$F_T + (1\,\text{kg})a = 9.8\,\text{N}$$
$$F_T = 9.8\,\text{N} - (1\,\text{kg})a$$

Set the two equations equal to each other to solve for the system's acceleration:

$$(2\,\text{kg})a + 6.86\,\text{N} = F_T = 9.8\,\text{N} - (1\,\text{kg})a$$
$$(3\,\text{kg})a + 6.86\,\text{N} = 9.8\,\text{N}$$
$$(3\,\text{kg})a = 2.94\,\text{N}$$
$$a = 0.98\,\text{m/s}^2$$

With this acceleration, an initial velocity of 0 meters per second (the block was at rest), and a time of 1.2 seconds, use the velocity-time formula to solve for the hanging block's final velocity.

$$v_f = v_i + at$$
$$= (0\,\text{m/s}) + \left(0.98\,\text{m/s}^2\right)(1.2\,\text{s})$$
$$= 0\,\text{m/s} + 1.176\,\text{m/s}$$
$$= 1.176\,\text{m/s}$$

Finally, substitute the velocity and mass into the kinetic energy formula to solve the problem:

$$K = \frac{1}{2}mv^2$$
$$= \frac{1}{2}(1\,\text{kg})(1.176\,\text{m/s})^2$$
$$= 0.7\,\text{J}$$

279. $\sqrt{3}v_E$

The net work performed on an object is equal to the object's change in kinetic energy. Given that each racer's initial velocity is 0 meters per second, this reduces to

$$W = \Delta K$$
$$= K_f - K_i$$
$$= \frac{1}{2}mv_f^2 - \frac{1}{2}mv_i^2$$
$$= \frac{1}{2}mv_f^2 - \frac{1}{2}m(0\,\text{m/s})^2$$
$$= \frac{1}{2}mv_f^2$$

In this case, work is proportional to the square of the final velocity. If work is tripled, so too is the square of the final velocity. Or, removing the square, if work is multiplied by three, the velocity is multiplied by the square root of three.

$$W_S = 3W_E$$
$$\frac{1}{2}mv_S^2 = 3\left(\frac{1}{2}mv_E^2\right)$$
$$\frac{1}{2}mv_S^2 = \frac{3}{2}mv_S^2$$
$$v_S^2 = 3v_E^2$$
$$v_S = \sqrt{3v_E^2}$$
$$= (\sqrt{3})\left(\sqrt{v_E^2}\right)$$
$$= \sqrt{3}v_E$$

280. 50 J

The work-energy theorem states that the net amount of work done on an object is equal to the object's final kinetic energy minus its initial kinetic energy. Both the engine and friction do work on the car; the sum of that work must be equal to the change in the car's kinetic energy.

$$W_{\text{engine}} + W_F = K_f - K_i$$
$$(500 \text{ J}) + W_F = (670 \text{ J}) - (220 \text{ J})$$
$$500 \text{ J} + W_F = 450 \text{ J}$$
$$W_F = -50 \text{ J}$$

Friction does 50 joules of work; the negative sign indicates that the force of friction points in the opposite direction from the car's motion — which is always the case with frictional forces.

ANSWERS 201–300

281. –312.8 J

First convert all the units given into "correct" SI form:

$$(480 \text{ g})\left(\frac{1 \text{ kg}}{1{,}000 \text{ g}}\right) = 0.48 \text{ kg}$$

$$\left(130 \, \frac{\text{km}}{\text{h}}\right)\left(\frac{1{,}000 \text{ m}}{1 \text{ km}}\right)\left(\frac{1 \text{ h}}{60 \text{ min}}\right)\left(\frac{1 \text{ min}}{60 \text{ s}}\right) = 36.1 \, \frac{\text{m}}{\text{s}}$$

The work-energy formula $W = \Delta K = K_f - K_i$, where K represents kinetic energy ($\frac{1}{2}mv^2$), states that work done on an object changes its kinetic energy. Use the formula to solve for the work done on the baseball:

$$W = K_f - K_i$$
$$= \frac{1}{2}mv_f^2 - \frac{1}{2}mv_i^2$$
$$= \frac{1}{2}m\left(v_f^2 - v_i^2\right)$$
$$= \frac{1}{2}(0.48 \text{ kg})\left[(0 \text{ m/s})^2 - (36.1 \text{ m/s})^2\right]$$
$$= \frac{1}{2}(0.48 \text{ kg})\left(0 \text{ m}^2/\text{s} - 1{,}303.2 \text{ m}^2/\text{s}^2\right)$$
$$= -312.8 \text{ J}$$

The negative sign indicates that the force used to stop the ball and the distance the ball moves while that force is exerted are in opposite directions, which is indeed the case.

282. 2.5 kJ

Use the velocity-time and velocity-displacement formulas to set up a system of equations with two unknown quantities: the acceleration and the initial velocity. Because the velocity doubles, use the substitution $v_f = 2v_i$.

$$v_f = v_i + at$$
$$2v_i = v_i + a(8 \text{ s})$$
$$v_i = (8 \text{ s})a$$
$$\frac{v_i}{(8 \text{ s})} = a$$

$$v_f^2 = v_i^2 + 2as$$

$$(2v_i)^2 = v_i^2 + 2a(48 \text{ m})$$

$$4v_i^2 = v_i^2 + 2a(48 \text{ m})$$

$$4v_i^2 = v_i^2 + (96 \text{ m})a$$

$$3v_i^2 = (96 \text{ m})a$$

$$\frac{v_i^2}{(32 \text{ m})} = a$$

Set the two equations equal to each other and solve for the initial velocity:

$$\frac{v_i}{8 \text{ s}} = a = \frac{v_i^2}{32 \text{ m}}$$

$$\frac{v_i}{8 \text{ s}} = \frac{v_i^2}{32 \text{ m}}$$

$$(32 \text{ m})v_i = (8 \text{ s})v_i^2$$

$$(32 \text{ m}) = (8 \text{ s})v_i$$

$$4 \text{ m/s} = v_i$$

Since $v_f = 2v_i$

$$v_f = 2v_i$$

$$= 2(4 \text{ m/s})$$

$$= 8 \text{ m/s}$$

By the work-energy theorem, the work done on the skier is equal to the skier's change in kinetic energy, so calculate the initial and final values using the formula for kinetic energy, $K = \frac{1}{2}mv^2$ (m is mass, and v is velocity).

$$K_i = \frac{1}{2}mv_i^2$$

$$= \frac{1}{2}(105 \text{ kg})(4 \text{ m/s})^2$$

$$= 840 \text{ J}$$

$$K_f = \frac{1}{2}mv_f^2$$

$$= \frac{1}{2}(105 \text{ kg})(8 \text{ m/s})^2$$

$$= 3,360 \text{ J}$$

$$W = K_f - K_i$$

$$= (3,360 \text{ J}) - (840 \text{ J})$$

$$= 2,520 \text{ J}$$

Finally, convert to kilojoules and round to two significant digits:

$$(2,520 \text{ J})\left(\frac{1 \text{ kJ}}{1,000 \text{ J}}\right) = 2.52 \text{ kJ} \approx 2.5 \text{ kJ}$$

283. 410 m

First deal with the necessary conversions and turn all the given data into "correct" units:

$$(1.1\,\text{km})\left(\frac{1{,}000\,\text{m}}{1\,\text{km}}\right)=1{,}100\,\text{m}$$

$$\left(18\,\frac{\text{km}}{\text{h}}\right)\left(\frac{1{,}000\,\text{m}}{1\,\text{km}}\right)\left(\frac{1\,\text{h}}{60\,\text{min}}\right)\left(\frac{1\,\text{min}}{60\,\text{s}}\right)=5\,\frac{\text{m}}{\text{s}}$$

The soldier's journey has two segments: One segment deals with a friction force related to a closed parachute, and the other segment deals with friction from an open parachute. The work-energy theorem states that the net amount of work done on an object is equal to the object's final kinetic energy minus its initial kinetic energy. Both gravity and friction perform work on the falling soldier. So, let d be the total distance the soldier falls, d_c be the distance the soldier falls with the parachute closed, and d_o be the distance the soldier falls with the parachute open. You can use the formulas for work ($W = Fd\cos\theta$, where F is force, d is distance, and θ is the angle between the two aforementioned vectors) and kinetic energy ($K = \frac{1}{2}mv^2$, where m is mass and v is velocity) to set up the following:

$$W_G - W_{Fc} - W_{Fo} = K_f - K_i$$

$$F_G d - F_{Fc} d_c - F_{Fo} d_o = \frac{1}{2}mv_f^2 - \frac{1}{2}mv_i^2$$

$$mgd - F_{Fc} d_c - F_{Fo} d_o = \frac{1}{2}m\left(v_f^2 - v_i^2\right)$$

$$(100\,\text{kg})\left(9.8\,\text{m/s}^2\right)(1{,}100\,\text{m}) - (70\,\text{N})d_c - (2{,}500\,\text{N})d_o = \frac{1}{2}(100\,\text{kg})\left[(5\,\text{m/s})^2 - (0\,\text{m/s})^2\right]$$

$$1.078 \times 10^6\,\text{J} - (70\,\text{N})d_c - (2{,}500\,\text{N})d_o = \frac{1}{2}(100\,\text{kg})\left(25\,\text{m}^2/\text{s}^2\right)$$

$$1.078 \times 10^6\,\text{J} - (70\,\text{N})d_c - (2{,}500\,\text{N})d_o = 1{,}250\,\text{J}$$

You have two unknowns, so you need a second equation. Fortunately, the way the variables are set up, the sum of the two unknown distances must equal the total distance the soldier falls:

$$d_c + d_o = 1{,}100\,\text{m}$$
$$d_c = 1{,}100\,\text{m} - d_o$$

Substitute this into the prior result.

$$1.078 \times 10^6\,\text{J} - (70\,\text{N})d_c - (2{,}500\,\text{N})d_o = 1{,}250\,\text{J}$$
$$1.078 \times 10^6\,\text{J} - (70\,\text{N})\left(1{,}100\,\text{m} - d_o\right) - (2{,}500\,\text{N})d_o = 1{,}250\,\text{J}$$
$$1.078 \times 10^6\,\text{J} - 7.7 \times 10^4\,\text{J} + (70\,\text{N})d_o - (2{,}500\,\text{N})d_o = 1{,}250\,\text{J}$$
$$1.001 \times 10^6\,\text{J} - (2{,}430\,\text{N})d_o = 1{,}250\,\text{J}$$
$$-(2{,}430\,\text{N})d_o = -9.9975 \times 10^5\,\text{J}$$
$$d_o = 411\,\text{m} \approx 410\,\text{m}$$

This is the distance of the soldier's fall during which the parachute is open; therefore, it must also be the height above the ground at which the soldier opens the chute.

284. 800 g

Use the formula for gravitational potential energy, $U_G = mgh$, where U_G is the potential energy due to gravity, m is an object's mass, g is the acceleration due to gravity near Earth's surface, and h is the object's height above Earth's surface. You know that the two potential energies are equal, so $U_{G1} = U_{G2}$, with this solution using the basketball as object 1 and the baseball as object 2:

$$U_{G1} = U_{G2}$$
$$m_1 g h_1 = m_2 g h_2$$
$$m_1 h_1 = m_2 h_2$$

Substitute the baseball's mass and the relationship $h_2 = 4h_1$ because the baseball (object 2) has a height four times that of the basketball (object 1).

$$m_1 h_1 = m_2 h_2$$
$$m_1 h_1 = (200 \text{ g})(4h_1)$$
$$m_1 h_1 = (800 \text{ g})h_1$$
$$m_1 = 800 \text{ g}$$

Note that it isn't necessary to convert the baseball's mass into the "correct" physics units of kilograms at the beginning because the final answer is requested in the same units.

285. 5.0 J

First convert the mass to "correct" units:

$$(200 \text{ g})\left(\frac{1 \text{ kg}}{1,000 \text{ g}}\right) = 0.2 \text{ kg}$$

Then use trigonometry to determine the tablet's change in height after Stan has finished walking 6 meters along the ramp:

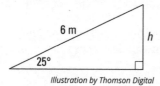

Illustration by Thomson Digital

$$\sin 25° = \frac{h}{6 \text{ m}}$$
$$0.423 = \frac{h}{6 \text{ m}}$$
$$2.54 \text{ m} = h$$

Then use the formula for gravitational potential energy, $U_G = mgh$, where U_G is the potential energy due to gravity, m is an object's mass, g is the acceleration due to gravity near the Earth's surface, and h is the object's height above Earth's surface.

$$
\begin{aligned}
U_G &= mgh \\
&= (0.2 \text{ kg})(9.8 \text{ m/s}^2)(2.54 \text{ m}) \\
&= 4.98 \\
&\approx 5.0 \text{ J}
\end{aligned}
$$

286. −160 J

If m_2 moves to the left, m_1 must also move to the left, down the ramp. To solve for the amount of "height" it loses doing so, use trigonometry:

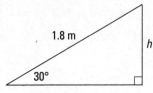

Illustration by Thomson Digital

$$\sin 30° = \frac{h}{1.8\,\text{m}}$$

$$0.5 = \frac{h}{1.8\,\text{m}}$$

$$0.9\,\text{m} = h$$

Now use the formula for gravitational potential energy, $U_G = mgh$, where U_G is the potential energy due to gravity, m is an object's mass, g is the acceleration due to gravity near Earth's surface, and h is the object's height relative to the starting position (h doesn't always need to refer to height above Earth's surface). Because m_1's height drops 0.9 meters, make sure to use a negative sign in the formula.

$$U_G = mgh$$

$$= (18\,\text{kg})\left(9.8\,\text{m/s}^2\right)(-0.9\,\text{m})$$

$$= -158.8\,\text{J}$$

$$\approx -160\,\text{J}$$

287. 0.20

Use the formula for gravitational potential energy ($U_G = mgh$, where U is potential energy, m is mass, g is the acceleration due to gravity near Earth's surface, and h is the height above a starting location) to calculate the distance m_1 loses vertically:

$$U_G = mgh$$

$$30\,\text{J} = (60\,\text{kg})\left(9.8\,\text{m/s}^2\right)h$$

$$30\,\text{J} = (588\,\text{N})h$$

$$0.051\,\text{m} = h$$

Then use trigonometry to determine the distance along the ramp that m_1 slides during that vertical loss:

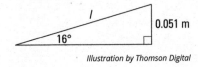

Illustration by Thomson Digital

$$\sin 16° = \frac{0.051 \text{ m}}{l}$$

$$0.276 = \frac{0.051 \text{ m}}{l}$$

$$0.276l = 0.051 \text{ m}$$

$$l = 0.185 \text{ m}$$

Starting from rest, the mass takes 6 seconds to travel that far, so use the displacement formula to calculate m_1's acceleration:

$$s = v_i t + \frac{1}{2} a t^2$$

$$0.185 \text{ m} = (0 \text{ m/s})(6 \text{ s}) + \frac{1}{2} a (6 \text{ s})^2$$

$$0.185 \text{ m} = 0 \text{ m} + \left(18 \text{ s}^2\right) a$$

$$0.185 \text{ m} = \left(18 \text{ s}^2\right) a$$

$$0.01 \text{ m/s}^2 = a$$

Use two free-body diagrams to analyze the forces and relate them to the acceleration you calculated:

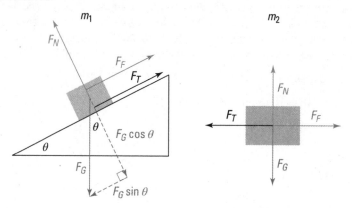

Illustration by Thomson Digital

Start with m_1 and use Newton's second law to develop a relationship between the tension in the string, F_T, and the coefficient of friction, μ. First, in the "vertical" direction:

$$F_{y,\text{net}} = \sum F_y$$

$$ma_y = F_N - F_G \cos\theta$$

$$ma_y = F_N - mg \cos\theta$$

$$(60 \text{ kg})\left(0 \text{ m/s}^2\right) = F_N - (60 \text{ kg})\left(9.8 \text{ m/s}^2\right)\cos 16°$$

$$(60 \text{ kg})\left(0 \text{ m/s}^2\right) = F_N - (60 \text{ kg})\left(9.8 \text{ m/s}^2\right)(0.961)$$

$$0 \text{ N} = F_N - 565.1 \text{ N}$$

$$565.1 \text{ N} = F_N$$

And in the horizontal direction:

$$F_{x,\text{net}} = \sum F_x$$
$$ma_x = F_G \sin\theta - F_T - F_F$$
$$ma_x = mg\sin\theta - F_T - \mu F_N$$
$$(60\,\text{kg})(0.01\,\text{m/s}^2) = (60\,\text{kg})(9.8\,\text{m/s}^2)\sin 16° - F_T - \mu(565.1\,\text{N})$$
$$(60\,\text{kg})(0.01\,\text{m/s}^2) = (60\,\text{kg})(9.8\,\text{m/s}^2)(0.276) - F_T - \mu(565.1\,\text{N})$$
$$0.6\text{N} = 162.3\,\text{N} - F_T - (565.1\,\text{N})\mu$$
$$-161.7\,\text{N} = -F_T - (565.1\,\text{N})\mu$$
$$-161.7\,\text{N} + (565.1\,\text{N})\mu = -F_T$$
$$161.7\,\text{N} - (565.1\,\text{N})\mu = F_T$$

Next work on m_2. As usual, start with the vertical direction:

$$F_{y,\text{net}} = \sum F_y$$
$$ma_y = F_N - F_G$$
$$ma_y = F_N - mg$$
$$(23\,\text{kg})(0\,\text{m/s}^2) = F_N - (23\,\text{kg})(9.8\,\text{m/s}^2)$$
$$0\,\text{N} = F_N - 225.4\,\text{N}$$
$$225.4\,\text{N} = F_N$$

And in the horizontal direction:

$$F_{x,\text{net}} = \sum F_x$$
$$ma_x = F_T - F_F$$
$$ma_x = F_T - \mu F_N$$
$$(23\,\text{kg})(0.01\,\text{m/s}^2) = F_T - \mu(225.4\,\text{N})$$
$$0.23\,\text{N} = F_T - (225.4\,\text{N})\mu$$
$$0.23\,\text{N} + (225.4\,\text{N})\mu = F_T$$

Set the two expressions for F_T equal to each other and solve for the acceleration:

$$161.7\,\text{N} - (565.1\,\text{N})\mu = F_T = 0.23\,\text{N} + (225.4\,\text{N})\mu$$
$$161.7\,\text{N} - (565.1\,\text{N})\mu = 0.23 + (225.4\,\text{N})\mu$$
$$161.47\,\text{N} - (565.1\,\text{N})\mu = (225.4\,\text{N})\mu$$
$$161.47\,\text{N} = (790.5\,\text{N})\mu$$
$$0.20 = \mu$$

288. 20 J

In the absence of friction, mechanical energy is conserved:

$$E_i = E_f$$
$$K_i + U_i = K_f + U_f$$

where K is kinetic energy and U is potential energy. The ball is released from rest, so its initial velocity is 0, meaning that its initial kinetic energy is also 0 ($K = \frac{1}{2}mv^2$). If the ground is designated with a value of 0 meters for h, then the final potential energy (gravitational) is also 0. Thus, the above equation simplifies to

$$K_i + U_i = K_f + U_f$$
$$0 + U_i = K_f + 0$$
$$U_i = K_f$$

The final kinetic energy is equal to the initial potential energy of 20 joules.

289.   **14.5 m/s**

In the absence of friction, mechanical energy is conserved:

$$E_i = E_f$$
$$K_i + U_i = K_f + U_f$$

where K is kinetic energy and U is potential energy. The football is initially at rest, and it has no kinetic energy at that point; its velocity is 0. If the height of the ground is designated as 0 meters, then the final gravitational potential energy is also 0. Thus, the above equation simplifies to

$$K_i + U_i = K_f + U_f$$
$$0 + U_i = K_f + 0$$
$$U_i = K_f$$

When the equations for kinetic and gravitational potential energy are substituted, this becomes

$$U_i = K_f$$
$$mgh_i = \frac{1}{2}mv_f^2$$

where m is the mass, h_i is the initial height, g is the acceleration due to gravity near Earth's surface, and v_f is the final velocity (in this case, when the height equals zero). Substitute the correct values after cancelling the m from both sides to solve for v_f:

$$mgh_i = \frac{1}{2}mv_f^2$$
$$gh_i = \frac{1}{2}v_f^2$$
$$\left(9.8 \text{ m/s}^2\right)(10.8 \text{ m}) = \frac{1}{2}v_f^2$$
$$105.84 \text{ m}^2/\text{s}^2 = \frac{1}{2}v_f^2$$
$$211.68 \text{ m}^2/\text{s}^2 = v_f^2$$
$$\sqrt{211.68 \text{ m}^2/\text{s}^2} = v_f$$
$$14.5 \text{ m/s} = v_f$$

290.　**13.9 m/s**

In the absence of friction, mechanical energy is conserved:

$$E_i = E_f$$
$$K_i + U_i = K_f + U_f$$

where K is kinetic energy and U is potential energy. Substituting the equations for kinetic and gravitational potential energy — $K = \frac{1}{2}mv^2$ and $U_G = mgh$, respectively, where m is mass, h is height, g is the acceleration due to gravity near Earth's surface, and v is velocity — turns the equation into

$$K_i + U_i = K_f + U_f$$
$$\frac{1}{2}mv_i^2 + mgh_1 = \frac{1}{2}mv_f^2 + mgh_f$$

At the top of the slide (the "initial" position), the velocity is 0 meters per second because the ball just makes it up there without stopping. When the slide ends, the ball has both potential and kinetic energy because the height is not 0 at the end of the slide — it's 80 centimeters, which you should convert into meters before doing any calculations: $(80 \text{ cm})\left(\dfrac{1\,\text{m}}{100 \text{ cm}}\right) = 0.8 \text{ m}$.

Cancel m from all terms in the energy equation and substitute the values of the heights and initial velocity to solve for the final velocity:

$$\frac{1}{2}mv_i^2 + mgh_1 = \frac{1}{2}mv_f^2 + mgh_f$$

$$\frac{1}{2}v_i^2 + gh_1 = \frac{1}{2}v_f^2 + gh_f$$

$$\frac{1}{2}(0\,\text{m/s})^2 + \left(9.8\,\text{m/s}^2\right)(10.6\,\text{m}) = \frac{1}{2}v_f^2 + \left(9.8\,\text{m/s}^2\right)(0.8\,\text{m})$$

$$0\,\text{m}^2/\text{s}^2 + 103.88\,\text{m}^2/\text{s}^2 = \frac{1}{2}v_f^2 + 7.84\,\text{m}^2/\text{s}^2$$

$$103.88\,\text{m}^2/\text{s}^2 = \frac{1}{2}v_f^2 + 7.84\,\text{m}^2/\text{s}^2$$

$$96.04\,\text{m}^2/\text{s}^2 = \frac{1}{2}v_f^2$$

$$192.08\,\text{m}^2/\text{s}^2 = v_f^2$$

$$\sqrt{192.08\,\text{m}^2/\text{s}^2} = v_f$$

$$13.9\,\text{m/s} = v_f$$

291. 54.9 m

You can use the horizontal distance and the time to calculate the horizontal component of the cannonball's velocity by using the displacement formula in the x direction. In the absence of air resistance, the acceleration of a projectile is always 0 meters per second squared in the x direction.

$$s_x = v_{ix}t + \frac{1}{2}a_xt^2$$

$$208 \text{ m} = v_{ix}(6.8 \text{ s}) + \frac{1}{2}\left(0 \text{ m/s}^2\right)(6.8 \text{ s})^2$$

$$208 \text{ m} = (6.8 \text{ s})v_{ix} + 0 \text{ m}$$

$$208 \text{ m} = (6.8 \text{ s})v_{ix}$$

$$30.59 \text{ m/s} = v_{ix}$$

The horizontal component of a vector (in this case, velocity) is equal to the product of the vector's magnitude and the cosine of the angle of elevation. Therefore,

$$v_x = v \cos \theta$$

$$30.59 \text{ m/s} = v \cos 47°$$

$$30.59 \text{ m/s} = v(0.682)$$

$$44.85 \text{ m/s} = v$$

Use your knowledge of projectile motion and the conservation of energy to solve for the cannonball's maximum height. A projectile has a vertical velocity of 0 when it reaches its apex, meaning that the only velocity involved at the highest point is the horizontal velocity. Because there is no horizontal acceleration, the initial horizontal velocity remains constant throughout the trip. So the cannonball's initial velocity is 44.85 meters per second, and its final velocity (at the highest point of its journey) is 30.59 meters per second.

In the absence of friction, mechanical energy is conserved:

$$E_i = E_f$$

$$K_i + U_i = K_f + U_f$$

$$\frac{1}{2}mv_i^2 + mgh_1 = \frac{1}{2}mv_f^2 + mgh_f$$

where m is mass, h is height, g is the acceleration due to gravity near Earth's surface, and v is velocity. If the starting location is designated as having a height of 0 meters, the energy equation becomes

$$\frac{1}{2}mv_i^2 + mgh_1 = \frac{1}{2}mv_f^2 + mgh_f$$

$$\frac{1}{2}v_i^2 + gh_1 = \frac{1}{2}v_f^2 + gh_f$$

$$\frac{1}{2}(44.85 \text{ m/s})^2 + \left(9.8 \text{ m/s}^2\right)(0 \text{ m}) = \frac{1}{2}(30.59 \text{ m/s})^2 + \left(9.8 \text{ m/s}^2\right)h_f$$

$$1{,}005.8 \text{ m}^2/\text{s}^2 + 0 \text{ m}^2/\text{s}^2 = 467.9 \text{ m}^2/\text{s}^2 + \left(9.8 \text{ m/s}^2\right)h_f$$

$$1{,}005.8 \text{ m}^2/\text{s}^2 = 467.9 \text{ m}^2/\text{s}^2 + \left(9.8 \text{ m/s}^2\right)h_f$$

$$537.9 \text{ m}^2/\text{s}^2 = \left(9.8 \text{ m/s}^2\right)h_f$$

$$54.9 \text{ m} = h_f$$

292. 95%

The key to starting the solution is to draw the free-body diagram and analyze the forces on the desk using Newton's second law:

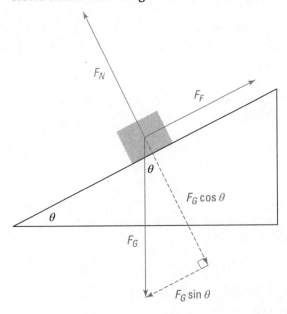

Illustration by Thomson Digital

In the y or vertical direction:

$$F_{y,\text{net}} = \sum F_y$$
$$ma_y = F_N - F_G \cos\theta$$
$$ma_y = F_N - mg \cos\theta$$
$$m\left(0\,\text{m/s}^2\right) = F_N - m\left(9.8\,\text{m/s}^2\right)\cos 3°$$
$$0\,\text{N} = F_N - m\left(9.8\,\text{m/s}^2\right)(0.9986)$$
$$0\,\text{N} = F_N - \left(9.786\,\text{m/s}^2\right)m$$
$$\left(9.786\,\text{m/s}^2\right)m = F_N$$

And in the x or horizontal direction, where "down the slope" is taken as positive in this solution:

$$F_{x,\text{net}} = \sum F_x$$
$$ma_x = F_G \sin\theta - F_F$$
$$ma_x = mg \sin\theta - \mu F_N$$
$$ma = m\left(9.8\,\text{m/s}^2\right)\sin 3° - (0.05)\left(\left(9.786\,\text{m/s}^2\right)m\right)$$
$$ma = m\left(9.8\,\text{m/s}^2\right)(0.0523) - \left(0.489\,\text{m/s}^2\right)m$$
$$ma = \left(0.513\,\text{m/s}^2\right)m - \left(0.489\,\text{m/s}^2\right)m$$
$$a = \left(0.513\,\text{m/s}^2\right) - \left(0.489\,\text{m/s}^2\right)$$
$$a = 0.024\,\text{m/s}^2$$

Substitute this result, along with the final velocity, into the velocity–displacement formula to solve for the distance that the desk rolls:

$$v_f^2 = v_i^2 + 2as$$

$$(2.3 \text{ m/s})^2 = (0 \text{ m/s})^2 + 2(0.024 \text{ m/s}^2) s$$

$$5.29 \text{ m}^2/\text{s}^2 = 0 \text{ m}^2/\text{s}^2 + (0.048 \text{ m/s}^2) s$$

$$5.29 \text{ m}^2/\text{s}^2 = (0.048 \text{ m/s}^2) s$$

$$110 \text{ m} = s$$

Now that you know the length of the ramp and its angle of elevation, you can calculate the height from which the desk starts:

$$\sin 3° = \frac{h}{110 \text{ m}}$$

$$0.0523 = \frac{h}{110 \text{ m}}$$

$$5.75 \text{ m} = h$$

Because the desk starts from rest, it has no initial kinetic energy — all its initial mechanical energy is potential. Therefore,

$$E_i = U_i + K_i$$

$$= mgh_i + 0 \text{ J}$$

$$= mgh_i$$

$$= m(9.8 \text{ m/s}^2)(5.75 \text{ m})$$

$$= (56.35 \text{ m}^2/\text{s}^2) m$$

where E is the total mechanical energy, U is the gravitational potential energy, and K is the kinetic energy.

When the desk reaches the end of the ramp, its height is 0, so its gravitational potential energy is 0, and all its energy is kinetic:

$$E_f = U_f + K_f$$

$$= 0 \text{ J} + \frac{1}{2} mv_f^2$$

$$= \frac{1}{2} mv_f^2$$

$$= \frac{1}{2} m(2.3 \text{ m/s})^2$$

$$= (2.645 \text{ m}^2/\text{s}^2) m$$

To find out what percentage of the initial amount of energy this final amount is, use the following setup, where x represents the percent:

$$E_f = \frac{x}{100} E_i$$

$$(2.645 \text{ m}^2/\text{s}^2) m = \frac{x}{100} ((56.35 \text{ m}^2/\text{s}^2) m)$$

$$2.645m = 0.5635mx$$

$$2.645 = 0.5635x$$

$$4.7 = x$$

So, the final amount of mechanical energy is 4.7 percent of the initial mechanical energy. The other $100 - 4.7 = 95.3 \approx 95$ percent is used to counteract the force of friction between the wheels and the ramp.

293. 50 J

A particle's total mechanical energy is the sum of the particle's kinetic energy and its potential energy. At the first moment mentioned in the problem, the particle has $37.5 + 12.5 = 50$ joules of mechanical energy. In the absence of friction (such as air resistance), this amount stays constant throughout the fall. Although the potential energy is 0 on the ground, and therefore all the mechanical energy is in the form of kinetic energy, the particle's total mechanical energy remains 50 joules.

294. 15.6 m/s

First choose a reference location for your gravitational potential energy calculations. (This solution designates point C as the location where $h = 0$ meters.) Because friction is not present, start with the standard conservation-of-mechanical-energy formula:

$$E_i = E_f$$
$$U_i + K_i = U_f + K_f$$

where E represents the total mechanical energy, U is the potential energy (entirely gravitational in this problem), and K is the kinetic energy. Then substitute the formulas for kinetic and potential energy:

$$U_i + K_i = U_f + K_f$$
$$mgh_i + \frac{1}{2}mv_i^2 = mgh_f + \frac{1}{2}mv_f^2$$
$$gh_i + \frac{1}{2}v_i^2 = gh_f + \frac{1}{2}v_f^2$$

where g is the acceleration due to gravity near Earth's surface, h is the height relative to your reference point, and v is the cart's velocity. The initial location (*i*) is point A, and the final location (*f*) is point D. If the height at C is 0 meters, then the height at A is 24 meters — $2r = 2(12\,\text{m}) = 24\,\text{m}$ — and the height at D is 12 meters. Substitute everything you know into the equation to solve for the velocity at point D:

$$gh_i + \frac{1}{2}v_i^2 = gh_f + \frac{1}{2}v_f^2$$
$$gh_A + \frac{1}{2}v_A^2 = gh_D + \frac{1}{2}v_D^2$$
$$\left(9.8\,\text{m/s}^2\right)(24\,\text{m}) + \frac{1}{2}(3\,\text{m/s})^2 = \left(9.8\,\text{m/s}^2\right)(12\,\text{m}) + \frac{1}{2}v_D^2$$
$$235.2\,\text{m}^2/\text{s}^2 + 4.5\,\text{m}^2/\text{s}^2 = 117.6\,\text{m}^2/\text{s}^2 + \frac{1}{2}v_D^2$$
$$239.7\,\text{m}^2/\text{s}^2 = 117.6\,\text{m}^2/\text{s}^2 + \frac{1}{2}v_D^2$$
$$122.1\,\text{m}^2/\text{s}^2 = \frac{1}{2}v_D^2$$
$$244.2\,\text{m}^2/\text{s}^2 = v_D^2$$
$$\sqrt{244.2\,\text{m}^2/\text{s}^2} = v_D$$
$$15.6\,\text{m/s} = v_D$$

295. $5mg$

At point C the cart is kept in a circular path by a centripetal force, which is the net result of the vertical forces on the cart: gravity (pointing down) and the normal force (pointing up). Centripetal force points toward the center of the circle, which is upward in this case. By Newton's second law:

$$\sum F_y = F_{y,\text{net}}$$
$$F_N - F_G = F_C$$
$$F_N - ma_G = ma_C$$
$$F_N - mg = m\frac{v^2}{r}$$
$$F_N = mg + \frac{mv^2}{r}$$

where F_c is the centripetal force, a_c is the centripetal acceleration, and v is the cart's velocity along the curved path. If v were allowed in the final solution, this would be the answer. Instead, you need to use the formula for the conservation of mechanical energy to find an expression equivalent to v (or v^2) in terms of m, g, and/or r. Because friction is not present, start with the standard conservation-of-mechanical-energy formula:

$$E_i = E_f$$
$$U_i + K_i = U_f + K_f$$

where E represents the total mechanical energy, U is the potential energy (entirely gravitational in this problem), and K is the kinetic energy. Then substitute the formulas for kinetic and potential energy:

$$U_i + K_i = U_f + K_f$$
$$mgh_i + \frac{1}{2}mv_i^2 = mgh_f + \frac{1}{2}mv_f^2$$
$$gh_i + \frac{1}{2}v_i^2 = gh_f + \frac{1}{2}v_f^2$$

where g is the acceleration due to gravity near Earth's surface, h is the height relative to your reference point, and v is the cart's velocity. The initial location (*i*) is point A, and the final location (*f*) is point C. If the height at C is designated as 0 meters, then the height at A is 2*r* meters. Substitute everything you know into the equation to solve for v_f^2:

$$gh_i + \frac{1}{2}v_i^2 = gh_f + \frac{1}{2}v_f^2$$
$$gh_A + \frac{1}{2}v_A^2 = gh_C + \frac{1}{2}v_C^2$$
$$g(2r) + \frac{1}{2}(0)^2 = g(0) + \frac{1}{2}v^2$$
$$2gr + 0 = 0 + \frac{1}{2}v^2$$
$$2gr = \frac{1}{2}v^2$$
$$4gr = v^2$$

Substitute this into your result for the normal force to develop the final answer.

$$F_N = mg + \frac{mv^2}{r}$$
$$= mg + \frac{m(4gr)}{r}$$
$$= mg + \frac{4mgr}{r}$$
$$= mg + 4mg$$
$$= 5mg$$

296. 15.2 m/s

First choose a reference location for your gravitational potential energy calculations. (This solution designates point B as the location where $h = 0$ meters.) Because friction is not present, start with the standard conservation-of-mechanical-energy formula:

$$E_i = E_f$$
$$U_i + K_i = U_f + K_f$$

where E represents the total mechanical energy, U is the potential energy (entirely gravitational in this problem), and K is the kinetic energy. Then substitute the formulas for kinetic and potential energy:

$$U_i + K_i = U_f + K_f$$
$$mgh_i + \frac{1}{2}mv_i^2 = mgh_f + \frac{1}{2}mv_f^2$$
$$gh_i + \frac{1}{2}v_i^2 = gh_f + \frac{1}{2}v_f^2$$

where g is the acceleration due to gravity near Earth's surface, h is the height relative to your reference point, and v is the cart's velocity. The initial location (i) is point C, and the final location (f) is point E. If the height at point B is 0 meters, then the height at point C is $-r$ meters and the height at point E is r meters. Substitute everything you know at this point into the equation to solve for the value of r:

$$gh_i + \frac{1}{2}v_i^2 = gh_f + \frac{1}{2}v_f^2$$
$$gh_C + \frac{1}{2}v_C^2 = gh_E + \frac{1}{2}v_E^2$$
$$\left(9.8 \text{ m/s}^2\right)(-r) + \frac{1}{2}(20 \text{ m/s})^2 = \left(9.8 \text{ m/s}^2\right)(r) + \frac{1}{2}(8 \text{ m/s})^2$$
$$-\left(9.8 \text{ m/s}^2\right)r + 200 \text{ m}^2/\text{s}^2 = \left(9.8 \text{ m/s}^2\right)r + 32 \text{ m}^2/\text{s}^2$$
$$200 \text{ m}^2/\text{s}^2 = \left(19.6 \text{ m/s}^2\right)r + 32 \text{ m}^2/\text{s}^2$$
$$168 \text{ m}^2/\text{s}^2 = \left(19.6 \text{ m/s}^2\right)r$$
$$8.57 \text{ m} = r$$

Use the conservation-of-mechanical-energy formula a second time, comparing point B with either C or E; now that you know the height, you can solve for velocity:

$$gh_i + \frac{1}{2}v_i^2 = gh_f + \frac{1}{2}v_f^2$$

$$gh_B + \frac{1}{2}v_B^2 = gh_C + \frac{1}{2}v_C^2$$

$$\left(9.8\,\text{m/s}^2\right)(0\,\text{m}) + \frac{1}{2}v_B^2 = \left(9.8\,\text{m/s}^2\right)(-8.57\,\text{m}) + \frac{1}{2}(20\,\text{m/s})^2$$

$$0\,\text{m}^2/\text{s}^2 + \frac{1}{2}v_B^2 = -83.99\,\text{m}^2/\text{s}^2 + 200\,\text{m}^2/\text{s}^2$$

$$\frac{1}{2}v_B^2 = 116.01\,\text{m}^2/\text{s}^2$$

$$v_B^2 = 232.02\,\text{m}^2/\text{s}^2$$

$$v_B = \sqrt{232.02\,\text{m}^2/\text{s}^2}$$

$$v_B = 15.2\,\text{m/s}$$

297. power

Power is most commonly seen in textbooks as the rate of change of work done on/by an object, but, because work is energy, power is also the rate of change of an object's energy.

298. 7.2×10^6 W

First convert the energy and time values into "correct" units:

$$(120\,\text{kWh})\left(\frac{1 \times 10^3\,\text{W}}{1\,\text{kW}}\right)\left(\frac{1\,\text{J/s}}{1\,\text{W}}\right)\left(\frac{60\,\text{min}}{1\,\text{h}}\right)\left(\frac{60\,\text{s}}{1\,\text{min}}\right) = 4.32 \times 10^8\,\text{J}$$

$$(1\,\text{min})\left(\frac{60\,\text{s}}{1\,\text{min}}\right) = 60\,\text{s}$$

Now solve the power formula, $\bar{P} = \dfrac{W}{t}$, where $\bar{P}$ is the average power delivered to an object, W is the average work done on the object, and t is the amount of time during which the work is performed:

$$\bar{P} = \frac{W}{t}$$

$$= \frac{4.32 \times 10^8\,\text{J}}{60\,\text{s}}$$

$$= 7.2 \times 10^6\,\text{W}$$

299. 1.2×10^4 W

First convert the speeds into the "correct" units:

$$\left(25\frac{\text{km}}{\text{h}}\right)\left(\frac{1,000\,\text{m}}{1\,\text{km}}\right)\left(\frac{1\,\text{h}}{60\,\text{min}}\right)\left(\frac{1\,\text{min}}{60\,\text{s}}\right) = 6.94\,\text{m/s}$$

$$\left(50\frac{\text{km}}{\text{h}}\right)\left(\frac{1,000\,\text{m}}{1\,\text{km}}\right)\left(\frac{1\,\text{h}}{60\,\text{min}}\right)\left(\frac{1\,\text{min}}{60\,\text{s}}\right) = 13.9\,\text{m/s}$$

In the absence of friction, the work-energy theorem states that the change in an object's kinetic energy is equal to the work done on it. Therefore,

$$W = K_f - K_i$$
$$= \frac{1}{2}mv_f^2 - \frac{1}{2}mv_i^2$$
$$= \frac{1}{2}(1{,}300\text{ kg})(13.9\text{ m/s})^2 - \frac{1}{2}(1{,}300\text{ kg})(6.94\text{ m/s})^2$$
$$= 1.256 \times 10^5\text{ J} - 3.13 \times 10^4\text{ J}$$
$$= 9.43 \times 10^4\text{ J}$$

Finally, use the power formula, $\bar{P} = \frac{W}{t}$, where $\bar{P}$ is the average power delivered to an object, W is the average work done on the object, and t is the amount of time during which the work is performed:

$$\bar{P} = \frac{W}{t}$$
$$= \frac{9.43 \times 10^4\text{ J}}{8\text{ s}}$$
$$= 1.2 \times 10^4\text{ W}$$

---

300. 8 s

The power formula that directly utilizes a value for time is $\bar{P} = \frac{W}{t}$, where $\bar{P}$ is the average power, W is the work performed, and t is the time during which the power acts. Work is the product of force and distance (which you're given), so you need to calculate the amount of force Matt uses to drag the stone. Start with a free-body diagram.

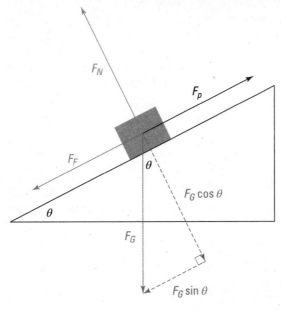

Illustration by Thomson Digital

(F_p is Matt's "pulling" force.) The stone moves up the ramp, so the force of friction — which always opposes motion — points down the ramp. This solution uses "up the hill" as the positive direction for the following calculations. The force form of Newton's second law in the vertical direction states that

$$F_{y,net} = \sum F_y$$
$$ma_y = F_N - F_G \cos \theta$$
$$ma_y = F_N - mg \cos \theta$$
$$m\left(0\,\text{m/s}^2\right) = F_N - mg \cos \theta$$
$$0\,\text{N} = F_N - mg \cos \theta$$
$$mg \cos \theta = F_N$$

Likewise, in the horizontal direction,

$$F_{x,net} = \sum F_x$$
$$ma_x = F_p - F_F - F_G \sin \theta$$
$$ma_x = F_p - \mu F_N - mg \sin \theta$$
$$ma_x = F_p - \mu(mg \cos \theta) - mg \sin \theta$$
$$ma_x = F_p - \mu mg \cos \theta - mg \sin \theta$$
$$(80\,\text{kg})\left(0.1\,\text{m/s}^2\right) = F_p - (0.4)(80\,\text{kg})\left(9.8\,\text{m/s}^2\right)\cos 25°$$
$$- (80\,\text{kg})\left(9.8\,\text{m/s}^2\right)\sin 25°$$
$$(80\,\text{kg})\left(0.1\,\text{m/s}^2\right) = F_p - (0.4)(80\,\text{kg})\left(9.8\,\text{m/s}^2\right)(0.906)$$
$$- (80\,\text{kg})\left(9.8\,\text{m/s}^2\right)(0.423)$$
$$8\,\text{N} = F_p - 284.1\,\text{N} - 331.6\,\text{N}$$
$$8\,\text{N} = F_p - 615.7\,\text{N}$$
$$623.7\,\text{N} = F_p$$

The stone moves in the exact direction as Matt's force of pulling, so the angle between the two is 0 degrees, making the work formula

$$W = Fd \cos \theta$$
$$W_p = F_p d \cos \theta$$
$$= (623.7\,\text{N})(11.2\,\text{m}) \cos 0°$$
$$= (623.7\,\text{N})(11.2\,\text{m})(1)$$
$$= 6{,}985.4\,\text{J}$$

Matt does 6,985.4 joules of work. To find how long Matt works, use the power formula:

$$\bar{P} = \frac{W}{t}$$
$$870\,\text{W} = \frac{6{,}985.4\,\text{J}}{t}$$
$$(870\,\text{W})t = 6{,}985.4\,\text{J}$$
$$t = 8\,\text{s}$$

301. velocity

The basic definition of power is "the rate at which work is performed." Because work is the product of force and distance, the equation can be manipulated as follows:

$$P = \frac{W}{t}$$
$$= \frac{Fd}{t}$$
$$= F\frac{d}{t}$$
$$= Fv$$

where P is power, F is force, d is distance, t is time, and v is velocity. This shows that a secondary definition of power is the product of force exerted on an object and the object's acquired velocity.

302. 20 m/s

Use the velocity form of the power equation, $\bar{P} = F\bar{v}$, where $\bar{P}$ is average power, F is force, and $\bar{v}$ is average velocity. For instantaneous situations — as in this problem — the average power *is* the power and the average velocity *is* the velocity:

$$P = Fv$$
$$2{,}000\text{ W} = (100\text{ N})v$$
$$20\text{ m/s} = v$$

303. 11.1 m/s

First convert the kilowatts of power into the "correct" units:

$$(13\text{ kW})\left(\frac{1 \times 10^3\text{ W}}{1\text{ kW}}\right) = 1.3 \times 10^4\text{ W}$$

Use the force form of Newton's second law to solve for the amount of force exerted on the speedboat:

$$F = ma$$
$$= (850\text{ kg})\left(1.6\text{ m/s}^2\right)$$
$$= 1{,}360\text{ N}$$

Then use the velocity form of the power equation, $\bar{P} = F\bar{v}$, where $\bar{P}$ is average power, F is force, and $\bar{v}$ is average velocity:

$$\bar{P} = F\bar{v}$$
$$\bar{P} = F\left(\frac{v_i + v_f}{2}\right)$$
$$1.3 \times 10^4\text{ W} = (1{,}360\text{ N})\left[\frac{(8\text{ m/s}) + v_f}{2}\right]$$
$$1.3 \times 10^4\text{ W} = (680\text{ N})\left((8\text{ m/s}) + v_f\right)$$
$$1.3 \times 10^4\text{ W} = 5{,}440\text{ W} + (680\text{ N})v_f$$
$$7{,}560\text{ W} = (680\text{ N})v_f$$
$$11.1\text{ m/s} = v_f$$

304. 4.56 W

First convert the car's mass into "correct" units:

$$(950 \text{ g})\left(\frac{1 \text{ kg}}{1,000 \text{ g}}\right) = 0.95 \text{ kg}$$

Let the car's initial velocity be v; therefore, its final velocity is $2v$. Combine those with the given time and acceleration and the velocity-time formula to solve for the values of the velocities:

$$v_f = v_i + at$$
$$2v = v + \left(0.8 \text{ m/s}^2\right)(5 \text{ s})$$
$$2v = v + 4 \text{ m/s}$$
$$v = 4 \text{ m/s}$$

So the car's initial speed is 4 meters per second; following the period of acceleration, its speed is 8 meters per second.

Now use the velocity form of the power equation, $\bar{P} = F\bar{v}$, where $\bar{P}$ is average power, F is force, and $\bar{v}$ is average velocity. You can compute the force using Newton's second law, $F = ma$:

$$F = ma$$
$$= (0.95 \text{ kg})\left(0.8 \text{ m/s}^2\right)$$
$$= 0.76 \text{ N}$$

Substitute this value and the velocities into the power formula to solve.

$$\bar{P} = F\bar{v}$$
$$\bar{P} = F\left(\frac{v_i + v_f}{2}\right)$$
$$= (0.76 \text{ N})\left[\frac{(4 \text{ m/s}) + (8 \text{ m/s})}{2}\right]$$
$$= (0.76 \text{ N})\left(\frac{12 \text{ m/s}}{2}\right)$$
$$= (0.76 \text{ N})(6 \text{ m/s})$$
$$= 4.56 \text{ W}$$

305. 2.1 m

Start with the velocity form of the power equation, $\bar{P} = F\bar{v}$, where $\bar{P}$ is average power, F is force, and $\bar{v}$ is average velocity.

$$\bar{P} = F\bar{v}$$
$$\bar{P} = F\left(\frac{v_i + v_f}{2}\right)$$
$$600 \text{ W} = F\left[\frac{(0 \text{ m/s}) + (0.9 \text{ m/s})}{2}\right]$$
$$600 \text{ W} = F(0.45 \text{ m/s})$$
$$1,333.3 \text{ N} = F$$

This is the amount of force Paul exerts pushing the cabinet. You now have all the pieces to fill in a free-body diagram and write out Newton's second law equations:

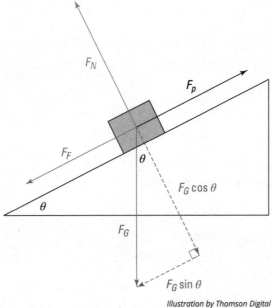

Illustration by Thomson Digital

Start with the vertical forces:

$$F_{y,net} = \sum F_y$$
$$ma_y = F_N - F_G \cos \theta$$
$$ma_y = F_N - mg \cos \theta$$
$$(400 \text{ kg}) \left(0 \text{ m/s}^2 \right) = F_N - (400 \text{ kg}) \left(9.8 \text{ m/s}^2 \right) \cos 7°$$
$$(400 \text{ kg}) \left(0 \text{ m/s}^2 \right) = F_N - (400 \text{ kg}) \left(9.8 \text{ m/s}^2 \right) (0.9925)$$
$$0 \text{ N} = F_N - 3{,}890.6 \text{ N}$$
$$3{,}890.6 = F_N$$

Then, for the horizontal forces,

$$F_{x,net} = \sum F_x$$
$$ma_x = F_P - F_G \sin \theta - F_F$$
$$ma_x = F_P - mg \sin \theta - \mu F_N$$
$$(400 \text{ kg})a = (1{,}333.3 \text{ N}) - (400 \text{ kg}) \left(9.8 \text{ m/s}^2 \right) \sin 7° - (0.2)(3{,}890.6 \text{ N})$$
$$(400 \text{ kg})a = (1{,}333.3 \text{ N}) - (400 \text{ kg}) \left(9.8 \text{ m/s}^2 \right) (0.122) - (0.2)(3{,}890.6 \text{ N})$$
$$(400 \text{ kg})a = 1{,}333.3 \text{ N} - 478.2 \text{ N} - 778.1 \text{ N}$$
$$(400 \text{ kg})a = 77 \text{ N}$$
$$a = 0.19 \text{ m/s}^2$$

Finally, use the cabinet's acceleration along with its initial and final velocities in the velocity-displacement formula to calculate the distance the cabinet traveled from the bottom to the top of the ramp — in other words, the length of the ramp.

$$v_f^2 = v_i^2 + 2as$$
$$(0.9\,\text{m/s})^2 = (0\,\text{m/s})^2 + 2(0.19\,\text{m/s}^2)\,s$$
$$0.81\,\text{m}^2/\text{s}^2 = 0\,\text{m}^2/\text{s}^2 + (0.38\,\text{m/s}^2)\,s$$
$$0.81\,\text{m}^2/\text{s}^2 = (0.38\,\text{m/s}^2)\,s$$
$$2.1\,\text{m} = s$$

306. time

Impulse is the product of force and time:

$$\mathbf{J} = \mathbf{F}\Delta t$$

Impulse = Force · Change in Time

307. 2J

Impulse is directly proportional to the change in time between the beginning and the end of a collision. Increasing that time by a factor of 2 also increases the impulse by a factor of 2.

308. – 20 N · s

Because the baseball is being slowed down — which indicates a negative acceleration — the force acting on the ball is also negative. Therefore, when using the impulse formula, $J = F\Delta t$, be sure to use the correct sign for F. The change in time (Δt) refers only to the collision itself; in this case, the collision occurs when the baseball slams into the catcher's glove.

$$J = F\Delta t$$
$$= (-400\,\text{N})(1.08\,\text{s} - 1.03\,\text{s})$$
$$= (-400\,\text{N})(0.05\,\text{s})$$
$$= -20\,\text{N} \cdot \text{s}$$

309. 3

Momentum is directly proportional to both mass and velocity. To triple momentum, you can triple either mass or velocity — but not both or else the momentum would increase by a factor of nine!

310. displacement

Because momentum is the product of mass and velocity, rearranging the equation yields the following:

$$p = mv$$
$$\frac{p}{m} = v$$

Multiplying that result by time, $vt = s$, which is displacement.

311.

$7.2 \times 10^3 \ \mathrm{kg} \times \mathrm{m/s}$

First, convert the velocities into units of meters per second:

$$\left(80\frac{\mathrm{km}}{\mathrm{h}}\right)\left(\frac{1{,}000\ \mathrm{m}}{1\ \mathrm{km}}\right)\left(\frac{1\ \mathrm{h}}{60\ \mathrm{min}}\right)\left(\frac{1\ \mathrm{min}}{60\ \mathrm{s}}\right) = 22.2\ \frac{\mathrm{m}}{\mathrm{s}}$$

$$\left(120\frac{\mathrm{km}}{\mathrm{h}}\right)\left(\frac{1{,}000\ \mathrm{m}}{1\ \mathrm{km}}\right)\left(\frac{1\ \mathrm{h}}{60\ \mathrm{min}}\right)\left(\frac{1\ \mathrm{min}}{60\ \mathrm{s}}\right) = 33.3\ \frac{\mathrm{m}}{\mathrm{s}}$$

Then use the momentum equation with a slight notational difference: $\Delta p = m\Delta v$, where Δp is the magnitude of the momentum change, m is the object's mass, and Δv is the change in its speed.

$$
\begin{aligned}
\Delta p &= m\Delta v \\
&= m\left(v_f - v_i\right) \\
&= (650\ \mathrm{kg})\,(33.3\ \mathrm{m/s} - 22.2\ \mathrm{m/s}) \\
&= 7.2 \times 10^3\ \mathrm{kg \cdot m/s}
\end{aligned}
$$

312.

3.08 ThKilos

Given mass and momentum, you can quickly solve for velocity using the momentum formula $(p = mv)$. For the first stage:

$$
\begin{aligned}
p &= mv \\
6.3 \times 10^5\ \mathrm{kg \cdot m/s} &= (2{,}350\ \mathrm{kg})v \\
268.1\ \mathrm{m/s} &= v
\end{aligned}
$$

This velocity is in the same direction as the momentum, 20 degrees south of east.

For the second stage:

$$
\begin{aligned}
p &= mv \\
1.01 \times 10^6\ \mathrm{kg \cdot m/s} &= (2{,}350\ \mathrm{kg})v \\
429.8\ \mathrm{m/s} &= v
\end{aligned}
$$

This velocity is in the same direction as the momentum, 80 degrees north of east.

The first stage lasts for 1 hour, or

$$(1\ \mathrm{h})\left(\frac{60\ \mathrm{min}}{1\ \mathrm{h}}\right)\left(\frac{60\ \mathrm{s}}{1\ \mathrm{min}}\right) = 3{,}600\ \mathrm{s}$$

From the definition of speed, $v = \dfrac{d}{t}$, where d is distance and t is time, you can solve for the distance traveled by the plane in the first stage:

$$
\begin{aligned}
v &= \frac{d}{t} \\
vt &= d \\
(268.1\ \mathrm{m/s})(3{,}600\ \mathrm{s}) &= d \\
9.652 \times 10^5\ \mathrm{m} &= d
\end{aligned}
$$

Likewise, for the second stage and its 2-hour duration (7,200 seconds),

$$v = \frac{d}{t}$$

$$vt = d$$

$$(429.8 \text{ m/s})(7{,}200 \text{ s}) = d$$

$$3.095 \times 10^6 \text{ m} = d$$

To calculate the net distance after the two stages, you have to break the two vectors (9.652×10^5 meters 20 degrees south of east; 3.095×10^6 meters 80 degrees north of east) into their components (s_x and s_y — horizontal and vertical, respectively). Remember that the angle is always measured relative to the positive x-axis, with positive angles being north of the east-west axis and negative angles being south of the east-west axis:

$$s_{x1} = s_1 \cos \theta$$
$$= \left(9.652 \times 10^5 \text{ m}\right) \cos\left(-20°\right)$$
$$= 9.07 \times 10^5 \text{ m}$$
$$s_{y1} = s_1 \sin \theta$$
$$= \left(9.652 \times 10^5 \text{ m}\right) \sin\left(-20°\right)$$
$$= -3.301 \times 10^5 \text{ m}$$
$$s_{x2} = s_2 \cos \theta$$
$$= \left(3.095 \times 10^6 \text{ m}\right) \cos 80°$$
$$= 5.374 \times 10^5 \text{ m}$$
$$s_{y2} = s_2 \sin \theta$$
$$= \left(3.095 \times 10^6 \text{m}\right) \sin 80°$$
$$= 3.048 \times 10^6 \text{ m}$$

Adding up the x- and y-components:

$$s_x = s_{x1} + s_{x2}$$
$$= 9.07 \times 10^5 \text{ m} + 5.374 \times 10^5 \text{ m}$$
$$= 1.444 \times 10^6 \text{ m}$$
$$s_y = s_{y1} + s_{y2}$$
$$= -3.301 \times 10^5 \text{ m} + 3.048 \times 10^6 \text{ m}$$
$$= 2.718 \times 10^6 \text{ m}$$

To find the net displacement from the individual components, use the Pythagorean theorem:

$$s^2 = s_x^2 + s_y^2$$
$$s = \sqrt{\left(1.444 \times 10^6 \text{ m}\right)^2 + \left(2.718 \times 10^6 \text{ m}\right)^2}$$
$$= \sqrt{2.085 \times 10^{12} \text{ m}^2 + 7.388 \times 10^{12} \text{ m}^2}$$
$$= \sqrt{9.473 \times 10^{12} \text{ m}^2}$$
$$= 3.08 \times 10^6 \text{ m}$$

This, in units of thousands of kilometers (abbreviated as ThKilos here), would equal:

$$\left(3.08 \times 10^6 \text{ m}\right)\left(\frac{1 \text{ km}}{1 \times 10^3 \text{ m}}\right)\left(\frac{1 \text{ ThKilos}}{1{,}000 \text{ km}}\right) = 3.08 \text{ ThKilos}$$

313. **mass**

The impulse–momentum theorem states that $J = \Delta p$, where J is the impulse imparted onto an object and Δp ("delta-p") is the object's resulting change in momentum. Because momentum is mass times velocity, you can substitute to solve for impulse divided by velocity:

$$J = \Delta p$$
$$J = m\Delta v$$
$$\frac{J}{\Delta v} = m$$

314. $\dfrac{50DW}{3gT}$

Start with the impulse–momentum theorem and substitute the definitions of impulse ($J = F\Delta t$) and momentum ($\Delta p = m\Delta v$):

$$J = \Delta p$$
$$F\Delta t = m\Delta v$$
$$F\Delta t = m\left(v_f - v_i\right)$$
$$F(1\,\text{s}) = m\left(v_f\,\text{m/s} - 0\,\text{m/s}\right)$$
$$F(1\,\text{s}) = m\left(v_f\,\text{m/s}\right)$$
$$F = m\left(v_f\,\text{m/s}^2\right)$$
$$F = mv_f\,\text{N}$$

where the subscripts i and f stand for "initial" and "final," respectively.

Remember that the t in the impulse formula is the length of time of the collision itself — not the length of time that the baseball travels afterward.

Although values aren't explicitly given for mass and velocity, you can find expressions for them in terms of the given variables by using two more definitions. First, Newton's second law:

$$F = ma$$
$$F_G = ma_G$$
$$\frac{F_G}{a_G} = m$$
$$\frac{W}{g} = m$$

Second, the definition of velocity:

$$v = \frac{d}{t}$$
$$= \frac{(D\,\text{km})}{(T\,\text{min})}$$
$$= \frac{D}{T}\,\text{km/min}$$

These aren't the desired units to match up with g, so convert the velocity before substituting it into the force relationship you derived earlier.

$$v = \left(\frac{D}{T}\frac{\text{km}}{\text{min}}\right)\left(\frac{1,000\,\text{m}}{1\,\text{km}}\right)\left(\frac{1\,\text{min}}{60\,\text{s}}\right) = \frac{1,000D}{60T}\frac{\text{m}}{\text{s}}$$

$$= \frac{50D}{3T}\,\text{m/s}$$

Therefore,

$$F = mv_f = mv$$

$$= \left(\frac{W}{g}\right)\left(\frac{50D}{3T}\right)$$

$$= \frac{50DW}{3gT}$$

315. 50 m/s

You can solve this by using one of two mental approaches. Option 1 is the impulse-momentum theorem: $F\Delta t = m\Delta v$.

If the force equals 0 newtons, then the left side — and therefore the right side — of that equation also equals 0. Because the mass doesn't equal 0, the change in velocity must equal 0; in other words, the velocity doesn't change from its original value of 50 meters per second.

Option 2 is just remembering Newton's first law: In the absence of any forces, an object maintains its current velocity. No force means no change in velocity.

316. $\Delta v_A = \frac{1}{2}\Delta v_B$

According to Newton's third law, the force that each ball exerts on the other has the same magnitude (different directions). Because the collision time is obviously the same for both, that means that the impulse given to A by B is the same magnitude as the impulse given to B by A. If both bowling balls experience the same impulse, they must both experience the same change in momentum (again, directions are opposite, but magnitudes are equal). Mass is inversely proportional to the change in velocity, so doubling an object's mass halves its change in velocity. Ball A has twice the mass of ball B, so it also has one-half the change in velocity of ball B.

317. −29.3 N

You're interested in the force experienced by the billiard ball, so you need to focus on the billiard ball's change in momentum. First, convert its mass into kilograms:

$$(200\,\text{g})\left(\frac{1\,\text{kg}}{1,000\,\text{g}}\right) = 0.2\,\text{kg}$$

Then, use the impulse-momentum theorem, remembering that delta always subtracts an initial value from a final value ($\Delta v = v_{\text{final}} - v_{\text{initial}}$). The initial velocity is positive because the ball is traveling in the positive x-direction, and the final velocity is negative because it's traveling opposite the positive x-direction:

$$J = \Delta p$$
$$F\Delta t = m\Delta v$$
$$F(0.15\,\text{s}) = (0.2\,\text{kg})(-6\,\text{m/s} - 16\,\text{m/s})$$
$$(0.15\,\text{s})F = (0.2\,\text{kg})(-22\,\text{m/s})$$
$$(0.15\,\text{s})F = -4.4\,\text{kg}\cdot\text{m/s}$$
$$F = -29.3\,\text{N}$$

318. kinetic energy

Although momentum is always conserved during a collision — whether elastic or inelastic — kinetic energy isn't conserved during inelastic collisions.

319. momentum

During a collision of two objects, each object exerts a force upon the other object. Newton's third law says those forces are equal in magnitude and opposite in direction. Because an equal force is exerted for an equal time on each object, the magnitude of the impulse is the same for each object. The impulse for each object is in an opposite direction, so the total impulse is zero. But impulse is just the change in momentum, so the change in momentum is zero. Momentum is always conserved during a collision.

320. $\frac{1}{2}v$

Initially, the stationary glob has no momentum ($v = 0$), so all the momentum in the "system" — composed of the two globs — is maintained by the moving glob. After the collision, that momentum now supports two globs — twice the mass it was originally propelling. Because momentum is the product of mass and velocity, the only way to keep the momentum constant is for the mass to be *multiplied* by a factor of 2 at the same time the velocity is *divided* by a factor of 2.

321. 180°

Although the situation is more of a reverse-collision than the types with which you're more familiar, momentum must still be conserved in the astronaut-wrench system. Regardless of the resulting speeds of the astronaut and the wrench, the only way that the final momentum of the astronaut-wrench system can equal 0 (which it was before the wrench was thrown; the astronaut was motionless, and 0 velocity means 0 momentum) is if the two objects move in opposite directions. Therefore, if the wrench is moving in the 0-degree direction, the astronaut must be moving in the 180-degree direction.

322. 1.5 m/s

The collision is perfectly inelastic, so objects A and B will stick together after the collision and have the same velocity. Mass and velocity are inversely related in the formula for momentum, which is conserved in collisions. The mass of the moving object is increased by a factor of 2 from before the collision to after the collision, so the velocity must decrease by a factor of 2.

If you're more comfortable working with the equation, let each object's mass be labeled something generic — like m — and solve the conservation of momentum equation for the final velocity:

$$p_i = p_f$$
$$m_A v_A + m_B v_B = (m_A + m_B)v_f$$
$$m(3\,\text{m/s}) + m(0\,\text{m/s}) = (m+m)v_f$$
$$(3\,\text{m/s})m = 2m v_f$$
$$3\,\text{m/s} = 2v_f$$
$$1.5\,\text{m/s} = v_f$$

323.   0.69 m/s

This is an example of a perfectly inelastic collision because the skater catches the snowball and doesn't let it bounce off; after the collision, only one object is moving. If you label the original direction of the snowball as positive and the skater as object 1, and then substitute values, you get

$$p_i = p_f$$
$$m_1 v_{i1} + m_2 v_{i2} = (m_1 + m_2)v_f$$
$$(32\,\text{kg})(0\,\text{m/s}) + (0.5\,\text{kg})(45\,\text{m/s}) = (32\,\text{kg} + 0.5\,\text{kg})v_f$$
$$22.5\,\text{kg}\cdot\text{m/s} = (32.5\,\text{kg})v_f$$
$$0.69\,\text{m/s} = v_f$$

324.   80 cm

Before jumping into any formulas, convert the bullet's speed into units of meters per second:

$$\left(400\,\frac{\text{km}}{\text{h}}\right)\left(\frac{1{,}000\,\text{m}}{1\,\text{km}}\right)\left(\frac{1\,\text{h}}{60\,\text{min}}\right)\left(\frac{1\,\text{min}}{60\,\text{s}}\right) = 111.1\,\frac{\text{m}}{\text{s}}$$

First, analyze the perfectly inelastic collision between the bullet and the rubber block. If you declare the bullet to be object 1 in the formula, then

$$p_i = p_f$$
$$m_1 v_{i1} + m_2 v_{i2} = (m_1 + m_2)v_f$$
$$(0.156\,\text{kg})(111.1\,\text{m/s}) + (4.25\,\text{kg})(0\,\text{m/s}) = (0.156\,\text{kg} + 4.25\,\text{kg})v_f$$
$$17.3\,\text{kg}\cdot\text{m/s} = (4.406\,\text{kg})v_f$$
$$3.93\,\text{m/s} = v_f$$

Therefore, the rubber block with an embedded bullet has a velocity of 3.93 meters per second immediately following the collision. It will then pull on its attached string, causing it to swing upward. You can determine its height using the principle of energy conservation: $U_i + K_i = U_f + K_f$, where U stands for potential energy and K stands for kinetic energy (and i and f stand for *initial* and *final*). Kinetic energy equals $\frac{1}{2}mv^2$, and in this case, the potential energy is entirely gravitational and equals mgh, where h is the height above/below the starting position. When the height is at its maximum, the velocity equals 0 (the object temporarily stops as it "turns around"). Now, solve for the final height of the block–bullet combo:

$$U_i + K_i = U_f + K_f$$

$$mgh_i + \frac{1}{2}mv_i^2 = mgh_f + \frac{1}{2}mv_f^2$$

$$gh_i + \frac{1}{2}v_i^2 = gh_f + \frac{1}{2}v_f^2$$

$$\left(9.8\,\text{m/s}^2\right)(0\,\text{m}) + \frac{1}{2}(3.93\,\text{m/s})^2 = \left(9.8\,\text{m/s}^2\right)h_f + \frac{1}{2}(0\,\text{m/s})^2$$

$$7.722\,\text{m}^2/\text{s}^2 = \left(9.8\,\text{m/s}^2\right)h_f$$

$$0.788\,\text{m} = h_f$$

As specified by the question, convert this result into centimeters, rounding to the nearest ten: $(0.788\,\text{m})\left(\dfrac{100\,\text{cm}}{1\,\text{m}}\right) = 78.8\,\text{cm} \approx 80\,\text{cm}.$

---

325.   20.4 m

This problem has three stages: energy conservation, momentum conservation, and projectile motion. First, use energy conservation to determine how fast Tarzan is moving before his collision with the watermelon. The rope is 6.5 meters long, so Tarzan is 6.5 meters lower than his starting point when at the bottom of the swing. By the conservation of energy equation:

$$U_i + K_i = U_f + K_f$$

$$mgh_i + \frac{1}{2}mv_i^2 = mgh_f + \frac{1}{2}mv_f^2$$

$$gh_i + \frac{1}{2}v_i^2 = gh_f + \frac{1}{2}v_f^2$$

$$\left(9.8\,\text{m/s}^2\right)(0\,\text{m}) + \frac{1}{2}(0\,\text{m/s})^2 = \left(9.8\,\text{m/s}^2\right)(-6.5\,\text{m}) + \frac{1}{2}v_f^2$$

$$0\,\text{m}^2/\text{s}^2 = -63.7\,\text{m}^2/\text{s}^2 + \frac{1}{2}v_f^2$$

$$63.7\,\text{m}^2/\text{s}^2 = \frac{1}{2}v_f^2$$

$$127.4\,\text{m}^2/\text{s}^2 = v_f^2$$

$$11.3\,\text{m/s} = v_f$$

So Tarzan is moving 11.3 meters per second just before colliding with the watermelon. Use the formula for conservation of momentum to calculate his speed upon letting go of the rope and holding onto the watermelon.

$$p_i = p_f$$

$$m_1 v_{i1} + m_2 v_{i2} = \left(m_1 + m_2\right)v_f$$

$$(108\,\text{kg})(11.3\,\text{m/s}) + (22\,\text{kg})(-6\,\text{m/s}) = (108\,\text{kg} + 22\,\text{kg})v_f$$

$$1{,}220.4\,\text{kg}\cdot\text{m/s} - 132\,\text{kg}\cdot\text{m/s} = (130\,\text{kg})v_f$$

$$1{,}088.4\,\text{kg}\cdot\text{m/s} = (130\,\text{kg})v_f$$

$$8.37\,\text{m/s} = v_f$$

Finally, use the kinematics of projectile motion to find out how much farther Tarzan and the watermelon travel horizontally.

$$\mathbf{s} = \left(s_x, s_y\right) = \left(s_x, -13.5\,\text{m}\right)$$
$$\mathbf{v} = \left(v_x, v_y\right) = \left(8.37\,\text{m/s}, 0\,\text{m/s}\right)$$
$$\mathbf{a} = \left(a_x, a_y\right) = \left(0\,\text{m/s}^2, -9.8\,\text{m/s}^2\right)$$

where $\mathbf{s}$, $\mathbf{v}$, and $\mathbf{a}$ are the position, velocity, and acceleration vectors, respectively, and the x- and y-subscripts indicate those vectors' horizontal and vertical components. The vertical component of displacement is –13.5 meters because Tarzan is already 6.5 meters below his 20-meter-high starting point as a result of reaching the bottom of the swing on the rope. Therefore, he only has 13.5 more meters to fall to hit the ground.

Use the formula for calculating displacement, $\mathbf{s} = \mathbf{v}_i t + \frac{1}{2}\mathbf{a}t^2$, to write out each component's equation. First, vertically:

$$s_y = v_y t + \frac{1}{2}a_y t^2$$
$$-13.5\,\text{m} = (0\,\text{m/s})t + \frac{1}{2}\left(-9.8\,\text{m/s}^2\right)t^2$$
$$-13.5\,\text{m} = \left(-4.9\,\text{m/s}^2\right)t^2$$
$$2.76\,\text{s}^2 = t^2$$
$$1.66\,\text{s} = t$$

Use this additional piece of information to help you complete the horizontal component's equation:

$$s_x = v_x t + \frac{1}{2}a_x t^2$$
$$= (8.37\,\text{m/s})(1.66\,\text{s}) + \frac{1}{2}\left(0\,\text{m/s}^2\right)(1.66\,\text{s})^2$$
$$= (8.37\,\text{m/s})(1.66\,\text{s})$$
$$= 13.9\,\text{m}$$

Add this to the 6.5 meters Tarzan has already moved horizontally away from the treehouse during his swing on the vine to find the total horizontal displacement: $13.9\,\text{m} + 6.5\,\text{m} = 20.4\,\text{m}$.

326. 0.2 m

The solution to this problem involves three steps: the brick's velocity before the collision, the brick-box combination's velocity immediately after the collision, and the distance traveled afterward. The first and last stages involve friction — a force — so drawing force diagrams helps you calculate the acceleration in each of those stages. Three forces act on the brick while on the ramp — gravitational (F_G), normal (F_N), and frictional (F_F):

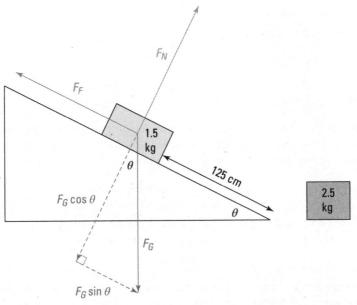

Illustration by Thomson Digital

The vectors are oriented with respect to the ramp's surface because you need to know the acceleration *in the direction of* that surface.

Taking horizontal to be the ramp's surface, there's no vertical net acceleration, and therefore, the net force in the vertical direction is 0 newtons, so $F_N - F_G \cos \theta = 0$, or, rearranging, $F_N = F_G \cos \theta = mg \cos \theta$. The brick does have an acceleration in the horizontal direction, so adding up the forces there results in $F_G \sin \theta - F_F = ma_{net}$. The mathematical definition of the force of friction ($F_f = \mu F_N$, where μ is the coefficient of friction) allows you to make a substitution and solve for the brick's acceleration:

$$F_G \sin \theta - F_F = ma_{net}$$
$$F_G \sin \theta - \mu F_N = ma_{net}$$
$$mg \sin \theta - \mu mg \cos \theta = ma_{net}$$
$$g \sin \theta - \mu g \cos \theta = a_{net}$$
$$\left(9.8 \text{ m/s}^2\right) \sin 20° - (0.08)\left(9.8 \text{ m/s}^2\right) \cos 20° =$$
$$\left(9.8 \text{ m/s}^2\right)(0.342) - (0.08)\left(9.8 \text{ m/s}^2\right)(0.94) =$$
$$3.352 \text{ m/s}^2 - 0.737 \text{ m/s}^2 =$$
$$2.62 \text{ m/s}^2 =$$

The brick's horizontal displacement is 125 centimeters, which you need to convert into meters: $(125 \text{ cm})\left(\dfrac{1 \text{ m}}{100 \text{ cm}}\right) = 1.25 \text{ m}$.

The brick starts from rest, so use the horizontal component of the velocity-displacement formula ($\mathbf{v}_f^2 = \mathbf{v}_i^2 + 2\mathbf{as}$) to solve for the brick's velocity after 1.25 meters:

$$v_{fx}^2 = v_{ix}^2 + 2a_x s_x$$
$$v_{fx} = \sqrt{v_{ix}^2 + 2a_x s_x}$$
$$= \sqrt{(0\text{ m/s})^2 + 2(2.62\text{ m/s}^2)(1.25\text{ m})}$$
$$= \sqrt{6.55\text{ m}^2/\text{s}^2}$$
$$= 2.56\text{ m/s}$$

You now have the brick's initial velocity heading into the collision with the box. Use the conservation of momentum formula to solve for the velocity of the brick-box combination after the collision. If you assign the brick as object 1, then

$$p_i = p_f$$
$$m_1 v_{i1} + m_2 v_{i2} = (m_1 + m_2)v_f$$
$$(1.5\text{ kg})(2.56\text{ m/s}) + (2.5\text{ kg})(0\text{ m/s}) = (1.5\text{ kg} + 2.5\text{ kg})v_f$$
$$3.84\text{ kg} \cdot \text{m/s} = (4\text{ kg})v_f$$
$$0.96\text{ m/s} = v_f$$

After the brick and box are attached to each other, and the situation is once again in the presence of friction (and other forces), you can draw a second force diagram:

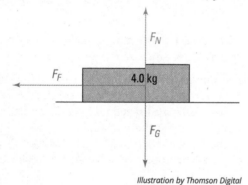

Illustration by Thomson Digital

Once again, there's no net acceleration in the vertical direction (and therefore no net force), so the sum of the vertical forces must equal 0:

$$F_N - F_G = 0$$
$$F_N = F_G$$
$$= mg$$

Only one force contributes to acceleration in the horizontal direction — friction — and it points in the negative direction (opposite to "forward" motion). Make substitutions to find the acceleration of the brick-box combo:

$$-F_F = ma_{net}$$
$$-\mu F_N = ma_{net}$$
$$-\mu mg = ma_{net}$$
$$-\mu g = a_{net}$$
$$-(0.2)(9.8\text{ m/s}^2) = a_{net}$$
$$-1.96\text{ m/s}^2 = a_{net}$$

You know the brick-box's initial velocity and its acceleration, so use the velocity-displacement formula ($v_f^2 = v_i^2 + 2\mathbf{as}$) in the horizontal direction to solve for the distance traveled before it stops (final velocity = 0 meters per second):

$$v_{fx}^2 = v_{ix}^2 + 2a_x s_x$$
$$(0 \text{ m/s})^2 = (0.96 \text{ m/s})^2 + 2(-1.96 \text{ m/s}^2) s_x$$
$$0 \text{ m}^2/\text{s}^2 = 0.922 \text{ m}^2/\text{s}^2 - (3.92 \text{ m/s}^2) s_x$$
$$-0.922 \text{ m}^2/\text{s}^2 = -(3.92 \text{ m/s}^2) s_x$$
$$0.2 \text{ m} = s_x$$

327. 8 m/s

Use the conservation of momentum formula where m is mass and v is velocity (the subscripts represent the values before [i] and after [f] the collision, respectively).

$$p_i = p_f$$
$$m_1 v_{i1} + m_2 v_{i2} = m_1 v_{f1} + m_2 v_{f2}$$
$$(42 \text{ kg})(15 \text{ m/s}) + (62 \text{ kg})(0 \text{ m/s}) = (42 \text{ kg})(3 \text{ m/s}) + (62 \text{ kg})v_{f2}$$
$$630 \text{ kg} \cdot \text{m/s} = 126 \text{ kg} \cdot \text{m/s} + (62 \text{ kg})v_{f2}$$
$$504 \text{ kg} \cdot \text{m/s} = (62 \text{ kg})v_{f2}$$
$$8.1 \text{ m/s} = v_{f2}$$
$$8 \text{ m/s} \approx v_{f2}$$

328. 1.7 m/s toward home plate

Use the formula for the conservation of momentum, $m_1 v_{i1} + m_2 v_{i2} = m_1 v_{f1} + m_2 v_{f2}$, where m is mass and v is velocity (the subscripts represent the values before [i] and after [f] the collision, respectively). Choose a direction for positive motion (this solution uses "toward home plate" as positive) to correctly place signs on the velocities. If the runner is designated as object 1, then

$$m_1 v_{i1} + m_2 v_{i2} = m_1 v_{f1} + m_2 v_{f2}$$
$$(91 \text{ kg})(5 \text{ m/s}) + (100 \text{ kg})(-2 \text{ m/s}) = (91 \text{ kg})v_{f1} + (100 \text{ kg})(1 \text{ m/s})$$
$$455 \text{ kg} \cdot \text{m/s} - 200 \text{ kg} \cdot \text{m/s} = (91 \text{ kg})v_{f1} + 100 \text{ kg} \cdot \text{m/s}$$
$$255 \text{ kg} \cdot \text{m/s} = (91 \text{ kg})v_{f1} + 100 \text{ kg} \cdot \text{m/s}$$
$$155 \text{ kg} \cdot \text{m/s} = (91 \text{ kg})v_{f1}$$
$$1.7 \text{ m/s} = v_{f1}$$

329. −1.3 m/s

Use the formula for the conservation of momentum, $m_1 v_{i1} + m_2 v_{i2} = m_1 v_{f1} + m_2 v_{f2}$, where m is mass and v is velocity (the subscripts represent the values before [i] and after [f] the collision, respectively). The problem dictates that Kate's initial direction of motion is the "positive" direction, so be sure to use negative values for Axel's velocities, as Axel is moving in the opposite direction both before and after the collision.

If Kate is designated as object 1, then

$$m_1 v_{i1} + m_2 v_{i2} = m_1 v_{f1} + m_2 v_{f2}$$
$$(45 \text{ kg})(4 \text{ m/s}) + (80 \text{ kg})(-3.2 \text{ m/s}) = (45 \text{ kg})v_{f1} + (80 \text{ kg})(-0.2 \text{ m/s})$$
$$180 \text{ kg} \cdot \text{m/s} - 256 \text{ kg} \cdot \text{m/s} = (45 \text{ kg})v_{f1} - 16 \text{ kg} \cdot \text{m/s}$$
$$-76 \text{ kg} \cdot \text{m/s} = (45 \text{ kg})v_{f1} - 16 \text{ kg} \cdot \text{m/s}$$
$$-60 \text{ kg} \cdot \text{m/s} = (45 \text{ kg})v_{f1}$$
$$-1.3 \text{ m/s} = v_{f1}$$

330. 6.9 m/s

Use the principle of energy conservation to start the solution to this problem.

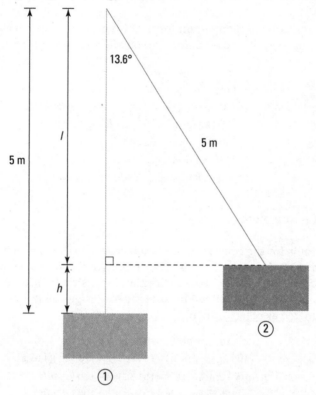

Illustration by Thomson Digital

By analyzing the amount of the wooden block's kinetic and potential energies at positions 1 and 2, you can solve for the velocity of the block at position 1 — which is, in turn, the block's final velocity after the collision with the rock.

By the conservation of energy, $U_1 + K_1 = U_2 + K_2$, where U stands for potential energy and K stands for kinetic energy. Kinetic energy always equals $\frac{1}{2}mv^2$, and in this case, the potential energy is entirely gravitational and equals mgh, where h is the height above/below the starting position.

To find the block's final value of h, use trigonometry:

$$\cos\theta = \frac{l}{5\,\text{m}}$$

$$\cos 13.6° = \frac{l}{5\,\text{m}}$$

$$0.972 = \frac{l}{5\,\text{m}}$$

$$4.86\,\text{m} = l$$

$$l + h = 5\,\text{m}$$

$$4.86\,\text{m} + h = 5\,\text{m}$$

$$h = 0.14\,\text{m}$$

Now substitute values into the energy conversation formula, being sure to identify the block's final velocity as 0 (it temporarily stops moving when it reaches its highest point):

$$U_1 + K_1 = U_2 + K_2$$

$$mgh_1 + \frac{1}{2}mv_1^2 = mgh_2 + \frac{1}{2}mv_2^2$$

$$gh_1 + \frac{1}{2}v_1^2 = gh_2 + \frac{1}{2}v_2^2$$

$$\left(9.8\,\text{m/s}^2\right)(0\,\text{m}) + \frac{1}{2}v_1^2 = \left(9.8\,\text{m/s}^2\right)(0.14\,\text{m}) + \frac{1}{2}(0\,\text{m/s})^2$$

$$\frac{1}{2}v_1^2 = 1.372\,\text{m}^2/\text{s}^2$$

$$v_1^2 = 2.744\,\text{m}^2/\text{s}^2$$

$$v_1 = 1.66\,\text{m/s}$$

You now know the block's final velocity after the rock strikes it. Use the momentum conservation formula to finish off the solution to find the rock's initial velocity. Let the mass of the rock equal M; therefore, the mass of the wooden block equals $5M$. This solution uses the rock as object 1:

$$p_i = p_f$$

$$m_1 v_{i1} + m_2 v_{i2} = m_1 v_{f1} + m_2 v_{f2}$$

$$M v_{i1} + 5M(0\,\text{m/s}) = M(-1.4\,\text{m/s}) + 5M(1.66\,\text{m/s})$$

$$v_{i1} = -1.4\,\text{m/s} + 5(1.66\,\text{m/s})$$

$$= -1.4\,\text{m/s} + 8.3\,\text{m/s}$$

$$= 6.9\,\text{m/s}$$

331. v

Whenever one item is stationary before an elastic collision, you can calculate the stationary object's final velocity using the formula $v_{f2} = \frac{2m_1 v_{i1}}{m_1 + m_2}$, where object 1 is the moving mass and object 2 is the stationary one, and the i and f stand for *initial* and *final*, respectively.

Because the masses are equal, $m_1 = m_2$, and using v as object 1's initial velocity:

$$v_{f2} = \frac{2m_1 v_{i1}}{m_1 + m_2}$$

$$= \frac{2m_1 v}{m_1 + m_1}$$

$$= \frac{2m_1 v}{2m_1}$$

$$= v$$

---

332.   0.043 kg

Whenever one item is stationary before an elastic collision, you can calculate the objects' final velocities using the formulas $v_{f1} = \dfrac{(m_1 - m_2) v_{i1}}{m_1 + m_2}$ and $v_{f2} = \dfrac{2m_1 v_{i1}}{m_1 + m_2}$, where object 1 is the moving mass and object 2 is the stationary one, and the i and f stand for *initial* and *final*, respectively.

You need to solve for the mass of the grape gum — the stationary object — so you're looking to solve for m_2 given v_{i1}, m_1, v_{f1}, and v_{f2}. Either equation will do to solve for m_2; this solution uses the formula for v_{f2} simply because it has one fewer variable substitution:

$$v_{f2} = \frac{2m_1 v_{i1}}{m_1 + m_2}$$

$$0.1\,\text{m/s} = \frac{2m_1 \,(1.2\,\text{m/s})}{m_1 + 1\,\text{kg}}$$

$$0.1\,\text{m/s} = \frac{(2.4\,\text{m/s})m_1}{m_1 + 1\,\text{kg}}$$

$$(0.1\,\text{m/s})m_1 + 0.1\,\text{kg} \cdot \text{m/s} = (2.4\,\text{m/s})m_1$$

$$0.1\,\text{kg} \cdot \text{m/s} = (2.3\,\text{m/s})m_1$$

$$0.043\,\text{kg} = m_1$$

---

333. $v_{bi} + v_{bf}$

Because this is an elastic collision, you can use both the conservation of momentum equation and the conservation of kinetic energy equation. Rearrange the components of each so that all the terms containing mass A are on the opposite side from all the terms containing mass B. For momentum:

$$m_a v_{ai} + m_b v_{bi} = m_a v_{af} + m_b v_{bf}$$

$$m_a v_{ai} - m_a v_{af} = m_b v_{bf} - m_b v_{bi}$$

$$m_a \left(v_{ai} - v_{af} \right) = m_b \left(v_{bf} - v_{bi} \right)$$

And for kinetic energy:

$$\frac{1}{2} m_a v_{ai}^2 + \frac{1}{2} m_b v_{bi}^2 = \frac{1}{2} m_a v_{af}^2 + \frac{1}{2} m_b v_{bf}^2$$

$$\frac{1}{2} m_a v_{ai}^2 - \frac{1}{2} m_a v_{af}^2 = \frac{1}{2} m_b v_{bf}^2 - \frac{1}{2} m_b v_{bi}^2$$

$$\frac{1}{2} m_a \left(v_{ai}^2 - v_{af}^2 \right) = \frac{1}{2} m_b \left(v_{bf}^2 - v_{bi}^2 \right)$$

$$m_a \left(v_{ai}^2 - v_{af}^2 \right) = m_b \left(v_{bf}^2 - v_{bi}^2 \right)$$

$$m_a \left(v_{ai} + v_{af} \right)\left(v_{ai} - v_{af} \right) = m_b \left(v_{bf} + v_{bi} \right)\left(v_{bf} - v_{bi} \right)$$

Finally, take your result from the kinetic energy formula and divide it by your result from the momentum formula:

$$\frac{m_a(v_{ai}+v_{af})(v_{ai}-v_{af})=m_b(v_{bf}+v_{bi})(v_{bf}-v_{bi})}{m_a(v_{ai}-v_{af})=m_b(v_{bf}-v_{bi})}$$

$$\frac{m_a(v_{ai}+v_{af})(v_{ai}-v_{af})}{m_a(v_{ai}-v_{af})}=\frac{m_b(v_{bf}+v_{bi})(v_{bf}-v_{bi})}{m_b(v_{bf}-v_{bi})}$$

$$v_{ai}+v_{af}=v_{bf}+v_{bi}$$

which is exactly the relationship you were asked to discover.

334.   8.7 kg . m/s

While attached to the rope, the ball undergoes circular motion and therefore experiences a centripetal force. The only force involved is the tension in the rope caused by the astronaut's arm, so set the equation for centripetal force equal to the 100 newtons of tension:

$$F_C = F_T$$

$$\frac{mv^2}{r}=100\,\text{N}$$

m is the mass of the rubber ball, v is its tangential velocity (the velocity with which it would leave the circular path if it no longer experienced the centripetal force), and r is the radius of its circular path. Substitute the mass and radius, remembering to convert the desired units in both instances:

$$(650\,\text{g})\left(\frac{1\,\text{kg}}{1{,}000\,\text{g}}\right)=0.65\,\text{kg}$$

$$(80\,\text{cm})\left(\frac{1\,\text{m}}{100\,\text{cm}}\right)=0.8\,\text{m}$$

$$\frac{mv^2}{r}=100\,\text{N}$$

$$\frac{(0.65\,\text{kg})v^2}{0.8\,\text{m}}=100\,\text{N}$$

$$v^2=123.08\,\text{m}^2/\text{s}^2$$

$$v=\sqrt{123.08\,\text{m}^2/\text{s}^2}$$

$$=11.1\,\text{m/s}$$

After the ball detaches from the rope, it will travel toward its inevitable collision at this velocity — its initial velocity relative to the elastic collision. Although you have the option of performing some algebraic acrobatics to solve two equations (momentum and kinetic energy conservation) for the two unknowns (the second ball's mass and final velocity), you can use a shortcut when the problem meets the following circumstances:

I. The collision is elastic.

II. The collision is head-on.

III. You know three of the four velocities.

The relationship $v_{i1} + v_{f1} = v_{i2} + v_{f2}$ is always true if the problem meets the first two conditions. If you designate the rubber ball as object 1, you can solve for the final velocity of the debris (v_{f2}). Using the direction of the ball's initial velocity as positive,

$$v_{i1} + v_{f1} = v_{i2} + v_{f2}$$
$$(11.1\,\text{m/s}) + (3\,\text{m/s}) = (4\,\text{m/s}) + v_{f2}$$
$$14.1\,\text{m/s} = 4\,\text{m/s} + v_{f2}$$
$$10.1\,\text{m/s} = v_{f2}$$

The debris continues to move in its original direction, increased to 10.1 meters per second. To obtain the mass of the debris, again let object 1 reference the rubber ball and object 2 reference the debris, and use the conservation of momentum equation (using the equation for conservation of kinetic energy would also work, but the squares cause a little extra arithmetic):

$$m_1 v_{i1} + m_2 v_{i2} = m_1 v_{f1} + m_2 v_{f2}$$
$$(0.650\,\text{kg})(11.1\,\text{m/s}) + m_2(4\,\text{m/s}) = (0.650\,\text{kg})(3\,\text{m/s}) + m_2(10.1\,\text{m/s})$$
$$7.22\,\text{kg}\cdot\text{m/s} + (4\,\text{m/s})m_2 = 1.95\,\text{kg}\cdot\text{m/s} + (10.1\,\text{m/s})m_2$$
$$5.27\,\text{kg}\cdot\text{m/s} = (6.1\,\text{m/s})m_2$$
$$0.86\,\text{kg} = m_2$$

Because momentum is the product of mass and velocity, the final momentum of the debris is:

$$p_f = mv_f$$
$$= (0.86\,\text{kg})(10.1\,\text{m/s})$$
$$= 8.7\,\text{kg}\cdot\text{m/s}$$

335. 19.1 m

Because the ramp is frictionless, the easiest way to calculate the duckpin ball's velocity when it leaves the ramp is to use the principle of energy conservation: $U_i + K_i = U_f + K_f$, where U is the potential energy (all gravitational in this problem), K is the kinetic energy, and the subscripts refer to the duckpin ball's initial (i) and final (f) positions. Kinetic energy always equals $\frac{1}{2}mv^2$, and in this case, the potential energy is entirely gravitational and equals mgh, where h is the height above/below the starting position. Use trigonometry to find the height that the ball will fall during its trip along the ramp:

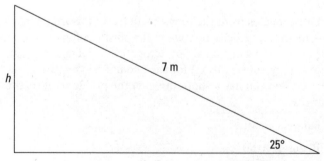

Illustration by Thomson Digital

$$\sin 25° = \frac{h}{7\,\text{m}}$$

$$0.423 = \frac{h}{7}$$

$$2.96\,\text{m} = h$$

So the height of the duckpin ball when it leaves the ramp is 2.96 meters below its starting height. Now, fill in all the known values in the energy conservation equation:

$$U_i + K_i = U_f + K_f$$

$$mgh_i + \frac{1}{2}mv_i^2 = mgh_f + \frac{1}{2}mv_f^2$$

$$gh_i + \frac{1}{2}v_i^2 = gh_f + \frac{1}{2}v_f^2$$

$$\left(9.8\,\text{m/s}^2\right)(0\,\text{m}) + \frac{1}{2}(0\,\text{m/s})^2 = \left(9.8\,\text{m/s}^2\right)(-2.96\,\text{m}) + \frac{1}{2}v_f^2$$

$$0\,\text{m}^2/\text{s}^2 = -29\,\text{m}^2/\text{s}^2 + \frac{1}{2}v_f^2$$

$$29\,\text{m}^2/\text{s}^2 = \frac{1}{2}v_f^2$$

$$58\text{m}^2/\text{s}^2 = v_f^2$$

$$\sqrt{58\,\text{m}^2/\text{s}^2} = v_f$$

$$7.62\,\text{m/s} = v_f$$

The relationship $v_{i1} + v_{f1} = v_{i2} + v_{f2}$ is always true in a collision between objects 1 and 2 if the collision is elastic and head–on. Because you now know the duckpin ball's initial velocity and the question gives you the initial and final velocities of the ten–pin ball, solving for the duckpin ball's final velocity is a simple matter of substitution. If you label the duckpin ball as object 1, keeping in mind that motion east is considered positive and motion west is considered negative,

$$v_{i1} + v_{f1} = v_{i2} + v_{f2}$$

$$(7.62\,\text{m/s}) + v_{f1} = (-10\,\text{m/s}) + (5\,\text{m/s})$$

$$7.62\,\text{m/s} + v_{f1} = -5\,\text{m/s}$$

$$v_{f1} = -12.62\,\text{m/s}$$

So the duckpin ball is indeed heading back along the ramp. To figure out how *far* along, first use energy conservation to find the maximum height the ball reaches (when the ball stops moving):

$$U_i + K_i = U_f + K_f$$

$$mgh_i + \frac{1}{2}mv_i^2 = mgh_f + \frac{1}{2}mv_f^2$$

$$gh_i + \frac{1}{2}v_i^2 = gh_f + \frac{1}{2}v_f^2$$

$$\left(9.8\,\text{m/s}^2\right)(0\,\text{m}) + \frac{1}{2}(-12.62\,\text{m/s})^2 = \left(9.8\,\text{m/s}^2\right)h_f + \frac{1}{2}(0\,\text{m/s})^2$$

$$79.63\text{m}^2/\text{s}^2 = \left(9.8\,\text{m/s}^2\right)h$$

$$79.63\,\text{m}^2/\text{s}^2 = \left(9.8\,\text{m/s}^2\right)h$$

$$8.1\,\text{m} = h$$

Examining the trigonometry once more,

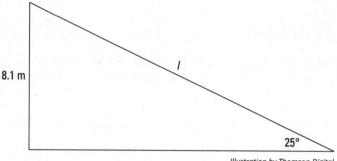

Illustration by Thomson Digital

$$\sin 25° = \frac{8.1\,\text{m}}{l}$$

$$0.423 = \frac{8.1\,\text{m}}{l}$$

$$0.423l = 8.1\,\text{m}$$

$$l = \frac{8.1\,\text{m}}{0.423}$$

$$= 19.1\,\text{m}$$

336. 1.9 m/s

You have three equations in your toolbox when solving elastic collisions. First, the conservation of momentum in the x-direction:

$$m_1 v_{i1} \cos \theta_{i1} + m_2 v_{i2} \cos \theta_{i2} = m_1 v_{f1} \cos \theta_{f1} + m_2 v_{f2} \cos \theta_{f2}$$

Next, the conservation of momentum in the y-direction:

$$m_1 v_{i1} \sin \theta_{i1} + m_2 v_{i2} \sin \theta_{i2} = m_1 v_{f1} \sin \theta_{f1} + m_2 v_{f2} \sin \theta_{f2}$$

And finally, the conservation of kinetic energy:

$$\frac{1}{2} m_1 v_{i1}^2 + \frac{1}{2} m_2 v_{i2}^2 = \frac{1}{2} m_1 v_{f1}^2 + \frac{1}{2} m_2 v_{f2}^2$$

which you can quickly simplify to $m_1 v_{i1}^2 + m_2 v_{i2}^2 = m_1 v_{f1}^2 + m_2 v_{f2}^2$.

If you know all but one of the velocities — as is the situation in this problem — the final equation is the easiest to use. Substitute the known values, including 0 meters per second for the initial velocity of the stationary, 2-kilogram particle. The following solution identifies the 1-kilogram particle as object 1:

$$m_1 v_{i1}^2 + m_2 v_{i2}^2 = m_1 v_{f1}^2 + m_2 v_{f2}^2$$

$$(1\,\text{kg})(2\,\text{m/s})^2 + (2\,\text{kg})(0\,\text{m/s})^2 = (1\,\text{kg})v_{f1}^2 + (2\,\text{kg})(0.5\,\text{m/s})^2$$

$$4\,\text{J} + 0\,\text{J} = (1\,\text{kg})v_{f1}^2 + 0.5\,\text{J}$$

$$3.5\,\text{J} = (1\,\text{kg})v_{f1}^2$$

$$\sqrt{3.5\,\text{m}^2/\text{s}^2} = v_{f1}$$

$$1.9\,\text{m/s} = v_{f1}$$

337. **30° south of east**

You have three equations in your toolbox when solving elastic collisions. First, the conservation of momentum in the x-direction:

$$m_1 v_{i1} \cos\theta_{i1} + m_2 v_{i2} \cos\theta_{i2} = m_1 v_{f1} \cos\theta_{f1} + m_2 v_{f2} \cos\theta_{f2}$$

Next, the conservation of momentum in the y-direction:

$$m_1 v_{i1} \sin\theta_{i1} + m_2 v_{i2} \sin\theta_{i2} = m_1 v_{f1} \sin\theta_{f1} + m_2 v_{f2} \sin\theta_{f2}$$

And finally, the conservation of kinetic energy:

$$\frac{1}{2} m_1 v_{i1}^2 + \frac{1}{2} m_2 v_{i2}^2 = \frac{1}{2} m_1 v_{f1}^2 + \frac{1}{2} m_2 v_{f2}^2$$

which you can quickly simplify to $m_1 v_{i1}^2 + m_2 v_{i2}^2 = m_1 v_{f1}^2 + m_2 v_{f2}^2$.

If you know all but one of the velocities — as is the situation in this problem — the final equation is the easiest to use. However, because the question wants to know the *direction*, the kinetic energy equation won't get you the answer in one step. You can use it to solve for the black bowling ball's speed and then use *that* value in one of the momentum conservation equations. So, start with the kinetic energy equation to solve for v_{f2} (assuming the black ball is object 2):

$$m_1 v_{i1}^2 + m_2 v_{i2}^2 = m_1 v_{f1}^2 + m_2 v_{f2}^2$$
$$(8 \text{ kg})(40 \text{ m/s})^2 + (8 \text{ kg})(0 \text{ m/s})^2 = (8 \text{ kg})(20 \text{ m/s})^2 + (8 \text{ kg}) v_{f2}^2$$
$$12{,}800 \text{ J} = 3{,}200 \text{ J} + (8 \text{ kg}) v_{f2}^2$$
$$9{,}600 \text{ J} = (8 \text{ kg}) v_{f2}^2$$
$$1{,}200 \text{ m}^2/\text{s}^2 = v_{f2}^2$$
$$\sqrt{1{,}200 \text{ m}^2/\text{s}^2} = v_{f2}$$
$$34.64 \text{ m/s} = v_{f2}$$

Then use one of the momentum equations to solve for θ_{f2}, the final direction of object 2 — the black bowling ball. This solution uses the y-components because using the x-components results in an ambiguous angle (this is sometimes the case when the entire initial velocity is in the x-direction because of the cyclical nature of angle measurements):

$$m_1 v_{i1} \sin\theta_{i1} + m_2 v_{i2} \sin\theta_{i2} = m_1 v_{f1} \sin\theta_{f1} + m_2 v_{f2} \sin\theta_{f2}$$
$$(8 \text{ kg})(40 \text{ m/s})\sin 0° + (8 \text{ kg})(0 \text{ m/s})\sin 0° = (8 \text{ kg})(20 \text{ m/s})\sin 60°$$
$$+ (8 \text{ kg})(34.64 \text{ m/s})\sin\theta_{f2}$$
$$0 \text{ kg} \cdot \text{m/s} = 138.57 \text{ kg} \cdot \text{m/s} + (277.12 \text{ kg} \cdot \text{m/s})\sin\theta_{f2}$$
$$-138.57 \text{ kg} \cdot \text{m/s} = (277.12 \text{ kg} \cdot \text{m/s})\sin\theta_{f2}$$
$$\frac{-138.57}{277.12} = \sin\theta_{f2}$$
$$-0.5 = \sin\theta_{f2}$$
$$\sin^{-1}(-0.5) = \theta_{f2}$$
$$-30° = \theta_{f2}$$

This is relative to the positive x-axis — or east — so this indicates that the direction is 30 degrees below, or south, of east.

338.　1.8 kg

You have three equations at your disposal when solving elastic collisions. First, the conservation of momentum in the x-direction:

$$m_1 v_{i1} \cos \theta_{i1} + m_2 v_{i2} \cos \theta_{i2} = m_1 v_{f1} \cos \theta_{f1} + m_2 v_{f2} \cos \theta_{f2}$$

Next, the conservation of momentum in the y-direction:

$$m_1 v_{i1} \sin \theta_{i1} + m_2 v_{i2} \sin \theta_{i2} = m_1 v_{f1} \sin \theta_{f1} + m_2 v_{f2} \sin \theta_{f2}$$

And finally, the conservation of kinetic energy:

$$\frac{1}{2} m_1 v_{i1}^2 + \frac{1}{2} m_2 v_{i2}^2 = \frac{1}{2} m_1 v_{f1}^2 + \frac{1}{2} m_2 v_{f2}^2$$

which you can quickly simplify to $m_1 v_{i1}^2 + m_2 v_{i2}^2 = m_1 v_{f1}^2 + m_2 v_{f2}^2$.

You know all values involved except for the final speed of block 1 and the mass of block 2. None of the three equations contains just one of these unknowns, so you need to use two equations. To avoid squares and square roots, this solution uses the two momentum equations. Block 1 is designated as object 1.

First, in the x-direction, solving for v_{f1} so as to leave m_2 — the value the question requested — in play:

$$m_1 v_{i1} \cos \theta_{i1} + m_2 v_{i2} \cos \theta_{i2} = m_1 v_{f1} \cos \theta_{f1} + m_2 v_{f2} \cos \theta_{f2}$$

$$(0.6 \text{ kg})(3 \text{ m/s}) \cos 0° + m_2(0 \text{ m/s}) \cos 0° = (0.6 \text{ kg}) v_{1f} \cos 45°$$
$$+ m_2(0.735 \text{ m/s}) \cos(-61°)$$

$$(1.8 \text{ kg} \cdot \text{m/s})(1) = (0.6 \text{ kg}) v_{f1}(0.707)$$
$$+ m_2(0.735 \text{ m/s})(0.485)$$

$$1.8 \text{ kg} \cdot \text{m/s} = (0.424 \text{ kg}) v_{f1} + (0.356 \text{ m/s}) m_2$$

$$1.8 \text{ kg} \cdot \text{m/s} - (0.356 \text{ m/s}) m_2 = (0.424 \text{ kg}) v_{f1}$$

$$\frac{1.8 \text{ kg} \cdot \text{m/s} - (0.356 \text{ m/s}) m_2}{0.424 \text{ kg}} = v_{f1}$$

Next, in the y-direction, again solving for v_{f1}:

$$m_1 v_{i1} \sin \theta_{i1} + m_2 v_{i2} \sin \theta_{i2} = m_1 v_{f1} \sin \theta_{f1} + m_2 v_{f2} \sin \theta_{f2}$$

$$(0.6 \text{ kg})(3 \text{ m/s}) \sin 0° + m_2(0 \text{ m/s}) \sin 0° = (0.6 \text{ kg}) v_{1f} \sin 45°$$
$$+ m_2(0.735 \text{ m/s}) \sin(-61°)$$

$$0 \text{ kg} \cdot \text{m/s} = (0.6 \text{ kg}) v_{f1}(0.707)$$
$$+ m_2(0.735 \text{ m/s})(-0.875)$$

$$0 \text{ kg} \cdot \text{m/s} = (0.424 \text{ kg}) v_{f1} - (0.643 \text{ m/s}) m_2$$

$$(0.643 \text{ m/s}) m_2 = (0.424 \text{ kg}) v_{f1}$$

$$\frac{(0.643 \text{ m/s}) m_2}{0.424 \text{ kg}} = v_{f1}$$

$$\left(1.52 \frac{\text{m/s}}{\text{kg}} \right) m_2 = v_{f1}$$

Substitute this into the result from the x-components:

$$\frac{1.8\ \text{kg}\cdot\text{m/s}-(0.356\ \text{m/s})m_2}{0.424\ \text{kg}}=v_{f1}$$

$$\frac{1.8\ \text{kg}\cdot\text{m/s}-(0.356\ \text{m/s})m_2}{0.424\ \text{kg}}=\left(1.52\frac{\text{m/s}}{\text{kg}}\right)m_2$$

$$1.8\ \text{kg}\cdot\text{m/s}-(0.356\ \text{m/s})m_2=(0.0644\ \text{m/s})m_2$$

$$1.8\ \text{kg}-0.356m_2=0.644m_2$$

$$1.8\ \text{kg}=m_2$$

---

339. 4.0 m/s

You have three equations in your toolbox when solving elastic collisions. First, the conservation of momentum in the x-direction:

$$m_1v_{i1}\cos\theta_{i1}+m_2v_{i2}\cos\theta_{i2}=m_1v_{f1}\cos\theta_{f1}+m_2v_{f2}\cos\theta_{f2}$$

Next, the conservation of momentum in the y-direction:

$$m_1v_{i1}\sin\theta_{i1}+m_2v_{i2}\sin\theta_{i2}=m_1v_{f1}\sin\theta_{f1}+m_2v_{f2}\sin\theta_{f2}$$

And finally, the conservation of kinetic energy:

$$\tfrac{1}{2}m_1v_{i1}^2+\tfrac{1}{2}m_2v_{i2}^2=\tfrac{1}{2}m_1v_{f1}^2+\tfrac{1}{2}m_2v_{f2}^2$$

which you can quickly simplify to $m_1v_{i1}^2+m_2v_{i2}^2=m_1v_{f1}^2+m_2v_{f2}^2$.

Because you're given all the angles — and you're missing two of the velocities — use the conservation of momentum formulas. First, for the x-components (using BB1 as object 1):

$$m_1v_{i1}\cos\theta_{i1}+m_2v_{i2}\cos\theta_{i2}=m_1v_{f1}\cos\theta_{f1}+m_2v_{f2}\cos\theta_{f2}$$

$$(0.475\ \text{kg})v_{i1}\cos0°+(0.182\ \text{kg})(0\ \text{m/s})\cos0°=(0.475\ \text{kg})(2.22\ \text{m/s})\cos20°$$

$$+(0.182\ \text{kg})v_{f2}\cos(-21.6°)$$

$$(0.475\ \text{kg})v_{i1}(1)=(1.055\ \text{kg}\cdot\text{m/s})(0.9397)$$

$$+(0.182\ \text{kg})v_{f2}(0.9298)$$

$$(0.475\ \text{kg})v_{i1}=0.991\ \text{kg}\cdot\text{m/s}+(0.169\ \text{kg})v_{f2}$$

Then, for the momentum's y-components:

$$m_1v_{i1}\sin\theta_{i1}+m_2v_{i2}\sin\theta_{i2}=m_1v_{f1}\sin\theta_{f1}+m_2v_{f2}\sin\theta_{f2}$$

$$(0.475\ \text{kg})v_{i1}\sin0°+(0.182\ \text{kg})(0\ \text{m/s})\sin0°=(0.475\ \text{kg})(2.22\ \text{m/s})\sin20°$$

$$+(0.182\ \text{kg})v_{f2}\sin(-21.6°)$$

$$(0.475\ \text{kg})v_{i1}(0)=(1.055\ \text{kg}\cdot\text{m/s})(0.342)$$

$$+(0.182\ \text{kg})v_{f2}(-0.3681)$$

$$0\ \text{kg}\cdot\text{m/s}=0.361\ \text{kg}\cdot\text{m/s}-(0.067\ \text{kg})v_{f2}$$

$$-0.361\ \text{kg}\cdot\text{m/s}=-(0.067\ \text{kg})v_{f2}$$

$$5.39\ \text{m/s}=v_{f2}$$

Substitute this result into the result from the *x*-components to solve for the speed of object 1 (BB1) before the collision:

$$(0.475 \text{ kg})v_{i1} = 0.991 \text{ kg} \cdot \text{m/s} + (0.169 \text{ kg})v_{f2}$$
$$(0.475 \text{ kg})v_{i1} = 0.991 \text{ kg} \cdot \text{m/s} + (0.169 \text{ kg})(5.39 \text{ m/s})$$
$$0.475v_{i1} = 0.991 \text{ m/s} + 0.911 \text{ m/s}$$
$$0.475v_{i1} = 1.902 \text{ m/s}$$
$$v_{i1} = 4.0 \text{ m/s}$$

340.   **0.7 m/s, −78°**

First, convert the masses into the desired units, as well as the ramp's height:

$$(620 \text{ g})\left(\frac{1 \text{ kg}}{1,000 \text{ g}}\right) = 0.62 \text{ g}$$

$$(65 \text{ cm})\left(\frac{1 \text{ m}}{100 \text{ cm}}\right) = 0.65 \text{ m}$$

The solution to this problem involves two steps: the ramp and the collision. The first requires the use of forces to determine the velocity the sliding block will have *before* the elastic collision, and the second uses conservation of momentum and energy to solve for the velocity of the stationary block *after* the collision. Use a force diagram to draw the three forces acting on the brick while on the ramp — gravitational (F_G), normal (F_N), and frictional (F_F):

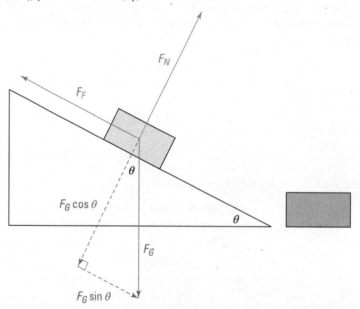

The vectors are oriented with respect to the ramp's surface because you need to know the acceleration *in the direction of* that surface.

Taking horizontal to be the ramp's surface, there's no vertical net acceleration, and therefore, the net force in the vertical direction is 0 newtons, so $F_N - F_G \cos\theta = 0$, or, rearranging, $F_N = F_G \cos\theta = mg \cos\theta$. The brick does have an acceleration in the horizontal direction, so adding up the forces there results in $F_G \sin\theta - F_F = ma_{net}$. The mathematical definition of the force of friction ($F_f = \mu F_N$, where μ is the coefficient of friction) allows you to make a substitution and solve for the brick's acceleration:

$$F_G \sin\theta - F_F = ma_{net}$$
$$F_G \sin\theta - \mu F_N = ma_{net}$$
$$mg \sin\theta - \mu mg \cos\theta = ma_{net}$$
$$g \sin\theta - \mu g \cos\theta = a_{net}$$
$$\left(9.8 \text{ m/s}^2\right) \sin 50° - (0.1)\left(9.8 \text{ m/s}^2\right) \cos 50° =$$
$$\left(9.8 \text{ m/s}^2\right)(0.766) - (0.1)\left(9.8 \text{ m/s}^2\right)(0.643) =$$
$$7.51 \text{ m/s}^2 - 0.63 \text{ m/s}^2 =$$
$$6.88 \text{ m/s}^2$$

Calculate the brick's displacement in the direction of this acceleration (along the ramp) by using trigonometry. You know the height from which the first block starts, and you know the ramp's angle:

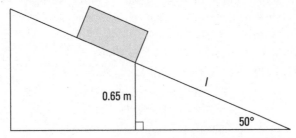

Illustration by Thomson Digital

$$\sin 50° = \frac{0.65 \text{ m}}{l}$$
$$l \sin 50° = 0.65 \text{ m}$$
$$l(0.766) = 0.65 \text{m}$$
$$l = \frac{0.65 \text{ m}}{0.766}$$
$$= 0.85 \text{ m}$$

The brick starts from rest, so use the horizontal component of the velocity-displacement formula $\left(\mathbf{v}_f^2 = \mathbf{v}_i^2 + 2\mathbf{as}\right)$ to solve for the brick's velocity after 0.85 meters:

$$v_{fx}^2 = v_{ix}^2 + 2a_x s_x$$
$$v_{fx} = \sqrt{v_{ix}^2 + 2a_x s_x}$$
$$= \sqrt{(0 \text{ m/s})^2 + 2\left(6.88 \text{ m/s}^2\right)(0.85 \text{ m})}$$
$$= \sqrt{0 \text{ m}^2/\text{s}^2 + 11.7 \text{ m}^2/\text{s}^2}$$
$$= \sqrt{11.7 \text{ m}^2/\text{s}^2}$$
$$= 3.42 \text{ m/s}$$

You now have the moving block's initial velocity heading into the collision with the stationary block. Use the conservation of momentum equations to solve for the final velocity of the brick-box combination after the collision. Let the originally stationary block be object 2, and, because the masses of both blocks are equal, let $m_2 = m_1$ to simplify the calculations.

The x-components of momentum conservation:

$$m_1 v_{i1} \cos \theta_{i1} + m_2 v_{i2} \cos \theta_{i2} = m_1 v_{f1} \cos \theta_{f1} + m_2 v_{f2} \cos \theta_{f2}$$
$$m_1 v_{i1} \cos \theta_{i1} + m_1 v_{i2} \cos \theta_{i2} = m_1 v_{f1} \cos \theta_{f1} + m_1 v_{f2} \cos \theta_{f2}$$
$$v_{i1} \cos \theta_{i1} + v_{i2} \cos \theta_{i2} = v_{f1} \cos \theta_{f1} + v_{f2} \cos \theta_{f2}$$
$$(3.42 \text{ m/s}) \cos 0° + (0 \text{ m/s}) \cos 0° = v_{f1} \cos 12° + v_{f2} \cos \theta_{f2}$$
$$(3.42 \text{ m/s})(1) + (0 \text{ m/s})(1) = v_{f1}(0.978) + v_{f2} \cos \theta_{f2}$$
$$3.42 \text{ m/s} = 0.978 v_{f1} + v_{f2} \cos \theta_{f2}$$
$$3.42 \text{ m/s} - 0.978 v_{f1} = v_{f2} \cos \theta_{f2}$$
$$\left(3.42 \text{ m/s} - 0.978 v_{f1}\right)^2 = \left(v_{f2} \cos \theta_{f2}\right)^2$$
$$11.6964 \text{ m}^2/\text{s}^2 - (6.6895 \text{ m/s}) v_{f1} + 0.9565 v_{f1}^2 = v_{f2}^2 \cos^2 \theta_{f2}$$

The y-components of momentum conservation:

$$m_1 v_{i1} \sin \theta_{i1} + m_2 v_{i2} \sin \theta_{i2} = m_1 v_{f1} \sin \theta_{f1} + m_2 v_{f2} \sin \theta_{f2}$$
$$m_{11} v_{i1} \sin \theta_{i1} + m_1 v_{i2} \sin \theta_{i2} = m_1 v_{f1} \sin \theta_{f1} + m_1 v_{f2} \sin \theta_{f2}$$
$$v_{i1} \sin \theta_{i1} + v_{i2} \sin \theta_{i2} = v_{f1} \sin \theta_{f1} + v_{f2} \sin \theta_{f2}$$
$$(3.42 \text{ m/s}) \sin 0° + (0 \text{ m/s}) \sin 0° = v_{f1} \sin 12° + v_{f2} \sin \theta_{f2}$$
$$(3.42 \text{ m/s})(0) + (0 \text{ m/s})(0) = v_{f1}(0.208) + v_{f2} \sin \theta_{f2}$$
$$0 \text{ m/s} = 0.208 v_{f1} + v_{f2} \sin \theta_{f2}$$
$$-0.208 v_{f1} = v_{f2} \sin \theta_{f2}$$
$$\left(-0.208 v_{f1}\right)^2 = \left(v_{f2} \sin \theta_{f2}\right)^2$$
$$0.0432 v_{f1}^2 = v_{f2}^2 \sin^2 \theta_{f2}$$

Now, combine your two resulting equations to make use of the trig identity:

$$v_{f2}^2 \cos^2 \theta_{f2} + v_{f2}^2 \sin^2 \theta_{f2} = 11.6964 \text{ m}^2/\text{s}^2 - (6.6895 \text{ m/s}) v_{f1} + 0.9565 v_{f1}^2 + 0.0432 v_{f1}^2$$
$$v_{f2}^2 \left(\cos^2 \theta_{f2} + \sin^2 \theta_{f2}\right) = 11.6964 \text{ m}^2/\text{s}^2 - (6.6895 \text{ m/s}) v_{f1} + 0.9997 v_{f1}^2$$
$$v_{f2}^2(1) = 0.9997 v_{f1}^2 - (6.6895 \text{ m/s}) v_{f1} + 11.6964 \text{ m}^2/\text{s}^2$$
$$v_{f2}^2 = 0.9997 v_{f1}^2 - (6.6895 \text{ m/s}) v_{f1} + 11.6964 \text{ m}^2/\text{s}^2$$

The conservation equation you haven't used yet is the kinetic energy equation, which also contains the unknown variables m_2 and v_{f1} — giving you two equations with only two unknowns now. The kinetic energy conservation formula yields (again making use of the fact that $m_2 = m_1$ to simplify the calculations):

$$m_1 v_{i1}^2 + m_2 v_{i2}^2 = m_1 v_{f1}^2 + m_2 v_{f2}^2$$
$$m_1 v_{i1}^2 + m_1 v_{i2}^2 = m_1 v_{f1}^2 + m_1 v_{f2}^2$$
$$v_{i1}^2 + v_{i2}^2 = v_{f1}^2 + v_{f2}^2$$
$$(3.42 \text{ m/s})^2 + (0 \text{ m/s})^2 = v_{f1}^2 + v_{f2}^2$$
$$11.6964 \text{ m}^2/\text{s}^2 = v_{f1}^2 + v_{f2}^2$$

Now, substitute the expression you obtained for v_{f2}^2:

$$11.6964 = v_{f1}^2 + v_{f2}^2$$
$$11.6964 \text{ m}^2/\text{s}^2 = v_{f1}^2 + \left(0.9997 v_{f1}^2 - (6.6895 \text{ m/s}) v_{f1} + 11.6964 \text{ m}^2/\text{s}^2\right)$$
$$11.6964 \text{ m}^2/\text{s}^2 = v_{f1}^2 + 0.9997 v_{f1}^2 - (6.6895 \text{ m/s}) v_{f1} + 11.6964 \text{ m}^2/\text{s}^2$$
$$11.6964 \text{ m}^2/\text{s}^2 = 1.9997 v_{f1}^2 - (6.6895 \text{ m/s}) v_{f1} + 11.6964 \text{ m}^2/\text{s}^2$$
$$0 \text{ m}^2/\text{s}^2 = 1.9997 v_{f1}^2 - (6.6895 \text{ m/s}) v_{f1}$$
$$(6.6895 \text{ m/s}) v_{f1} = 1.9997 v_{f1}^2$$
$$(6.6895 \text{ m/s}) = 1.9997 v_{f1}$$
$$3.345 \text{ m/s} = v_{f1}$$

Back in your algebraic simplifications in the momentum equations, you found two relationships between v_{f1} and v_{f2}: $3.42 \text{ m/s} - 0.978 v_{f1} = v_{f2} \cos \theta_{f2}$ and $-0.208 v_{f1} = v_{f2} \sin \theta_{f2}$. Substitute 3.345 meters per second for v_{f1} and divide the former by the latter to solve for θ_{f2}:

$$3.42 \text{ m/s} - (0.978)(3.345 \text{ m/s}) = v_{f2} \cos \theta_{f2}$$
$$3.42 \text{ m/s} - 3.27 \text{ m/s} = v_{f2} \cos \theta_{f2}$$
$$0.15 \text{ m/s} = v_{f2} \cos \theta_{f2}$$
$$(-0.208)(3.345 \text{ m/s}) = v_{f2} \sin \theta_{f2}$$
$$-0.696 \text{ m/s} = v_{f2} \sin \theta_{f2}$$

$$\frac{-0.696 \text{ m/s} = v_{f2} \sin \theta_{f2}}{0.15 \text{ m/s} = v_{f2} \cos \theta_{f2}} \Rightarrow \frac{-0.696}{0.15} = \frac{v_{f2} \sin \theta_{f2}}{v_{f2} \cos \theta_{f2}}$$

$$\frac{-0.696}{0.15} = \frac{\sin \theta_{f2}}{\cos \theta_{f2}} = \tan \theta_{f2}$$

$$-4.64 = \tan \theta_{f2}$$
$$\tan^{-1}(-4.64) = \theta_{f2}$$
$$-77.8° = \theta_{f2}$$

Finally (finally!), substitute this into either $0.15 \text{ m/s} = v_{f2} \cos \theta_{f2}$ or $-0.696 \text{ m/s} = v_{f2} \sin \theta_{f2}$ to solve for v_{f2}:

$$0.15 \text{ m/s} = v_{f2} \cos \theta_{f2}$$
$$0.15 \text{ m/s} = v_{f2} \cos (-77.8°)$$
$$0.15 \text{ m/s} = v_{f2}(0.211)$$
$$0.71 \text{ m/s} = v_{f2}$$

which, rounded to the nearest tenth, is 0.7 meters per second, and to the nearest degree is −78 degrees.

341. 0.6 m/s

A linear quantity is always the product of its angular counterpart and a radius. In the case of velocity, $v = r\omega$, where v is the linear velocity, r is the radius of the circular route being traveled, and ω is the angular velocity.

$$v = r\omega$$
$$= (0.2 \text{ m})(3 \text{ rad/s})$$
$$= 0.6 \text{ m/s}$$

(A radian is a dimensionless unit that vanishes whenever another unit such as meters is around to take its place.)

342. **3 cm**

A linear quantity is always the product of its angular counterpart and a radius. In the case of acceleration, $a = r\alpha$, where a is the linear acceleration, r is the radius of the circular route being traveled, and α is the angular acceleration.

$$a = r\alpha$$
$$0.18 \text{ m/s}^2 = r\left(7 \text{ rad/s}^2\right)$$
$$0.026 \text{ m} = r$$

Convert into the requested units and round:

$$(0.026 \text{ m})\left(\frac{100 \text{ cm}}{1 \text{ m}}\right) = 2.6 \text{ cm} \approx 3 \text{ cm}$$

343. **6.4 m**

First, convert the mass into "correct" units:

$$(12 \text{ g})\left(\frac{1 \text{ kg}}{1,000 \text{ g}}\right) = 0.012 \text{ kg}$$

At the bottom of a vertical circle, the force of tension *opposes* the force of gravity when summing the forces to produce centripetal (circular) motion. Maximum tension occurs here. Draw a free-body diagram of the ball at the bottom of the circle:

Illustration by Thomson Digital

Newton's second law states that the sum of the forces must equal the product of the mass times the net acceleration, which, in this case, is centripetal acceleration:

$$\sum F = F_{\text{net}}$$
$$F_T - F_G = ma_{\text{net}}$$
$$F_T - mg = ma_c$$
$$F_T - mg = m\frac{v^2}{r}$$

Make use of the fact that $v = r\omega$, where v is the linear velocity, r is the radius of the circular path, and ω is the angular velocity, as given in the problem.

$$F_T - mg = m\frac{v^2}{r}$$

$$F_T - mg = m\frac{(r\omega)^2}{r}$$

$$F_T - mg = m\frac{r^2\omega^2}{r}$$

$$F_T - mg = mr\omega^2$$

$$(31\,\text{N}) - (0.012\,\text{kg})(9.8\,\text{m/s}^2) = (0.012\,\text{kg})r(20\,\text{rad/s})^2$$

$$31\,\text{N} - 0.118\,\text{N} = (4.8\,\text{kg/s}^2)r$$

$$30.882\,\text{N} = (4.8\,\text{kg/s}^2)r$$

$$6.4\,\text{m} = r$$

344. $1.1\,\text{m/s}^2$

The formula for centripetal acceleration using an object's *angular* velocity is $a_c = \omega^2 r$, where ω is an object's angular velocity and r is the radius of the circle. Substitute the known values:

$$a_c = \omega^2 r$$
$$= (0.3\,\text{rad/s})^2(12\,\text{m})$$
$$= (.09\,\text{rad}^2/\text{s}^2)(12\,\text{m})$$
$$= 1.08\,\text{m/s}^2$$

(A radian is a dimensionless unit that vanishes whenever another unit such as meters is around to take its place.)

345. 12 m

First, adjust the time to read in units of seconds only. One minute is equal to 60 seconds, making the total time equal to $(60\,\text{s}) + (12\,\text{s}) = 72\,\text{s}$.

Then calculate the Ferris wheel's angular velocity. Any complete rotation has 2π radians, and the wheel completes *two* rotations in 72 seconds, so the orbital velocity is

$$\omega = \frac{\Delta\theta}{\Delta t}$$
$$= \frac{2(2\pi\,\text{rad})}{72\,\text{s}}$$
$$= \frac{4\pi\,\text{rad}}{72\,\text{s}}$$
$$= 0.175\,\text{rad/s}$$

Finally, use the angular velocity form of the centripetal acceleration equation, $a_c = \omega^2 r$, where ω is an object's angular velocity and r is the radius of the circle, to solve for the Ferris wheel's radius.

$$a_c = \omega^2 r$$
$$0.38\,\text{m/s}^2 = (0.175\,\text{rad/s})^2\,r$$
$$0.38\,\text{m/s}^2 = (0.0306\,\text{rad}^2/\text{s}^2)r$$
$$12.4\,\text{m} = r$$

(A radian is a dimensionless unit that vanishes whenever another unit such as meters is around to take its place.)

346.   0.54 rad/s

If any complete revolution has 2π radians, then 12 complete revolutions have $12(2\pi) = 24\pi$ radians. Radians are the unit of angular displacement, θ, and $\omega = \frac{\Delta\theta}{\Delta t}$, where ω is angular velocity and t is time. So if the carousel rider covers 24π radians in 140 seconds, the angular velocity is

$$\omega = \frac{\Delta\theta}{\Delta t}$$
$$= \frac{24\pi \text{ rad}}{140 \text{ s}}$$
$$= 0.54 \text{ rad/s}$$

347. 3.5×10^{-4} m/s

First, convert the length of the minute hand into the "correct" units of meters:

$$(20 \text{ cm})\left(\frac{1 \text{ m}}{100 \text{ cm}}\right) = 0.2 \text{ m}$$

Then determine the angular velocity of a minute hand, which makes 24 revolutions in the span of a day (1 revolution per hour). That's a total of 48π radians of angular displacement because every revolution has 2π radians:

$$(24 \text{ rev})\left(\frac{2\pi \text{ rad}}{1 \text{ rev}}\right) = 48\pi \text{ rad}$$

Now calculate how many seconds are in 1 day:

$$(1 \text{ day})\left(\frac{24 \text{ h}}{1 \text{ day}}\right)\left(\frac{60 \text{ min}}{1 \text{ h}}\right)\left(\frac{60 \text{ s}}{1 \text{ min}}\right) = 86,400 \text{ s}$$

So the minute hand covers 48π radians in 86,400 seconds, which equals an angular velocity of:

$$\omega = \frac{\Delta\theta}{\Delta t}$$
$$= \frac{48\pi \text{ rad}}{86,400 \text{ s}}$$
$$= 1.745 \times 10^{-3} \text{ rad/s}$$

To convert between linear and angular velocities, use the equation $v = r\omega$, where v is the linear velocity, ω is the angular velocity, and r is the radius of the circle.

$$v = r\omega$$
$$= (0.2 \text{ m})\left(1.745 \times 10^{-3} \text{ rad/s}\right)$$
$$= 3.5 \times 10^{-4} \text{ m/s}$$

(A radian is a dimensionless unit that vanishes whenever another unit such as meters is around to take its place.)

348. 400 m

First, convert the radius into "correct" units:

$$(2.3 \text{ cm})\left(\frac{1 \text{ m}}{100 \text{ cm}}\right) = 0.023 \text{ m}$$

Use the angular displacement formula along with a time of 60 seconds (1 minute) to calculate the number of radians the gear rotates through during that time:

$$\theta = \omega_i t + \frac{1}{2}\alpha t^2$$
$$= (0 \text{ rad/s})(60 \text{ s}) + \frac{1}{2}\left(4.8 \text{ rad/s}^2\right)(60 \text{ s})^2$$
$$= 0 \text{ rad} + 8{,}640 \text{ rad}$$
$$= 8{,}640 \text{ rad}$$

Then use the conversion between angular displacement and linear displacement, $s = r\theta$, where s is the linear displacement, θ is the angular displacement, and r is the radius of the circular path traveled.

$$s = r\theta$$
$$= (0.023 \text{ m})(8{,}640 \text{ rad})$$
$$= 198.7 \text{ m}$$

To calculate the distance traveled in the second minute, first calculate the distance traveled in the first 2 minutes:

$$\theta = \omega_i t + \frac{1}{2}\alpha t^2$$
$$= (0 \text{ rad/s})(120 \text{ s}) + \frac{1}{2}\left(4.8 \text{ rad/s}^2\right)(120 \text{ s})^2$$
$$= 0 \text{ rad} + 34{,}560 \text{ rad}$$
$$= 34{,}560 \text{ rad}$$

Convert into a linear displacement:

$$s = r\theta$$
$$= (0.023 \text{ m})(34{,}560 \text{ rad})$$
$$= 794.9 \text{ m}$$

From this amount, subtract the displacement from the first 60 seconds ($t = 0$ s to $t = 60$ s) to find the displacement between $t = 60$ s and $t = 120$ s:

$$s_{60-120} = s_{0-120} - s_{0-60}$$
$$= (794.9 \text{ m}) - (198.7 \text{ m})$$
$$= 596.2 \text{ m}$$

This amount is $(596.2 \text{ m}) - (198.7 \text{ m}) = 397.5 \text{ m} \approx 400 \text{ m}$ more than the amount traveled during the first minute.

349. angular acceleration

Just as force always points in the same direction as linear acceleration, torque points in the same direction as angular acceleration.

350.   6 N·m

First, convert the given length into meters:

$$(3.8 \text{ cm})\left(\frac{1\,\text{m}}{100\,\text{cm}}\right) = 0.038 \text{ m}$$

Then use the formula for torque, $\tau = rF\sin\theta$, where F is the force exerted, r is the distance from the center of rotation to the point where the force is exerted, and θ is the angle between the two vectors. The angle here is 90 degrees because the force is exerted tangentially to the lid:

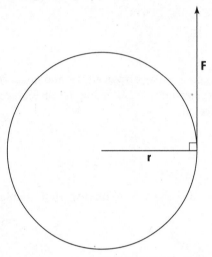

Illustration by Thomson Digital

Therefore,

$$\begin{aligned}
\tau &= rF\sin\theta \\
&= (0.038 \text{ m})(150 \text{ N})\sin 90° \\
&= (0.038 \text{ m})(150 \text{ N})(1) \\
&= 5.7\text{N}\cdot\text{m} \\
&\approx 6\text{N}\cdot\text{m}
\end{aligned}$$

351.   0.3 N·m

First, convert the mass into "correct" units:

$$(78 \text{ g})\left(\frac{1\,\text{kg}}{1{,}000\,\text{g}}\right) = 0.078 \text{ kg}$$

Use the formula for torque, $\tau = rF\sin\theta$, where F is the force exerted, r is the distance from the center of rotation to the point where the force is exerted, and θ is the angle between the two vectors.

In this problem, the string is the pivot arm, so $r = 2.8$ meters. The force exerted on it at the point of contact with the pendulum is the force of gravity on the pendulum: the weight of the pendulum. The angle between the two vectors can be seen as either 8 degrees or 172 degrees:

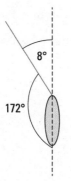

8°

172°

Illustration by Thomson Digital

Either number works in the formula, so you can choose to focus on the 8 degree angle. The problem asks you to find the *maximum* torque — for angles less than 90 degrees, the larger the angle, the larger the sine of the angle will be. So the maximum torque occurs at the maximum angle: 8 degrees. Therefore, the maximum torque is

$$\tau = rF \sin \theta$$
$$= rmg \sin \theta$$
$$= (2.8 \text{ m})(0.078 \text{ kg})(9.8 \text{ m/s}^2)\sin 8°$$
$$= (2.8 \text{ m})(0.078 \text{ kg})(9.8 \text{ m/s}^2)(0.139)$$
$$= 0.3 \text{ N} \cdot \text{m}$$

352.   **22 N · m; counterclockwise**

Three forces are at work, so you need to calculate three torques.

For τ_1 — the torque caused by F_1 — the force is located 90 centimeters (half the length of the bar) from the pivot. Convert this into meters:

$$(90 \text{ cm})\left(\frac{1 \text{ m}}{100 \text{ cm}}\right) = 0.9 \text{ m}$$

The angle between the "radius" and the force is 50 degrees (or 130 degrees):

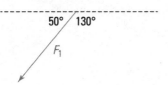

50°   130°

F_1

Illustration by Thomson Digital

You have all the information you need to solve the torque formula, $\tau = rF \sin \theta$, where F is the force exerted, r is the distance from the center of rotation to the point where the force is exerted, and θ is the angle between the two vectors. (The following calculations use the convention that a force directed *counterclockwise* yields a *positive* torque.)

$$\tau_1 = r_1 F_1 \sin \theta_1$$
$$= (0.9 \text{ m})(50 \text{ N}) \sin 50°$$
$$= (0.9 \text{ m})(50 \text{ N})(0.766)$$
$$= 34.47 \text{ N} \cdot \text{m}$$

F_1 is a counterclockwise force, so τ_1 is a positive 34.47 newton-meters.

Now continue with τ_2: The force is located 40 centimeters from the pivot. Convert this into meters:

$$(40\,\text{cm})\left(\frac{1\,\text{m}}{100\,\text{cm}}\right) = 0.4\,\text{m}$$

The angle between the "radius" and the force is $90 - 25 = 65$ (or 115 degrees):

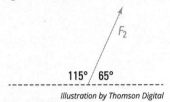

115° / 65°

Illustration by Thomson Digital

$$\begin{aligned}
\tau_2 &= r_2 F_2 \sin\theta_2 \\
&= (0.4\,\text{m})(40\,\text{N}) \sin 65° \\
&= (0.4\,\text{m})(40\,\text{N})(0.906) \\
&= 14.5\,\text{N}\cdot\text{m}
\end{aligned}$$

F_2 is a counterclockwise force, so τ_2 is a positive 14.5 newton-meters.

Finally, calculate τ_2. The force is again 0.9 meters away from the pivot, but the angle is 90 degrees:

Illustration by Thomson Digital

$$\begin{aligned}
\tau_2 &= r_2 F_2 \sin\theta_2 \\
&= (0.9\,\text{m})(30\,\text{N}) \sin 90° \\
&= (0.9\,\text{m})(30\,\text{N})(1) \\
&= 27\,\text{N}\cdot\text{m}
\end{aligned}$$

F_3 is a *clockwise* force, so τ_3 is a *negative* 27 newton-meters.

Add the three torques to find the net torque:

$$\begin{aligned}
\tau_{\text{net}} &= \tau_1 + \tau_2 + \tau_3 \\
&= (34.47\,\text{N}\cdot\text{m}) + (14.5\,\text{N}\cdot\text{m}) + (-27\,\text{N}\cdot\text{m}) \\
&= 34.47\,\text{N}\cdot\text{m} + 14.5\,\text{N}\cdot\text{m} - 27\,\text{N}\cdot\text{m} \\
&= 21.97\,\text{N}\cdot\text{m} \approx 22\,\text{N}\cdot\text{m}
\end{aligned}$$

This result is positive, meaning that the net force producing it is in the counterclockwise direction; therefore, the bar will rotate counterclockwise.

353. torque

A system is in rotational equilibrium when the sum of the torques — the net torque — is 0 newton-meters. (Because $\tau = I\alpha$ — where I is inertia and α is angular acceleration — if torque is 0, then angular acceleration is also 0, and vice versa.)

354. angular velocity

Rotational equilibrium exists when the net torque on an object is equal to 0. If the torque equals 0, so does the angular acceleration. Even with a nonzero value, a constant angular velocity means that the angular acceleration is 0 because the angular velocity isn't changing.

355. $250 \, \text{N} \cdot \text{m}$

Start with a free-body drawing of the forces on the skier:

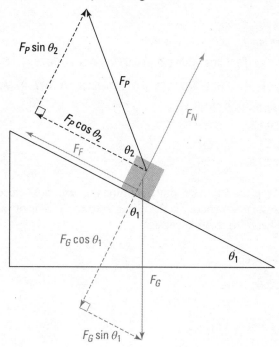

Illustration by Thomson Digital

F_G is the force of gravity, F_N is the normal force, F_F is the force of static friction, and F_P is the force of the ground pushing back against the ski poles (by Newton's third law, if the pole pushes on Earth, Earth pushes on the pole — and thereby the "attached" skier — with an equal magnitude of force). Friction points *up* the hill because gravity would naturally move the skier *down* the hill if the ski poles weren't present, and friction always opposes the direction of motion.

The net force in the horizontal direction, the net force in the vertical direction, and the torque about any point on the ladder must all equal 0 if the skier is to remain in equilibrium — translational and rotational. Use Newton's second law to analyze the forces, first in the vertical (y) direction:

$$F_{y,\text{net}} = \sum F_y$$
$$= F_N - F_G \cos \theta_1 + F_P \sin \theta_2$$
$$= F_N - mg \cos 22° + F_P \sin \theta_2$$
$$0\,\text{N} = F_N - (95\,\text{kg})(9.8\,\text{m/s}^2)\cos 22° + F_P \sin 45°$$
$$0\,\text{N} = F_N - (95\,\text{kg})(9.8\,\text{m/s}^2)(0.927) + F_P(0.707)$$
$$0\,\text{N} = F_N - 863\,\text{N} + 0.707 F_P$$
$$863\,\text{N} - 0.707 F_P = F_N$$

Then look at the horizontal (x) direction, where you can substitute the previous result:

$$F_{x,\text{net}} = \sum F_x$$
$$= F_G \sin \theta_1 - F_F - F_P \cos \theta_2$$
$$= mg \sin \theta_1 - \mu F_N - F_P \cos \theta_2$$
$$0\,\text{N} = (95\,\text{kg})(9.8\,\text{m/s}^2)\sin 22° - (0.08)(863\,\text{N} - 0.707 F_P) - F_P \cos 45°$$
$$0\,\text{N} = (95\,\text{kg})(9.8\,\text{m/s}^2)(0.375) - (0.08)(863\,\text{N} - 0.707 F_P) - F_P(0.707)$$
$$0\,\text{N} = 349.1\,\text{N} - 69.0\,\text{N} + 0.0566 F_P - 0.707 F_P$$
$$0\,\text{N} = 280.1\,\text{N} - 0.6504 F_P$$
$$0.6504 F_P = 280.1\,\text{N}$$
$$F_P = 430.7\,\text{N}$$

To find the torque about the shoulder joint, examine the arm more closely and use the torque formula $\tau = rF \sin \theta$ — where r is the distance from the axis of rotation to the force, F, and θ is the angle between the two vectors.

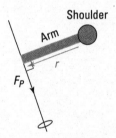

Illustration by Thomson Digital

The question states that the arm is perpendicular to the pole, so the angle between the arm (r) and the force along the pole (F_P) is 90 degrees:

$$\tau = rF \sin \theta$$
$$= rF_P \sin \theta$$
$$= (0.58\,\text{m})(430.7\,\text{N}) \sin 90°$$
$$= (0.58\,\text{m})(430.7\,\text{N})(1)$$
$$= 250\,\text{N} \cdot \text{m}$$

356. 8,900 N

First, draw a free-body diagram of the bodybuilder's arm:

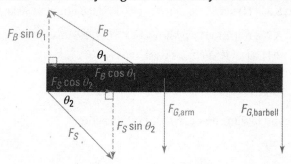

Illustration by Thomson Digital

$F_{G,\text{arm}}$ is the force due to the gravitational attraction between Earth and the arm, $F_{G,\text{barbell}}$ is the force due to the gravitational attraction between Earth and the barbell, F_B is the force of tension in the biceps muscle, and F_S is the force exerted by the shoulder joint. The net force in the horizontal direction, the net force in the vertical direction, and the torque about any point on the arm must all equal 0 if the arm is to remain in equilibrium. With the previous diagrams at your disposal, write out each equation, starting by analyzing the torque about an axis of rotation on the arm. Any point suffices, but the easiest calculations result from choosing an axis where a force is being exerted. The following solution uses the shoulder joint as the axis of rotation. Using the sign convention that clockwise-oriented forces produce negative torques and vice versa, the result follows:

$$\tau_{\text{net}} = \sum \tau$$
$$= \tau_B + \left(-\tau_{G,\text{arm}}\right) + \left(-\tau_{G,\text{barbell}}\right)$$
$$= \tau_B - \tau_{G,\text{arm}} - \tau_{G,\text{barbell}}$$
$$= r_B F_B \sin\theta_B - r_{G,\text{arm}} F_{G,\text{arm}} \sin\theta_{G,\text{arm}} - r_{G,\text{barbell}} F_{G,\text{barbell}} \sin\theta_{G,\text{barbell}}$$
$$= r_B F_B \sin\theta_B - r_{G,\text{arm}} m_{\text{arm}} g \sin\theta_{G,\text{arm}} - r_{G,\text{barbell}} m_{\text{barbell}} g \sin\theta_{G,\text{barbell}}$$
$$0\,\text{N}\cdot\text{m} = \left(\frac{L}{4}\right) F_B \sin 8° - \left(\frac{L}{2}\right)(3.5\ \text{kg})\left(9.8\ \text{m/s}^2\right)\sin 90°$$
$$- (L)(30\ \text{kg})\left(9.8\ \text{m/s}^2\right)\sin 90°$$
$$0 = \left(\frac{L}{4}\right) F_B (0.139) - \left(\frac{L}{2}\right)(3.5\ \text{kg})\left(9.8\ \text{m/s}^2\right)(1) - (L)(30\ \text{kg})\left(9.8\ \text{m/s}^2\right)(1)$$
$$0 = 0.03475 L F_B - (17.15\ \text{N})L - (294\ \text{N})L$$
$$0 = 0.03475 F_B - 17.15\ \text{N} - 294\ \text{N}$$
$$0 = 0.03475 F_B - 311.15\ \text{N}$$
$$311.15\ \text{N} = 0.03475 F_B$$
$$8,954\ \text{N} = F_B$$

Substitute this result into Newton's second law, first in the vertical (*y*) direction:

$$F_{y,\text{net}} = \sum F_y$$
$$= F_B \sin \theta_1 - F_S \sin \theta_2 - F_{G,\text{arm}} - F_{G,\text{barbell}}$$
$$= F_B \sin \theta_1 - F_S \sin \theta_2 - m_{\text{arm}}\, g - m_{\text{barbell}}\, g$$
$$0\,\text{N} = (8{,}954\,\text{N}) \sin 8° - F_S \sin \theta_2 - (3.5\,\text{kg})\left(9.8\,\text{m/s}^2\right) - (30\,\text{kg})\left(9.8\,\text{m/s}^2\right)$$
$$0 = (8{,}954\,\text{N})(0.139) - F_S \sin \theta_2 - (3.5\,\text{kg})\left(9.8\,\text{m/s}^2\right) - (30\,\text{kg})\left(9.8\,\text{m/s}^2\right)$$
$$0 = 1{,}244.6\,\text{N} - F_S \sin \theta_2 - 34.3\,\text{N} - 294\,\text{N}$$
$$0 = 916.3\,\text{N} - F_S \sin \theta_2$$
$$F_S \sin \theta_2 = 916.3\,\text{N}$$

Then use the second law in the horizontal (x) direction:

$$F_{x,\text{net}} = \sum F_x$$
$$= F_S \cos \theta_2 - F_B \cos \theta_1$$
$$0\,\text{N} = F_S \cos \theta_2 - (8{,}954\,\text{N}) \cos 8°$$
$$0\,\text{N} = F_S \cos \theta_2 - (8{,}954\,\text{N})(0.99)$$
$$0\,\text{N} = F_S \cos \theta_2 - 8{,}864.5\,\text{N}$$
$$8{,}864.5\,\text{N} = F_S \cos \theta_2$$

Divide the previous two results by one another to solve for the unknown angle:

$$\frac{F_S \sin \theta = 916.3\,\text{N}}{F_S \cos \theta = 8{,}864.5\,\text{N}} \Rightarrow \frac{F_S \sin \theta}{F_S \cos \theta} = \frac{916.3\,\text{N}}{8{,}864.5\,\text{N}}$$
$$\frac{\sin \theta}{\cos \theta} = 0.103$$
$$\tan \theta = 0.103$$
$$\theta = \tan^{-1}(0.103)$$
$$\theta = 5.9°$$

Finally, substitute this angle into one of the two equations you just used to solve for the amount of force exerted by the shoulder:

$$916.3\,\text{N} = F_S \sin \theta$$
$$916.3\,\text{N} = F_S \sin 5.9°$$
$$916.3\,\text{N} = F_S(0.103)$$
$$8{,}896\,\text{N} = F_S \approx 8{,}900\,\text{N}$$

357. 40 N

The net torque on an object in rotational equilibrium always equals 0 newton-meters. Therefore, the two forces on opposite sides of the axis of rotation must create equal and opposite torques to cancel out for a 0 sum. Use the force form of the torque formula ($\tau = rF \sin \theta$, where r is the distance from the axis of rotation to the force F and θ is the angle between the two vectors) to set the two torques equal to each other. In both cases, the angle between the r and F vectors is 90 degrees, so:

$$\tau_1 = \tau_2$$
$$r_1 F_1 \sin \theta_1 = r_2 F_2 \sin \theta_2$$
$$(0.8\,\text{m})(25\,\text{N}) \sin 90° = (0.5\,\text{m}) F_2 \sin 90°$$
$$(0.8\,\text{m})(25\,\text{N})(1) = (0.5\,\text{m}) F_2 (1)$$
$$20\,\text{N} \cdot \text{m} = (0.5\,\text{m}) F_2$$
$$40\,\text{N} = F_2$$

358. 0.75 meters

First, convert person A's distance into "correct" units:

$$(50\,\text{cm})\left(\frac{1\,\text{m}}{100\,\text{cm}}\right) = 0.5\,\text{m}$$

Then choose a point as the axis of rotation. Any point along the platform suffices, but the easiest calculation results from choosing an axis where a force is being exerted. Because you can ignore the platform's mass, four forces are acting on the platform:

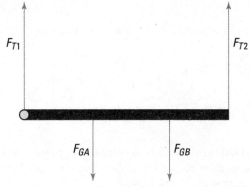

Illustration by Thomson Digital

F_{T1} is the force of tension from rope 1, F_{T2} is the force of tension from rope 2, F_{GA} is the force of gravitational attraction between Earth and person A, and F_{GB} is the force of gravitational attraction between Earth and person B. The yellow dot indicates the chosen axis of rotation for the solution presented here. Because the two gravitational forces tend to rotate the platform clockwise around the axis of rotation, the sign convention is to designate them as negative torques; the torque by the counterclockwise-pointing F_{T2} is designated positive. F_{T1} is located at the point of rotation, so it can't produce any torque (if $\tau = rF$ and $r = 0$, then $\tau = 0$ as well). All the forces act perpendicular (90 degrees) to the distances from the axis of rotation, and the entire system is in rotational (as well as translational) equilibrium, so the net torque on the system must equal 0 newton-meters. Start the solution by summing up the torques to calculate person B's distance from rope 1:

$$\tau_{net} = \sum \tau$$
$$= (-\tau_{GA}) + (-\tau_{GB}) + \tau_{T2}$$
$$= -\tau_{GA} - \tau_{GB} + \tau_{T2}$$
$$= -r_A F_{GA} \sin \theta_A - r_B F_{GB} \sin \theta_B + r_2 F_{T2} \sin \theta_2$$
$$= -r_A m_A g \sin \theta_A - r_B m_B g \sin \theta_B + r_2 F_{T2} \sin \theta_2$$
$$0\,\text{N}\cdot\text{m} = -(0.5\,\text{m})(100\,\text{kg})\left(9.8\,\text{m/s}^2\right)\sin 90° - r_B(60\,\text{kg})\left(9.8\,\text{m/s}^2\right)\sin 90°$$
$$+ (2.5\,\text{m})(608\,\text{N})\sin 90°$$
$$0\,\text{N}\cdot\text{m} = -(0.5\,\text{m})(100\,\text{kg})\left(9.8\,\text{m/s}^2\right)(1) - r_B(60\,\text{kg})\left(9.8\,\text{m/s}^2\right)(1) + (2.5\,\text{m})(608\,\text{N})(1)$$
$$0\,\text{N}\cdot\text{m} = -490\,\text{N}\cdot\text{m} - (588\,\text{N})r_B + 1{,}520\,\text{N}\cdot\text{m}$$
$$0\,\text{N}\cdot\text{m} = 1{,}030\,\text{N}\cdot\text{m} - (588\,\text{N})r_B - 1{,}030\,\text{N}\cdot\text{m} = -(588\,\text{N})r_B$$
$$1.75\,\text{m} = r_B$$

Therefore, person B is 1.75 meters away from the axis of rotation — rope 1 — and must be 2.5 – 1.75 = 0.75 meters away from rope 2.

---

359. **5,140 N**

First, convert the two given lengths into "correct" units:

$$(154\,\text{cm})\left(\frac{1\,\text{m}}{100\,\text{cm}}\right) = 1.54\,\text{m}$$

$$(30\,\text{cm})\left(\frac{1\,\text{m}}{100\,\text{cm}}\right) = 0.3\,\text{m}$$

Start by choosing a location for the axis of rotation. Any point along the platform suffices, but the easiest calculation comes from choosing an axis where a force is being exerted. Five forces are acting on the platform:

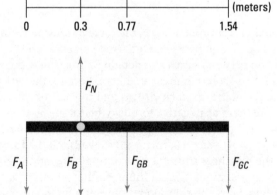

Illustration by Thomson Digital

F_A is the force provided by the nail at A, F_B is the force provided by the nail at B, F_N is the normal force supporting the weight of the system, F_{GC} is the force of gravitational attraction between Earth and the crate, and F_{GB} is the force of gravitational attraction between Earth and the horizontal bar. (You can approximate the entire mass of the bar at its middle because the mass is evenly distributed.) The yellow dot indicates the axis of rotation for the solution presented here; it was chosen in order to have F_N and F_B located at the point of rotation. Therefore, they can't produce any torque (if $\tau = rF$ and $r = 0$, then $\tau = 0$ as well), and their values cancel out of the calculation.

Because the two gravitational forces tend to rotate the platform clockwise around the axis of rotation, the sign convention is to designate them as negative torques; you can designate the torque by the counterclockwise-pointing F_A as positive. All the forces act perpendicular (90 degrees) to the distances from the axis of rotation, and the entire system is in rotational (as well as translational) equilibrium, so the net torque on the system must equal 0 newton-meters. Add up the torques to calculate the force exerted at pole A, making sure to use the distances relative to the axis of rotation (the point at pole B):

The distance of the bar mass is given by $\dfrac{1.54 \text{ m}}{2} - 0.3 \text{ m} = 0.47 \text{ m}$. The distance of the crate is given by $1.54 \text{ m} - 0.3 \text{ m} = 1.24 \text{ m}$.

$$
\begin{aligned}
\tau_{\text{net}} &= \sum \tau \\
&= \left(-\tau_{GB}\right) + \left(-\tau_{GC}\right) + \tau_A \\
&= -\tau_{GB} - \tau_{GC} + \tau_A \\
&= -r_B F_{GB} \sin\theta_B - r_C F_{GC} \sin\theta_C + r_A F_A \sin\theta_A \\
&= -r_B m_B g \sin\theta_B - r_C m_C g \sin\theta_C + r_A F_A \sin\theta_A
\end{aligned}
$$

$$
0 \text{ N} \cdot \text{m} = -(0.47 \text{ m})(18 \text{ kg})\left(9.8 \text{ m/s}^2\right)\sin 90°
$$
$$
- (1.24 \text{ m})(120 \text{ kg})\left(9.8 \text{ m/s}^2\right)\sin 90° + (0.3 \text{ m})F_A \sin 90°
$$

$$
0 \text{ N} \cdot \text{m} = -(0.47 \text{ m})(18 \text{ kg})\left(9.8 \text{ m/s}^2\right)(1)
$$
$$
- (1.24 \text{ m})(120 \text{ kg})\left(9.8 \text{ m/s}^2\right)(1) + (0.3 \text{ m})F_A(1)
$$

$$
0 \text{ N} \cdot \text{m} = -83 \text{ N} \cdot \text{m} - 1{,}458 \text{ N} \cdot \text{m} + (0.3 \text{ m})F_A
$$
$$
0 \text{ N} \cdot \text{m} = -1{,}541 \text{ N} \cdot \text{m} + (0.3 \text{ m})F_A
$$
$$
1{,}541 \text{ N} \cdot \text{m} = (0.3 \text{ m})F_A
$$
$$
5{,}137 \text{ N} = F_A \approx 5{,}140 \text{ N}
$$

360.   3.2 m

Choose a location for the axis of rotation. Any point along the platform suffices, but the easiest calculation comes from choosing an axis where a force is being exerted. According to the free-body diagram, these forces are acting on the beam:

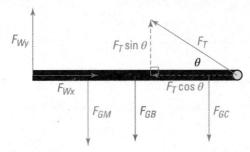

Illustration by Thomson Digital

F_{Wy} is the force the brick wall exerts in the vertical direction, F_{Wx} is the force the brick wall exerts in the horizontal direction (which is equivalent to $4F_{Wy}$ in this situation), F_T is the force of tension in the rope, F_{GM} is the force of gravitational attraction between Earth and the worker, F_{GC} is the force of gravitational attraction between Earth and the crate, and F_{GB} is the force of gravitational attraction between Earth and the horizontal bar. (You can approximate the entire mass of the bar at its middle because the mass is evenly distributed.)

The solution that follows places the axis of rotation on the right end of the beam where the rope is attached. (Because the distance r between the force of tension F and axis of rotation is 0 meters, the torque caused by the tension is 0 newton-meters $\tau = rF \sin \theta$; you can ignore it in the tension equation you use later in this solution.)

To start, calculate the angle of the rope relative to the beam using trigonometry:

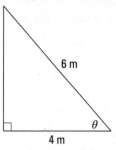

4 m

Illustration by Thomson Digital

$$\cos \theta = \frac{4 \text{ m}}{6 \text{ m}}$$

$$\cos \theta = 0.667$$

$$\theta = \cos^{-1}(0.667)$$

$$\theta = 48.2°$$

Then, using that fact that the system is in equilibrium (the net force and net torque must both equal 0), write the force form of Newton's second law for the forces acting on the beam in the vertical direction:

$$F_{y,\text{net}} = \sum_y F_y$$

$$0 \text{ N} = F_{Wy} - F_{GM} - F_{GB} - F_{GC} + F_T \sin \theta$$

$$0 \text{ N} = F_{Wy} - m_M g - m_B g - m_C g + F_T \sin \theta$$

$$0 \text{ N} = F_{Wy} - (90 \text{ kg})(9.8 \text{ m/s}^2) - (30 \text{ kg})(9.8 \text{ m/s}^2)$$
$$- (100 \text{ kg})(9.8 \text{ m/s}^2) + F_T \sin 48.2°$$

$$0 \text{ N} = F_{Wy} - 882 \text{ N} - 294 \text{ N} - 980 \text{ N} + 0.745 F_T$$

$$0 \text{ N} = F_{Wy} - 2,156 \text{ N} + 0.745 F_T$$

$$-0.745 F_T = F_{Wy} - 2,156 \text{ N}$$

$$F_T = -1.34 F_{Wy} + 2,894 \text{ N}$$

Next, do the same for the forces acting in the horizontal direction:

$$F_{x,\text{net}} = \sum_x F_x$$

$$0 \text{ N} = F_{Wx} - F_T \cos \theta$$

$$0 \text{ N} = 4 F_{Wy} - F_T \cos 48.2°$$

$$0 \text{ N} = 4 F_{Wy} - 0.667 F_T$$

$$0.667 F_T = 4 F_{Wy}$$

$$F_T = 6 F_{Wy}$$

Set the previous two equations equal to each other to solve for F_{Wy}:

$$-1.34F_{Wy} + 2{,}894\text{ N} = F_T = 6F_{Wy}$$
$$-1.34F_{Wy} + 2{,}894\text{ N} = 6F_{Wy}$$
$$2{,}894\text{ N} = 7.34F_{Wy}$$
$$394.3\text{ N} = F_{Wy}$$

Therefore,

$$F_{Wx} = 4F_{Wy}$$
$$= 4(394.3\text{ N})$$
$$= 1{,}577.2\text{ N}$$

Finally, add up the torques (this solution uses the standard convention that forces tending to produce clockwise rotation result in negative torques), and solve for the only unknown: the distance between the worker and the axis of rotation. All the gravitational forces act perpendicularly to the beam, so those angles are each 90 degrees. Note that the horizontal force from the wall is parallel to the beam and, therefore, has an angle of 0 degrees relative to the beam.

$$\tau_{net} = \sum \tau$$
$$= \left(-\tau_{Wx}\right) + \left(-\tau_{Wy}\right) + \tau_{GM} + \tau_{GB} + \tau_{GC}$$
$$= -\tau_{Wx} - \tau_{Wy} + \tau_{GM} + \tau_{GB} + \tau_{GC}$$
$$= -r_{Wx}F_{Wx}\sin\theta_{Wx} - r_{Wy}F_{Wy}\sin\theta_{Wy} + r_M F_{GM}\sin\theta_M$$
$$\quad + r_B F_{GB}\sin\theta_B + r_C F_{GC}\sin\theta_C$$
$$= -r_{Wx}F_{Wx}\sin\theta_{Wx} - r_{Wy}F_{Wy}\sin\theta_{Wy} + r_M m_M g\sin\theta_M$$
$$\quad + r_B m_B g\sin\theta_B + r_C m_C g\sin\theta_C$$
$$0\text{ N}\cdot\text{m} = -(4\text{ m})(1{,}577.2\text{ N})\sin 0° - (4\text{ m})(394.3\text{ N})\sin 90°$$
$$\quad + r_M(90\text{ kg})\left(9.8\text{ m/s}^2\right)\sin 90° + (2\text{ m})(30\text{ kg})\left(9.8\text{ m/s}^2\right)\sin 90°$$
$$\quad + (0.25\text{ m})(100\text{ kg})\left(9.8\text{ m/s}^2\right)\sin 90°$$
$$0\text{ N}\cdot\text{m} = -(4\text{ m})(1{,}577.2\text{ N})(0) - (4\text{ m})(394.3\text{ N})(1)$$
$$\quad + r_M(90\text{ kg})\left(9.8\text{ m/s}^2\right)(1) + (2\text{ m})(30\text{ kg})\left(9.8\text{ m/s}^2\right)(1)$$
$$\quad + (0.25\text{ m})(100\text{ kg})\left(9.8\text{ m/s}^2\right)(1)$$
$$0\text{ N}\cdot\text{m} = 0\text{ N}\cdot\text{m} - 1{,}577.2\text{ N}\cdot\text{m} + (882\text{ N})r_M + 588\text{ N}\cdot\text{m} + 245\text{ N}\cdot\text{m}$$
$$0\text{ N}\cdot\text{m} = -744.2\text{ N}\cdot\text{m} + (882\text{ N})r_M$$
$$744.2\text{ N}\cdot\text{m} = (882\text{ N})r_M$$
$$0.844\text{ m} = r_M$$

With respect to the question asked, the worker must walk $(4\text{ m}) - (0.844\text{ m}) = 3.156\text{ m} \approx 3.2\text{ m}$ away from the wall before the rope snaps.

361. 91 kg

First, draw the forces acting on the ladder:

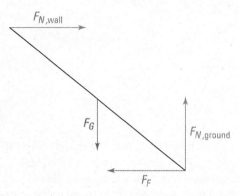

Illustration by Thomson Digital

F_G is the force of gravity acting on the ladder (when the mass is uniformly distributed in an object, the force is located in the middle of the object), $F_{N,\text{wall}}$ is the normal force exerted by the wall, $F_{N,\text{ground}}$ is the normal force exerted by the ground, and F_F is the force of static friction. (No static friction works on the *wall* because it's designated "smooth.") Friction always opposes motion, so — by using common sense that the ladder would move "right" if it slipped — the force of friction here must point "left."

You have enough data that you have the option of solving for the ladder's mass using either Newton's second law in the vertical and horizontal directions (two simpler equations) or the more complex torque equation. The solution presented here uses the latter method because it requires only one equation, and it uses the point of contact between the ladder and the ground as the axis of rotation. Draw a picture of the situation to make sure you use the correct angles with each force:

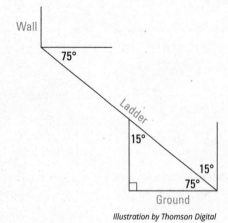

Illustration by Thomson Digital

$$\tau_{net} = \sum \tau$$
$$= \left(-\tau_{N,wall} \right) + \tau_G$$
$$= -\tau_{N,wall} + \tau_G$$
$$= -r_{N,wall} F_{N,wall} \sin \theta_{N,wall} + r_G F_G \sin \theta_A$$
$$= -r_{N,wall} F_{N,wall} \sin \theta_{N,wall} + r_G mg \sin \theta_G$$
$$0\,N\cdot m = -(2.8\,m)(120\,N)\sin 75° + (1.4\,m)m\left(9.8m/s^2\right)\sin 15°$$
$$0\,N\cdot m = -(2.8\,m)(120\,N)(0.966) + (1.4\,m)m\left(9.8m/s^2\right)(0.259)$$
$$0\,N\cdot m = -324.6\,N\cdot m + \left(3.55\,m^2/s^2\right)m$$
$$324.6\,N\cdot m = \left(3.55\,m^2/s^2\right)m$$
$$91.4\,kg = m$$

362.   3.1 m

First, draw the forces acting on the ladder:

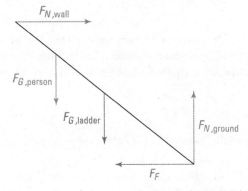

Illustration by Thomson Digital

$F_{G,ladder}$ is the force of gravity acting on the ladder (when the mass is uniformly distributed in an object, the force is located in the middle of the object), $F_{G,person}$ is the force of gravity acting on Joey, $F_{N,wall}$ is the normal force exerted by the wall, $F_{N,ground}$ is the normal force exerted by the ground, and F_F is the force of static friction. Friction always opposes motion, so by using common sense that the ladder would move "right" if it slipped, the force of friction here must point "left."

The net force in the horizontal direction, the net force in the vertical direction, and the torque about any point on the ladder must all equal 0 if the ladder is to remain in equilibrium (not slip). Use Newton's second law to analyze the forces, first in the vertical (y) direction:

$$F_{y,net} = \sum_y F_y$$
$$= F_{N,ground} - F_{G,ladder} - F_{G,person}$$
$$= F_{N,ground} - m_{ladder}g - m_{person}g$$
$$0\,N = F_{N,ground} - (35\,kg)\left(9.8\,m/s^2\right) - (95\,kg)\left(9.8\,m/s^2\right)$$
$$0\,N = F_{N,ground} - 343\,N - 931\,N$$
$$0\,N = F_{N,ground} - 1,274\,N$$
$$1,274\,N = F_{N,ground}$$

Then analyze the torque about an axis of rotation on the ladder. Any point along the platform suffices, but the easiest calculation comes from choosing an axis where a force is being exerted. You need to analyze the trigonometry of the scenario to obtain the correct angles:

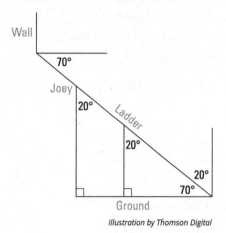

Illustration by Thomson Digital

The following solution uses the point of contact between the ladder and the wall as the axis of rotation. The length of the ladder is labeled l. Given the sign convention that clockwise-oriented forces produce negative torques and vice versa, the result is the following:

$$\tau_{net} = \sum \tau$$

$$= \left(-\tau_{G,person}\right) + \left(-\tau_{G,ladder}\right) + \left(-\tau_F\right) + \tau_{N,ground}$$

$$= -\tau_{G,person} - \tau_{G,ladder} - \tau_F + \tau_{N,ground}$$

$$= -r_{G,person} F_{G,person} \sin \theta_{G,person} - r_{G,ladder} F_{G,ladder} \sin \theta_{G,ladder}$$
$$- r_F F_F \sin \theta_F + r_{N,ground} F_{N,ground} \sin \theta_{N,ground}$$

$$= -r_{person} m_{person} g \sin \theta_{person} - r_{ladder} m_{ladder} g \sin \theta_{ladder}$$
$$- r_F \mu F_{N,ground} \sin \theta_F + r_{N,ground} F_{N_1 ground} \sin \theta_{N,ground}$$

$$0\,\text{N} \cdot \text{m} = -(1\,\text{m})(95\,\text{kg})\left(9.8\,\text{m/s}^2\right)\sin 20° - \frac{l}{2}(35\,\text{kg})\left(9.8\,\text{m/s}^2\right)\sin 20°$$
$$- l(0.23)(1{,}274\,\text{N})\sin 70° + l(1{,}274\,\text{N})\sin 20°$$

$$0\,\text{N} \cdot \text{m} = -(1\,\text{m})(95\,\text{kg})\left(9.8\,\text{m/s}^2\right)(0.342) - \frac{l}{2}(35\,\text{kg})\left(9.8\,\text{m/s}^2\right)(0.342)$$
$$- l(0.23)(1{,}274\,\text{N})(0.94) + l(1{,}274\,\text{N})(0.342)$$

$$0\,\text{N} \cdot \text{m} = -318.4\,\text{N} \cdot \text{m} - (58.65\,\text{N})l - (275.44\,\text{N})l + (435.71\,\text{N})l$$

$$0\,\text{N} \cdot \text{m} = -318.4\,\text{N} \cdot \text{m} + (101.62\text{N})l$$

$$318.4\,\text{N} \cdot \text{m} = (101.62\,\text{N})l$$

$$3.1\,\text{m} = l$$

363. 210 N

First, draw the forces acting on the sign:

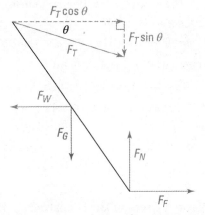

Illustration by Thomson Digital

F_G is the force of gravity acting on the sign, F_W is the force exerted by the wind, F_T is the force of tension in the rope, F_N is the normal force exerted by the ground, and F_F is the force of static friction between the sign and the ground. Friction always opposes motion, so if the wind is pushing the sign "left," then the force of friction must point "right."

The net force in the horizontal direction, the net force in the vertical direction, and the torque about any point on the sign must all equal 0 if the sign is to remain in equilibrium (not slip). Start by analyzing the torque about an axis of rotation on the ladder. Any point along the platform suffices, but the easiest calculation results from choosing an axis where a force is being exerted. You need to analyze the trigonometry of the scenario to obtain the correct angles:

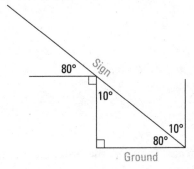

Illustration by Thomson Digital

The following solution uses the point of contact between the sign and the rope as the axis of rotation to eliminate the variable θ from the equation (forces exerted at the point of rotation provide no torque). Given the sign convention that clockwise-oriented forces produce negative torques and vice versa, the result is the following:

$$\tau_{net} = \sum \tau$$
$$= (-\tau_W) + (-\tau_G) + \tau_F + \tau_N$$
$$= -\tau_W - \tau_G + \tau_F + \tau_N$$
$$= -r_W F_W \sin\theta_W - r_G F_G \sin\theta_G + r_F F_F \sin\theta_F + r_N F_N \sin\theta_N$$
$$= -r_W F_W \sin\theta_W - r_G mg \sin\theta_G + r_F \mu F_N \sin\theta_F + r_N F_N \sin\theta_N$$
$$0\,\text{N} \cdot \text{m} = -(7\,\text{m})(180\,\text{N})\sin 80° - (7\,\text{m})(6\,\text{kg})(9.8\,\text{m/s}^2)\sin 10°$$
$$+ (14\,\text{m})(0.24)F_N \sin 80° + (14\,\text{m})F_N \sin 10°$$
$$0\,\text{N} \cdot \text{m} = -(7\,\text{m})(180\,\text{N})(0.985) - (7\,\text{m})(6\,\text{kg})(9.8\,\text{m/s}^2)(0.174)$$
$$+ (14\,\text{m})(0.24)F_N (0.985) + (14\,\text{m})F_N (0.174)$$
$$0\,\text{N} \cdot \text{m} = -1,241.1\,\text{N} \cdot \text{m} - 71.6\,\text{N} \cdot \text{m} + (3.31\,\text{m})F_N + (2.44\,\text{m})F_N$$
$$0\,\text{N} \cdot \text{m} = -1,312.7\,\text{N} \cdot \text{m} + (5.75\,\text{m})F_N$$
$$1,312.7\,\text{N} \cdot \text{m} = (5.75\,\text{m})F_N$$
$$228.3\,\text{N} = F_N$$

Now use this result along with Newton's second law to analyze the forces, first in the vertical (y) direction (*Note:* The variable θ in these calculations is the angle that the *rope* makes with the ground, not the one that the *sign* makes with the ground):

$$F_{y,net} = \sum F_y$$
$$= F_N - F_G - F_T \sin\theta$$
$$= F_N - mg - F_T \sin\theta$$
$$0\,\text{N} = (228.3\,\text{N}) - (6\,\text{kg})(9.8\,\text{m/s}^2) - F_T \sin\theta$$
$$0\,\text{N} = 228.3\,\text{N} - 58.8\,\text{N} - F_T \sin\theta$$
$$0\,\text{N} = 169.5\,\text{N} - F_T \sin\theta$$
$$-169.5\,\text{N} = -F_T \sin\theta$$
$$169.5\,\text{N} = F_T \sin\theta$$

Then look at the horizontal (x) direction:

$$F_{x,net} = \sum F_x$$
$$= -F_W + F_F + F_T \cos\theta$$
$$= -F_W + \mu F_N + F_T \cos\theta$$
$$0\,\text{N} = -(180\,\text{N}) + (0.24)(228.3\,\text{N}) + F_T \cos\theta$$
$$0\,\text{N} = -180\,\text{N} + 54.8\,\text{N} + F_T \cos\theta$$
$$0\,\text{N} = -125.2\,\text{N} + F_T \cos\theta$$
$$125.2\,\text{N} = F_T \cos\theta$$

Now divide the result from the vertical forces by the result from the horizontal forces:

$$\frac{169.5\,\text{N} = F_T \sin\theta}{125.2\,\text{N} = F_T \cos\theta} \Rightarrow$$
$$\frac{169.5\,\text{N}}{125.2\,\text{N}} = \frac{F_T \sin\theta}{F_T \cos\theta}$$
$$1.35 = \frac{\sin\theta}{\cos\theta} = \tan\theta$$
$$\tan^{-1}(1.35) = \theta$$
$$53.5° = \theta$$

Substitute this back into one of the force results to solve for the tension force in the rope:

$$125.2 \, \text{N} = F_T \cos\theta$$
$$125.2 \, \text{N} = F_T \cos 53.5°$$
$$125.2 \, \text{N} = F_T (0.595)$$
$$125.2 \, \text{N} = 0.595 F_T$$
$$210 \, \text{N} = F_T$$

364.   ground; 0.2

First, figure out where Lenny is on the ladder. If the ladder has 15 steps, then you have 14 gaps "on" the ladder, and each one is $\dfrac{3.4 \, \text{m}}{14} = 0.243$ m wide. The 13th step is 12 gaps from the bottom rung, so it must be $12(0.243 \, \text{m}) = 2.92$ m from the bottom rung. Now draw all the forces acting on the ladder:

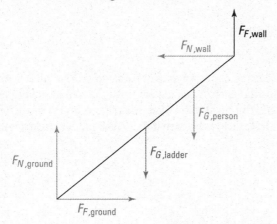

Illustration by Thomson Digital

$F_{G,\text{ladder}}$ is the force of gravity acting on the ladder (when the mass is uniformly distributed in an object, the force is located in the middle of the object), $F_{G,\text{person}}$ is the force of gravity acting on Lenny, $F_{N,\text{wall}}$ is the normal force exerted by the wall, $F_{F,\text{wall}}$ is the force of static friction between the ladder and the wall, $F_{N,\text{ground}}$ is the normal force exerted by the ground, and $F_{F,\text{ground}}$ is the force of static friction between the ladder and the ground. Friction always opposes motion, so by using common sense that the ladder would move "left" and "down" if it slipped, the forces of friction here must point "right" and "up."

The net force in the horizontal direction, the net force in the vertical direction, and the torque about any point on the ladder must all equal 0 if the ladder is to remain in equilibrium (not slip). Use Newton's second law to analyze the forces, first in the vertical (y) direction:

$$F_{y,\text{net}} = \sum F_y$$
$$= F_{N,\text{ground}} - F_{G,\text{ladder}} - F_{G,\text{person}} + F_{F,\text{wall}}$$
$$= F_{N,\text{ground}} - m_{\text{ladder}}\, g - m_{\text{person}}\, g + \mu_{\text{wall}}\, F_{N,\text{wall}}$$
$$0\,\text{N} = (820\,\text{N}) - (26\,\text{kg})\left(9.8\,\text{m/s}^2\right)$$
$$- (110\,\text{kg})\left(9.8\,\text{m/s}^2\right) + \mu_{\text{wall}}F_{N,\text{wall}}$$
$$0\,\text{N} = 820\,\text{N} - 254.8\,\text{N} - 1{,}078\,\text{N} + \mu_{\text{wall}}F_{N,\text{wall}}$$
$$0\,\text{N} = -512.8\,\text{N} + \mu_{\text{wall}}\, F_{N,\text{wall}}$$
$$512.8\,\text{N} = \mu_{\text{wall}}F_{N,\text{wall}}$$
$$\frac{512.8\,\text{N}}{\mu_{\text{wall}}} = F_{N,\text{wall}}$$

Then look at the horizontal (x) direction:

$$F_{x,\text{net}} = \sum F_x$$
$$= F_{F,\text{ground}} - F_{N,\text{wall}}$$
$$= \mu_{\text{ground}}\, F_{N,\text{ground}} - F_{N,\text{wall}}$$
$$0\,\text{N} = \mu_{\text{ground}}\,(820\,\text{N}) - F_{N,\text{wall}}$$
$$F_{N,\text{wall}} = (820\,\text{N})\mu_{\text{ground}}$$

Finally, analyze the torque about an axis of rotation on the ladder. Any point along the platform suffices, but the easiest calculation results from choosing an axis where a force is being exerted. You need to analyze the trigonometry of the scenario to obtain the correct angles. First, tackle the actual angle the ladder makes with the ground:

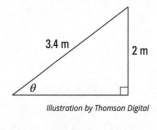

Illustration by Thomson Digital

$$\sin\theta = \frac{2\,\text{m}}{3.4\,\text{m}}$$
$$\sin\theta = 0.588$$
$$\theta = \sin^{-1}(0.588)$$
$$\theta = 36°$$

Thus, the angles relative to the forces are

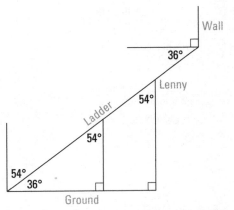

Wall

36°

Lenny

54°

Ladder

54°

54°

36°

Ground

Illustration by Thomson Digital

The following solution uses the point of contact between the ladder and the ground as the axis of rotation. Given the sign convention that clockwise-oriented forces produce negative torques and vice versa, the result is the following:

$$\tau_{net} = \sum \tau$$

$$= \left(-\tau_{G,person}\right) + \left(-\tau_{G,ladder}\right) + \tau_{N,wall} + \tau_{F,wall}$$

$$= -\tau_{G,person} - \tau_{G,ladder} + \tau_{N,wall} + \tau_{F,wall}$$

$$= -r_{G,person}F_{G,person}\sin\theta_{G,person} - r_{G,ladder}F_{G,ladder}\sin\theta_{G,ladder}$$

$$+ r_{N,wall}F_{N,wall}\sin\theta_{N,wall} + r_{F,wall}F_{F,wall}\sin\theta_{F,wall}$$

$$= -r_{person}m_{person}g\sin\theta_{person} - r_{ladder}m_{ladder}g\sin\theta_{ladder}$$

$$+ r_{N,wall}F_{N,wall}\sin\theta_{N,wall} + r_{F,wall}\mu_{wall}F_{N,wall}\sin\theta_{F,wall}$$

$$0\,\text{N}\cdot\text{m} = -(2.92\,\text{m})(110\,\text{kg})\left(9.8\,\text{m/s}^2\right)\sin 54°$$

$$-(1.7\,\text{m})(26\,\text{kg})\left(9.8\,\text{m/s}^2\right)\sin 54°$$

$$+(3.4\,\text{m})F_{N,wall}\sin 36° + (3.4\,\text{m})\mu_{wall}F_{N,wall}\sin 126°$$

$$0\,\text{N}\cdot\text{m} = -(2.92\,\text{m})(110\,\text{kg})\left(9.8\,\text{m/s}^2\right)(0.809)$$

$$-(1.7\,\text{m})(26\,\text{kg})\left(9.8\,\text{m/s}^2\right)(0.809)$$

$$+(3.4\,\text{m})F_{N,wall}(0.588) + (3.4\,\text{m})\mu_{wall}F_{N,wall}(0.809)$$

$$0\,\text{N}\cdot\text{m} = -2,546.5\,\text{N}\cdot\text{m} - 350.4\,\text{N}\cdot\text{m} + (2\,\text{m})F_{N,wall} + (2.75\,\text{m})\mu_{wall}F_{N,wall}$$

$$0\,\text{N}\cdot\text{m} = -2,896.9\,\text{N}\cdot\text{m} + (2\,\text{m})F_{N,wall} + (2.75\,\text{m})\mu_{wall}F_{N,wall}$$

$$2,896.9\,\text{N}\cdot\text{m} = (2\,\text{m})F_{N,wall} + (2.75\,\text{m})\mu_{wall}F_{N,wall}$$

$$2,896.9\,\text{N}\cdot\text{m} = F_{N,wall}\left((2\,\text{m}) + (2.75\,\text{m})\mu_{wall}\right)$$

$$\frac{2,896.9\,\text{N}}{\left(2 + 2.75\mu_{wall}\right)} = F_{N,wall}$$

Then set this equation equal to the result from the vertical force analysis — which has the same two unknowns — to solve for the value of μ_{wall}:

$$\frac{512.8\ \text{N}}{\mu_{\text{wall}}} = F_{N,\text{wall}} = \frac{2{,}896.9\ \text{N}}{\left(2 + 2.75\ \mu_{\text{wall}}\right)}$$

$$\frac{512.8\ \text{N}}{\mu_{\text{wall}}} = \frac{2{,}896.9\ \text{N}}{\left(2 + 2.75\ \mu_{\text{wall}}\right)}$$

$$(512.8\,\text{N})\,(2 + 2.275\,\mu_{\text{wall}}) = (2{,}896.9\ \text{N})\,\mu_{\text{wall}}$$

$$1{,}025.6\ \text{N} + (1{,}410.2\ \text{N})\mu_{\text{wall}} = (2{,}896.9\ \text{N})\mu_{\text{wall}}$$

$$1{,}025.6\ \text{N} = (1{,}486.7\ \text{N})\mu_{\text{wall}}$$

$$0.69 = \mu_{\text{wall}}$$

Therefore,

$$F_{N,\text{wall}} = \frac{512.8\ \text{N}}{\mu_{\text{wall}}}$$

$$= \frac{512.8\ \text{N}}{0.69}$$

$$= 743.2\ \text{N}$$

Substitute this value into the result from the horizontal force analysis to solve for the value of μ_{ground}:

$$F_{N,\text{wall}} = (820\ \text{N})\mu_{\text{ground}}$$

$$743.2\ \text{N} = (820\ \text{N})\mu_{\text{ground}}$$

$$0.91 = \mu_{\text{ground}}$$

So the coefficient of static friction on the ground is larger than that on the wall, by a difference of:

$$0.91 - 0.69 = 0.22 \approx 0.2$$

365.   0.24 m/s²

Use the conversion formula common in all linear-to-angular conversions. In the case of acceleration, $a = r\alpha$, where a is the linear, or tangential acceleration; r is the radius of the rotating object; and α is the angular acceleration.

$$a = r\alpha$$

$$= (0.12\ \text{m})\left(2\ \text{rad/s}^2\right)$$

$$= 0.24\ \text{m/s}^2$$

(A radian is a dimensionless unit that vanishes whenever another unit such as meters is around to take its place.)

366.   64 g

Use the angular acceleration form of the torque formula, $\tau = mr^2\,\alpha$, where m is the mass, r is radius of the circular path, and α is the angular acceleration. Before substituting values, convert the disc's diameter and calculate the length of the radius:

$$(16\ \text{cm})\left(\frac{1\ \text{m}}{100\ \text{cm}}\right) = 0.16\ \text{m}$$

Radius is half of the diameter, so:

$$r = \frac{1}{2}(0.16\ \text{m})$$

$$= 0.08\ \text{m}$$

Now substitute the known values to solve for the ant's mass:

$$\tau = mr^2\alpha$$
$$0.02\,\text{N}\cdot\text{m} = m(0.08\,\text{m})^2\left(49\,\text{rad/s}^2\right)$$
$$0.02\,\text{N}\cdot\text{m} = \left(0.314\text{m}^2/\text{s}^2\right)m$$
$$0.064\,\text{kg} = m$$

Finally, convert into the required units:

$$(0.064\,\text{kg})\left(\frac{1{,}000\,\text{g}}{1\,\text{kg}}\right) = 64\,\text{g}$$

---

367. $1.4\times10^4\,\text{N}\cdot\text{m}$

To calculate the torque with the formula $a\tau = mr^2\alpha$, you need Jennifer's mass (m), the Ferris wheel's radius (r), and the wheel's angular acceleration (α).

Use the fact that "weight" is the vernacular term used for the force of Earth's gravity on an object. Therefore,

$$W = F_G$$
$$W = ma_G$$
$$W = mg$$
$$480\,\text{N} = m\left(9.8\text{m/s}^2\right)$$
$$48.98\,\text{kg} = m$$

To find the wheel's radius, use geometry: one complete circular path is called a circumference, and its formula is $C = 2\pi r$, where r is the radius of the circle involved. The Ferris wheel makes four revolutions — four circumferences — and the total length of these revolutions is 260 meters. Therefore,

$$260\,\text{m} = 4C = 4(2\pi r)$$
$$260\,\text{m} = 8\pi r$$
$$10.35\,\text{m} = r$$

To get the final piece for the torque formula, use the angular velocity displacement formula to solve for the angular acceleration: $\omega_f^2 = \omega_i^2 + 2\alpha\theta$, where ω_i and ω_f are the initial and final angular velocities, respectively; α is the angular acceleration; and θ is the angular displacement. Any circular revolution has 2π radians of angular displacement, so four revolutions is equal to $4(2\pi) = 8\pi$ radians of angular displacement.

$$\omega_f^2 = \omega_i^2 + 2\alpha\theta$$
$$(12\,\text{rad/s})^2 = (3\,\text{rad/s})^2 + 2\alpha(8\pi\,\text{rad})$$
$$144\,\text{rad}^2/\text{s}^2 = 9\,\text{rad}^2/\text{s}^2 + (50.63\,\text{rad})\alpha$$
$$135\,\text{rad}^2/\text{s}^2 = (50.63\,\text{rad})\alpha$$
$$2.67\,\text{rad/s}^2 = \alpha$$

Therefore, the torque Jennifer experiences is

$$\tau = mr^2\alpha$$
$$= (48.98\,\text{kg})(10.35\,\text{m})^2(2.67\,\text{rad/s})$$
$$= 1.4\times10^4\,\text{N}\cdot\text{m}$$

368.   mass

You can calculate a particle's moment of inertia using this formula:

$$I = mr^2$$

I represents the moment of inertia, m represents the particle's mass, and r is the distance from the particle to the axis on which it spins.

369. $14\ \text{kg} \cdot \text{m}^2$

The system's total moment of inertia is the sum of the individual moments of inertia:

$$I_{\text{total}} = I_{\text{black}} + I_{\text{silver}}$$

The direction of rotation has no impact on the moment of inertia.

Use the moment of inertia formula on each ball bearing, and substitute the values into the previous equation. You can use the particle version of the formula $I = mr^2$, where m represents the particle's mass and r is the distance from the particle to the axis on which it spins, because the bearings have small volumes relative to the distance between them and the pole:

$$\begin{aligned}
I_{\text{black}} &= m_{\text{black}} r_{\text{black}}^2 \\
&= (2.8\ \text{kg})(1.8\ \text{m})^2 \\
&= 9.07\ \text{kg} \cdot \text{m}^2
\end{aligned}$$

$$\begin{aligned}
I_{\text{silver}} &= m_{\text{silver}} r_{\text{silver}}^2 \\
&= (3.3\ \text{kg})(1.2\ \text{m})^2 \\
&= 4.75\ \text{kg} \cdot \text{m}^2
\end{aligned}$$

Therefore,

$$\begin{aligned}
I_{\text{total}} &= I_{\text{black}} + I_{\text{silver}} \\
&= \left(9.07\ \text{kg} \cdot \text{m}^2\right) + \left(4.75\ \text{kg} \cdot \text{m}^2\right) \\
&= 13.82\ \text{kg} \cdot \text{m}^2 \approx 14\ \text{kg} \cdot \text{m}^2
\end{aligned}$$

370. $82\ \text{N} \cdot \text{m}$

Start with the forces acting on the black ball bearing by analyzing a free-body diagram.

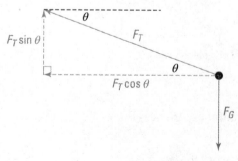

Illustration by Thomson Digital

Use the force form of Newton's second law in the vertical direction:

$$F_{y,net} = \sum F_y$$
$$ma_y = F_T \sin\theta - F_G$$
$$ma_y = F_T \sin\theta - mg$$
$$(2.8\text{ kg})(0\text{ m/s}^2) = (260\text{ N})\sin\theta - (2.8\text{ kg})(9.8\text{ m/s}^2)$$
$$0\text{ N} = (260\text{ N})\sin\theta - 27.44\text{ N}$$
$$27.44\text{ N} = (260\text{ N})\sin\theta$$
$$0.106 = \sin\theta$$
$$\sin^{-1}(0.106) = \theta$$
$$6.08° = \theta$$

Therefore, the angle between the string and the force of gravity providing the torque is $90° + 6.08° = 96.08°$ (its supplement, $83.92°$, is also appropriate for use).

Next, use trigonometry to find an expression for the distance between the bearing at the axis of rotation, the pole, labeled r here:

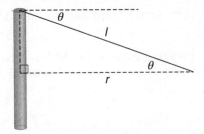

Illustration by Thomson Digital

$$\cos\theta = \frac{r}{l}$$
$$l\cos\theta = r$$

Substitute this result into the moment of inertia formula, $I = mr^2$, where m represents the particle's mass and r is the distance from the particle to the axis on which it spins, to solve for l, the length of the string:

$$I = mr^2$$
$$I = m(l\cos\theta)^2$$
$$I = ml^2\cos^2\theta$$
$$25\text{ kg}\cdot\text{m}^2 = (2.8\text{ kg})l^2(\cos 6.08°)^2$$
$$25\text{ kg}\cdot\text{m}^2 = (2.8\text{ kg})l^2(0.994)^2$$
$$25\text{ kg}\cdot\text{m}^2 = (2.77\text{ kg})l^2$$
$$9.025\text{ m}^2 = l^2$$
$$3\text{ m} = l$$

Armed with the distance from the pole to the force, as well as the angle between the distance and force vectors, solve the torque formula:

$$\tau = lF\sin\theta$$
$$= lF_G\sin\theta$$
$$= lmg\sin\theta$$
$$= (3\,\mathrm{m})(2.8\,\mathrm{kg})\left(9.8\,\mathrm{m/s^2}\right)\sin 96.08°$$
$$= (3\,\mathrm{m})(2.8\,\mathrm{kg})\left(9.8\,\mathrm{m/s^2}\right)(0.994)$$
$$= 81.8\,\mathrm{N\cdot m}$$

371. $3\alpha_i$

The amount of torque a force produces is directly proportional to that force's magnitude. Torque is also directly proportional to the resulting angular acceleration. Therefore, multiplying the force by 3 also multiplies the torque by 3 and, thereby, the angular acceleration by 3 as well.

372. 53 N

Torque has two main formulas: one in terms of force and the other in terms of angular acceleration:

$\tau = rF\sin\theta$ (r is the distance from center of rotation, F is the force that produces the torque, and θ is the angle between the two aforementioned vectors)

and

$\tau = I\alpha$ (I is the moment of inertia and α is the angular acceleration produced by the torque).

The angle is 90 degrees in this problem because the force is exerted *tangentially* to the wheel. Therefore,

$$rF\sin\theta = \tau = I\alpha$$
$$rF\sin\theta = I\alpha$$
$$(0.43\,\mathrm{m})F_F\sin 90° = \left(12\,\mathrm{kg\cdot m^2}\right)\left(1.9\,\mathrm{rad/s^2}\right)$$
$$(0.43\,\mathrm{m})F_F(1) = \left(12\,\mathrm{kg\cdot m^2}\right)\left(1.9\,\mathrm{rad/s^2}\right)$$
$$(0.43\,\mathrm{m})F_F = 22.8\,\mathrm{N\cdot m}$$
$$F_F = 53\,\mathrm{N}$$

373. $5\,\mathrm{kg\cdot m^2}$

Use the torque formula, $\tau = rF\sin\theta$, to solve for the pulley's radius. Because the force pulling down is tangential to the radius, $\theta = 90$ degrees:

$$\tau = rF\sin\theta$$
$$12\,\mathrm{N\cdot m} = r(20\,\mathrm{N})\sin 90°$$
$$12\,\mathrm{N\cdot m} = r(20\,\mathrm{N})(1)$$
$$12\,\mathrm{N\cdot m} = (20\,\mathrm{N})r$$
$$0.6\,\mathrm{m} = r$$

Then use the formula for a solid cylinder's moment of inertia (a disk such as a pulley is geometrically just a very short cylinder), $I = \frac{1}{2}mr^2$, where m is the pulley's mass and r is its radius:

$$I = \frac{1}{2}mr^2$$
$$= \frac{1}{2}(28\text{ kg})(0.6\text{ m})^2$$
$$= 5.04\text{ kg}\cdot\text{m}^2 \approx 5\text{ kg}\cdot\text{m}^2$$

---

374. 9.5 rad/s^2

First, convert the radius into "correct" units:

$$(12\text{ cm})\left(\frac{1\text{ m}}{100\text{ cm}}\right) = 0.12\text{ m}$$

When a pulley isn't frictionless, you must account for the torque provided by the moving string that will rotate it. Whereas in frictionless pulley problems, where the tension in a string may be seen as the same magnitude throughout the string, when a pulley rotates, the tensions must be different or else there would be no net force to provide the torque to rotate the pulley.

Therefore, four unknowns are present: *three* tension forces and the system's acceleration. You need four equations to solve the problem. Start by drawing free-body diagrams for the three masses and the pulley:

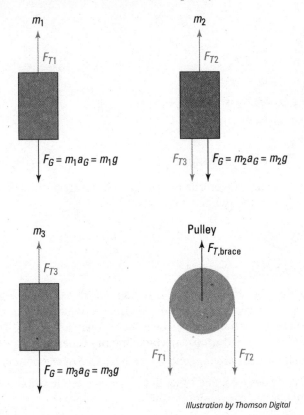

Illustration by Thomson Digital

The following solution designates the "positive" direction as that in which masses 2 and 3 fall and mass 1 rises.

Write Newton's law for each of the bodies. For m_1:

$$F_{y,net} = \sum F_y$$
$$m_1 a = F_{T1} - F_G$$
$$m_1 a = F_{T1} - m_1 g$$
$$(50\,\text{kg})a = F_{T1} - (50\,\text{kg})\left(9.8\,\text{m/s}^2\right)$$
$$(50\,\text{kg})a = F_{T1} - 490\,\text{N}$$
$$(50\,\text{kg})a + 490\,\text{N} = F_{T1}$$

For m_2:

$$F_{y,net} = \sum F_y$$
$$m_2 a = F_{T3} + F_G - F_{T2}$$
$$m_2 a = F_{T3} + m_2 g - F_{T2}$$
$$(25\,\text{kg})a = F_{T3} + (25\,\text{kg})\left(9.8\,\text{m/s}^2\right) - F_{T2}$$
$$(25\,\text{kg})a = F_{T3} + 245\,\text{N} - F_{T2}$$
$$(25\,\text{kg})a - 245\,\text{N} = F_{T3} - F_{T2}$$

For m_3:

$$F_{y,net} = \sum F_y$$
$$m_3 a = F_G - F_{T3}$$
$$m_3 a = m_3 g - F_{T3}$$
$$(40\,\text{kg})a = (40\,\text{kg})\left(9.8\,\text{m/s}^2\right) - F_{T3}$$
$$(40\,\text{kg})a = 392\,\text{N} - F_{T3}$$
$$F_{T3} + (40\,\text{kg})a = 392\,\text{N}$$
$$F_{T3} = 392\,\text{N} - (40\,\text{kg})a$$

And for the pulley, focus only on the forces that will create a torque around the pulley's axis of rotation — the two tension forces:

$$F_{net} = \sum F$$
$$= \left(-F_{T1}\right) + F_{T2}$$
$$= -F_{T1} + F_{T2} = F_{T2} - F_{T1}$$

You need to use this force in the torque formula $\tau = rF \sin\theta$. Torque is also the product of inertia and angular acceleration ($\tau = I\alpha$); set the two equations equal to one another to develop a numerical relationship between the two tension forces. The next steps make three significant substitutions: 1) The moment of inertia for a solid cylinder is $\frac{1}{2}mr^2$; 2) the tension forces are tangential to the surface of the pulley, so the angle between r and F is 90 degrees; and 3) tangential/linear acceleration and angular acceleration are related by the conversion formula $a = r\alpha$, or $\frac{a}{r} = \alpha$:

$$rF \sin \theta = \tau = I\alpha$$
$$rF \sin \theta = I\alpha$$
$$r\left(F_{T2} - F_{T1}\right)\sin \theta = \left(\frac{1}{2}m_{\text{pulley}}r^2\right)\left(\frac{a}{r}\right)$$
$$r\left(F_{T2} - F_{T1}\right)\sin \theta = \frac{1}{2}m_{\text{pulley}}ra$$
$$\left(F_{T2} - F_{T1}\right)\sin \theta = \frac{1}{2}m_{\text{pulley}}a$$
$$\left(F_{T2} - F_{T1}\right)\sin 90° = \frac{1}{2}(28 \text{ kg})a$$
$$\left(F_{T2} - F_{T1}\right)(1) = \frac{1}{2}(28 \text{ kg})a$$
$$F_{T2} - F_{T1} = (14 \text{ kg})a$$

To solve this complex system of equations, first substitute the result for m_3 into the one for m_2:

$$(25 \text{ kg})a - 245 \text{ N} = F_{T3} - F_{T2}$$
$$(25 \text{ kg})a - 245 \text{ N} = (392 \text{ N} - (40 \text{ kg})a) - F_{T2}$$
$$(25 \text{ kg})a - 245 \text{ N} = 392 \text{ N} - (40 \text{ kg})a - F_{T2}$$
$$(65 \text{ kg})a - 637 \text{ N} = -F_{T2}$$
$$-(65 \text{ kg})a + 637 \text{ N} = F_{T2}$$

Then substitute this result, as well as the one from m_1, into the result from the pulley:

$$F_{T2} - F_{T1} = (14 \text{ kg})a$$
$$(-(65 \text{ kg})a + 637 \text{ N}) - ((50 \text{ kg})a + 490 \text{ N}) = (14 \text{ kg})a$$
$$-(65 \text{ kg})a + 637 \text{ N} - (50 \text{ kg})a - 490 \text{ N} = (14 \text{ kg})a$$
$$-(115 \text{ kg})a + 147 \text{ N} = (14 \text{ kg})a$$
$$147 \text{ N} = (129 \text{ kg})a$$
$$1.14 \text{ m/s}^2 = a$$

Finally, use the conversion relationship between angular and linear/tangential acceleration to solve for the former:

$$a = r\alpha$$
$$1.14 \text{ m/s}^2 = (0.12 \text{ m})\alpha$$
$$9.5 \text{ rad/s}^2 = \alpha$$

375. B; 2 times

A combination of the angular work formula $W = \tau\theta$ — where τ is torque and θ is angular displacement — and the force torque formula $\tau = rF$ — where r is the radius of the wheel and F is the tangential force exerted — results in the following relationship:

$$W = \tau\theta$$
$$W = (rF)\theta$$
$$W = rF\theta$$
$$\frac{W}{F\theta} = r$$

Therefore, the radius is indirectly proportional to angular displacement, the fancy term for how much something rotates. If radius is multiplied by 2, then angular displacement is divided by 2. Wheel A, which has the larger radius, has the smaller displacement. Wheel B rotates twice as much as does wheel A.

376. 0.52 J

First, convert the centimeter lengths into meters:

$$(9 \text{ cm})\left(\frac{1 \text{ m}}{100 \text{ cm}}\right) = 0.09 \text{ m}$$

$$(130 \text{ cm})\left(\frac{1 \text{ m}}{100 \text{ cm}}\right) = 1.3 \text{ m}$$

Next, use the angular velocity displacement formula to calculate the necessary angular acceleration required to change the motor's angular velocity within 130 centimeters. Before substituting, convert that linear displacement into an angular one:

$$s = r\theta$$
$$1.3 \text{ m} = (0.09 \text{m})\theta$$
$$14.4 \text{ rad} = \theta$$
$$\omega_f^2 = \omega_i^2 + 2\alpha\theta$$
$$(12 \text{ rad/s})^2 = (0 \text{ rad/s})^2 + 2\alpha(14.4 \text{ rad})$$
$$144 \text{ rad}^2/\text{s}^2 = 0 \text{ rad}^2/\text{s}^2 + (28.8 \text{ rad})\alpha$$
$$144 \text{ rad}^2/\text{s}^2 = (28.8 \text{ rad})\alpha$$
$$5 \text{ rad/s}^2 = \alpha$$

The formula for a solid cylinder's moment of inertia is $I = \frac{1}{2}mr^2$, and one of the formulas for torque is $\tau = I\alpha$. Substitute the motor's mass, radius, and recently computed angular acceleration to solve for the torque:

$$\tau = I\alpha$$
$$= \left(\frac{1}{2}mr^2\right)\alpha$$
$$= \frac{1}{2}mr^2\alpha$$
$$= \frac{1}{2}(1.8 \text{ kg})(0.09 \text{ m})^2\left(5 \text{ rad/s}^2\right)$$
$$= 0.036 \text{ N}\cdot\text{m}$$

Finally, use the work formula for rotational motion $W = \tau\theta$, where τ is torque and θ is angular displacement:

$$W = \tau\theta$$
$$= (0.036 \text{ N}\cdot\text{m})(14.4 \text{ rad})$$
$$= 0.52 \text{ J}$$

Remember that a joule and a newton–meter are equivalent in terms of their base units:

$$\frac{\text{kg}\cdot\text{m}^2}{\text{s}^2} = \text{kg}\cdot\frac{\text{m}}{\text{s}^2}\cdot\text{m} = \text{N}\cdot\text{m}$$

or

$$\frac{\text{kg}\cdot\text{m}^2}{\text{s}^2} = \text{kg}\cdot\left(\frac{\text{m}}{\text{s}}\right)^2 = \text{J}$$

377. $3.1\,\text{m/s}^2$

First, convert the energy into the "correct" units of joules:

$$(6.8\,\text{kJ})\left(\frac{1{,}000\,\text{J}}{1\,\text{kJ}}\right) = 6{,}800\,\text{J}$$

Start with the angular work formula. $W = \tau\theta$, where τ is torque and θ is angular displacement, and substitute the inertia torque formula $\tau = I\alpha$, where I is the tire's inertia $(I = \frac{1}{2}mr^2)$ and α is the angular acceleration:

$$\begin{aligned} W &= \tau\theta \\ &= (I\alpha)\theta \\ &= \left[\left(\frac{1}{2}mr^2\right)\alpha\right]\theta \\ &= \frac{1}{2}mr^2\alpha\theta \end{aligned}$$

Then use the linear-to-angular conversions for the acceleration ($a = r\alpha$, or $\frac{a}{r} = \alpha$) and displacement ($s = r\theta$, or $\frac{s}{r} = \theta$) to replace all the rotational variables with their tangential equivalents:

$$\begin{aligned} W &= \frac{1}{2}mr^2\alpha\theta \\ &= \frac{1}{2}mr^2\left(\frac{a}{r}\right)\left(\frac{s}{r}\right) \\ &= \frac{mr^2as}{2r^2} \\ &= \frac{mas}{2} \\ &= \frac{1}{2}mas \end{aligned}$$

Substitute the given values for work, mass, and displacement to solve for the tangential acceleration.

$$\begin{aligned} W &= \frac{1}{2}mas \\ 6{,}800\,\text{J} &= \frac{1}{2}(11\,\text{kg})a(400\,\text{m}) \\ 6{,}800\,\text{J} &= (2{,}200\,\text{kg}\cdot\text{m})a \\ 3.1\,\text{m/s}^2 &= a \end{aligned}$$

378.   700 m

First, convert the energy into "correct" units:

$$(2{,}790\,\text{kJ})\left(\frac{1\times10^3\,\text{J}}{1\,\text{kJ}}\right) = 2.79\times10^6\,\text{J}$$

The angular work formula, $W = \tau\theta$, where τ is the applied torque and θ is the angular displacement, requires you to solve for the torque that the adult has applied accelerating the merry-go-round at a rate of 0.4 radians per second squared. The inertia torque formula $\tau = I\alpha$ states that torque is the product of the total moment of inertia (merry-go-round and children) and the angular acceleration of the merry-go-round/children system. To find the total moment of inertia, add the moments of the outer ring (I_{or}), the inner ring (I_{ir}), the 80-kilogram child (I_{ic}), and the 90-kilogram child (I_{oc}). The 90-kilogram child is the same distance from the axis of rotation as the outer ring — 4.8 meters — and the 80-kilogram child is the same distance from the axis of rotation as the inner ring — 80 percent of 4.8 meters, or $(0.8)(4.8\text{ m}) = 3.84\text{ m}$.

$$
\begin{aligned}
I_{total} &= I_{or} + I_{oc} + I_{ir} + I_{ic} \\
&= m_{or}r_{or}^2 + m_{oc}r_{or}^2 + m_{ir}r_{ir}^2 + m_{ic}r_{ir}^2 \\
&= (180\text{ kg})(4.8\text{ m})^2 + (90\text{ kg})(4.8\text{ m})^2 \\
&\quad + (144\text{ kg})(3.84\text{ m})^2 + (80\text{ kg})(3.84\text{ m})^2 \\
&= 4{,}147\text{ kg}\cdot\text{m}^2 + 2{,}074\text{ kg}\cdot\text{m}^2 + 2{,}123\text{ kg}\cdot\text{m}^2 + 1{,}180\text{ kg}\cdot\text{m}^2 \\
&= 9{,}524\text{ kg}\cdot\text{m}^2
\end{aligned}
$$

Therefore,

$$
\begin{aligned}
\tau &= I\alpha \\
&= \left(9{,}524\text{ kg}\cdot\text{m}^2\right)\left(0.4\text{ rad/s}^2\right) \\
&= 3{,}809.6\text{ N}\cdot\text{m}
\end{aligned}
$$

Now use the angular work formula, $W = \tau\theta$, to solve for θ:

$$
\begin{aligned}
W &= \tau\theta \\
2.79\times10^6\text{ J} &= (3{,}809.6\text{ N}\cdot\text{m})\theta \\
732.4\text{ rad} &= \theta
\end{aligned}
$$

Remember that a joule and a newton-meter are equivalent in terms of their base units:

$$
\frac{\text{kg}\cdot\text{m}^2}{\text{s}^2} = \text{kg}\cdot\frac{\text{m}}{\text{s}^2}\cdot\text{m} = \text{N}\cdot\text{m}
$$

or

$$
\frac{\text{kg}\cdot\text{m}^2}{\text{s}^2} = \text{kg}\cdot\left(\frac{\text{m}}{\text{s}}\right)^2 = \text{J}
$$

Finally, to convert this into a linear distance, use the angular linear conversion formula, $s = r\theta$, where s is the linear distance and r is the distance from the axis of rotation. So for the 80-kilogram child sitting on the inner ring:

$$
\begin{aligned}
s_{ic} &= r_{ir}\theta \\
&= (3.84\text{ m})(732.4\text{ rad}) \\
&= 2{,}812.4\text{ m}
\end{aligned}
$$

And for the 90-kilogram child sitting on the outer ring:

$$
\begin{aligned}
s_{oc} &= r_{or}\theta \\
&= (4.8\text{ m})(732.4\text{ rad}) \\
&= 3{,}515.5\text{ m}
\end{aligned}
$$

So the 90-kilogram child traveled $(3{,}515.5\text{ m}) - (2{,}812.4\text{ m}) = 703.1\text{ m}$ farther. Rounded to the nearest 10, that's 700 meters.

379. $\frac{1}{4}$

The only type of kinetic energy the album has is rotational (the whole album isn't "translating" — moving from one location to another), so its total kinetic energy is $K = \frac{1}{2}I\omega^2$. I represents the object's moment of inertia, and ω is its angular velocity. So kinetic energy is proportional to the square of angular velocity. Therefore, if you multiply the angular velocity by $\frac{1}{2}$, you multiply the rotational kinetic energy — and, therefore, the total kinetic energy in this example — by $\left(\frac{1}{2}\right)^2 = \frac{1}{4}$.

380. **6.1 m**

Use the formula for rotational kinetic energy, $K = \frac{1}{2}I\omega^2$, where I is the moment of inertia and ω is the angular velocity. The moment of inertia for a hollow sphere is $I = \frac{2}{3}mr^2$ (m is the mass of the sphere, and r is its radius). Combine the two formulas into one equation to solve for the globe's radius:

$$K = \frac{1}{2}I\omega^2$$
$$K = \frac{1}{2}\left(\frac{2}{3}mr^2\right)\omega^2$$
$$K = \frac{1}{3}mr^2\omega^2$$
$$12\,\text{J} = \frac{1}{3}(5.2\,\text{kg})r^2(2.7\,\text{rad/s})^2$$
$$12\,\text{J} = \left(12.64\,\text{kg/s}^2\right)r^2$$
$$0.949\,\text{m}^2 = r^2$$
$$0.974\,\text{m} = r$$

The length of the "outside" of a circle is called circumference; the length of the longest string that can fit around the globe and not overlap itself would have the "circumference" of the sphere at its widest point (this is called a Great Circle in cartography) — where the radius is equal to 0.974 meters. So substitute this radius into the formula for circumference to get the final answer:

$$C = 2\pi r$$
$$= 2\pi(0.974\,\text{m})$$
$$= 6.1\,\text{m}$$

381. **18 rad/s**

First, convert the two important distances for use later. The radius of the wheel r_w is

$$(15\,\text{cm})\left(\frac{1\,\text{m}}{100\,\text{cm}}\right) = 0.15\,\text{m}$$

And the radius of the block's contact point is

$$(18\,\text{cm})\left(\frac{1\,\text{m}}{100\,\text{cm}}\right) = 0.18\,\text{m}$$

Use the conservation of energy principle to calculate the velocity of the falling block just as it makes contact with the wheel's platform (this solution uses the original location of the platform as the "zero" height and takes "up" as the positive direction of motion). The mass of the block is represented by m_B.

$$U_i + K_i = U_f + K_f$$

$$m_B g h_i + \frac{1}{2} m_B v_i^2 = m_B g h_i + \frac{1}{2} m_B v_i^2$$

$$g h_i + \frac{1}{2} v_i^2 = g h_f + \frac{1}{2} v_f^2$$

$$\left(9.8 \text{ m/s}^2\right)(1.2 \text{ m}) + \frac{1}{2}(0 \text{ m/s})^2 = \left(9.8 \text{ m/s}^2\right)(0 \text{ m}) + \frac{1}{2} v_f^2$$

$$11.76 \text{ m}^2/\text{s}^2 + 0 \text{ m}^2/\text{s}^2 = 0 \text{ m}^2/\text{s}^2 + \frac{1}{2} v_f^2$$

$$11.76 \text{ m}^2/\text{s}^2 = \frac{1}{2} v_f^2$$

$$23.52 \text{ m}^2/\text{s}^2 = v_f^2$$

$$\sqrt{23.52 \text{ m}^2/\text{s}^2} = v_f$$

$$-4.85 \text{ m/s} = v_f$$

(The block is falling; therefore, its velocity is negative, to fit in the indicated reference system when "up" is positive.)

In an elastic collision, both angular momentum and kinetic energy are conserved. Use conservation of angular momentum about the axis running through the center of the wheel. The moment of inertia of the block is

$$I_B = m_B r_B^2$$
$$I_B = (1.8 \text{ kg})(0.18 \text{ m})^2$$
$$I_B = 0.05832 \text{ kg} \cdot \text{m}^2$$

where $r_b = 0.18$ m is the distance between the block and the center of the wheel. Because the wheel is a solid disk, the moment of inertia of the wheel is

$$I_W = \frac{1}{2} m_W r_W^2$$
$$I_W = \frac{1}{2}(10 \text{ kg})(0.15 \text{ m})^2$$
$$I_W = 0.1125 \text{ kg} \cdot \text{m}^2$$

The angular momentum equation yields

$$L_i = L_f$$

$$I_B \omega_{iB} + I_W \omega_{iW} = I_B \omega_{fB} + I_W \omega_W$$

$$\left(m_B r_B^2\right)(v_f / r_B) + I_W (0 \text{ rad/s}) = I_B \omega_{fB} + I_W \omega_{fW}$$

$$m_B r_B v_f + 0 = I_B \omega_{fB} + I_W \omega_{fW}$$

$$\frac{m_B r_B v_f - I_W \omega_{fW}}{I_B} = \omega_{fB}$$

The kinetic energy of the disc is $\frac{1}{2}I_w\omega_w^2$. The kinetic energy equation also results in an equation relating the two final angular velocities:

$$K_i = K_f$$

$$\frac{1}{2}m_B v_{iB}^2 + \frac{1}{2}I_W\omega_{iW}^2 = \frac{1}{2}m_B v_{fB}^2 + \frac{1}{2}I_W\omega_{fW}^2$$

$$m_B v_{iB}^2 + I_W\omega_{iW}^2 = m_B v_{fB}^2 + I_W\omega_{WW}^2$$

$$m_B v_f^2 + I_W\omega_{iW}^2 = m_B\omega_{fB}^2 r_B^2 + I_W\omega_{fW}^2$$

$$m_B v_f^2 + I_W\omega_{iW}^2 = I_B\omega_{fB}^2 + I_W\omega_{fW}^2$$

$$m_B v_f^2 + I_W(0 \text{ rad/s})^2 = I_B\omega_{fB}^2 + I_W\omega_{fW}^2$$

$$m_B v_f^2 + 0 = I_B\omega_{fB}^2 + I_W\omega_{fW}^2$$

$$m_B v_f^2 = I_B\omega_{fB}^2 + I_W\omega_{fW}^2$$

Now substitute the earlier result from angular momentum conservation into the energy equation:

$$m_B v_f^2 = I_B\omega_{fB}^2 + I_W\omega_{fW}^2$$

$$m_B v_f^2 = I_B\left(\frac{m_B r_B v_f - I_W\omega_{fW}}{I_B}\right)^2 + I_W\omega_{fW}^2$$

$$m_B v_f^2 = \frac{(m_B r_B v_f - I_W\omega_{fW})^2}{I_B} + I_W\omega_{TW}^2$$

$$m_B v_f^2 = \frac{m_B^2 r_B^2 v_f^2}{I_B} - \frac{2m_B r_B v_f I_W\omega_{fW}}{I_B} + \frac{I_W^2\omega_{fW}^2}{I_B} + I_W\omega_{fW}^2$$

$$m_B v_f^2 = \frac{m_B^2 r_B^2 v_f^2}{m_B r_B^2} - \frac{2m_B r_B v_f I_W\omega_{fW}}{m_B r_B^2} + \frac{I_W^2\omega_{fW}^2}{I_B} + I_W\omega_{fW}^2$$

$$m_B v_f^2 = m_B v_f^2 - \frac{2v_f I_W\omega_{fW}}{r_B} + \frac{I_W^2\omega_{fW}^2}{I_B} + I_W\omega_{fW}^2$$

$$0 = -\frac{2v_f I_W\omega_{fW}}{r_B} + \frac{I_W^2\omega_{fW}^2}{I_B} + I_W\omega_{fW}^2$$

$$\frac{2v_f}{r_B} = \frac{I_W\omega_{fW}}{I_B} + \omega_{fW}$$

$$\frac{2(-4.85 \text{ m/s})}{0.18 \text{ m}} = \frac{(0.1125 \text{ kg} \cdot \text{m}^2)\omega_{fW}}{0.05832 \text{ kg} \cdot \text{m}^2} + \omega_{fW}$$

$$-53.89 \text{ rad/s} = 1.929\omega_{fW} + \omega_{fW}$$

$$-53.89 \text{ rad/s} = 2.929\omega_{fW}$$

$$-18 \text{ rad/s} = \omega_{FW}$$

The negative sign just indicates that the wheel spins counterclockwise and isn't needed to find the magnitude of the angular velocity.

---

382. SS, SC, HS, HC

"Without slipping" means two things: 1. The objects roll — and, therefore, have rotational kinetic energy — and 2. the friction between the objects and the ramp providing the torque for the rotation does no work, meaning that mechanical energy is conserved.

All four objects start with the same amount of energy because they all started from the same height (gravitational potential energy) and at the same velocity (kinetic energy). By the time they reach the bottom of the ramp (where potential energy is 0 relative to the starting location), all that energy is distributed between *rotational* kinetic energy (spinning) and *translational* kinetic energy (sliding).

An object's rotational kinetic energy is proportional to its moment of inertia. So the larger the inertia, the larger the rotational kinetic energy. The formulas for inertia for the four shapes listed are

Solid cylinder: $\frac{1}{2}mr^2$

Hollow cylinder: mr^2

Solid sphere: $\frac{2}{5}mr^2$

Hollow sphere: $\frac{2}{3}mr^2$

The hollow cylinder is the only one without a fraction less than one attached to it, so it's the largest. Two-fifths is the smallest fraction — 0.4 as a decimal — followed by one-half (0.5) and two-thirds (approximately 0.67).

383. 60 J

"Without slipping" means two things: 1. The bowling ball rolls — and, therefore, has rotational kinetic energy — and 2. the friction between the ball's point of contact with the ramp providing the torque for the rotation does no work — meaning that mechanical energy is conserved.

Start with the formula for rotational kinetic energy, $K = \frac{1}{2}I\omega^2$. Then substitute both the moment of inertia of a solid sphere —$I = \frac{2}{5}mr^2$— and the conversion between translational (also known as linear or tangential) velocity and angular velocity — $\omega = \frac{v}{r}$, where ω is the angular velocity and v is the translational velocity:

$$K = \frac{1}{2}I\omega^2$$
$$= \frac{1}{2}\left(\frac{2}{5}mr^2\right)\left(\frac{v}{r}\right)^2$$
$$= \frac{1}{5}mr^2\left(\frac{v^2}{r^2}\right)$$
$$= \frac{1}{5}mv^2$$

Finally, substitute the known values:

$$K = \frac{1}{5}mv^2$$
$$= \frac{1}{5}(12\,\text{kg})(5\,\text{m/s})^2$$
$$= 60\,\text{J}$$

"Without slipping" means two things: 1. The wheel ball rolls — and, therefore, has rotational kinetic energy — and 2. the friction between the wheel's point of contact with the ramp providing the torque for the rotation does no work, meaning that mechanical energy is conserved.

Use the formula for the conservation of mechanical energy to solve for the height the wheel lost during its motion (the following solution uses the height at the end of the motion as the "zero" point). Substitute the formula for the moment of inertia of a solid disc ($I = \frac{1}{2}mr^2$), as well as the conversion between angular and translational/tangential velocities ($\omega = \frac{v}{r}$) along the way:

$$U_i + K_j = U_f + K_f$$

$$mgh_i + \frac{1}{2}I\omega_i^2 + \frac{1}{2}mv_i^2 = mgh_f + \frac{1}{2}I\omega_f^2 + \frac{1}{2}mv_f^2$$

$$mgh_i + \frac{1}{2}\left(\frac{1}{2}mr^2\right)\left(\frac{v_i}{r}\right)^2 + \frac{1}{2}mv_i^2 = mgh_f + \frac{1}{2}\left(\frac{1}{2}mr^2\right)\left(\frac{v_f}{r}\right)^2 + \frac{1}{2}mv_f^2$$

$$mgh_i + \frac{1}{4}mr^2\left(\frac{v_i^2}{r^2}\right) + \frac{1}{2}mv_i^2 = mgh_f + \frac{1}{4}mr^2\left(\frac{v_f^2}{r^2}\right) + \frac{1}{2}mv_f^2$$

$$mgh_i + \frac{1}{4}mv_i^2 + \frac{1}{2}mv_i^2 = mgh_f + \frac{1}{4}mv_f^2 + \frac{1}{2}mv_f^2$$

$$mgh_i + \frac{3}{4}mv_i^2 = mgh_f + \frac{3}{4}mv_f^2$$

$$gh_i + \frac{3}{4}v_i^2 = gh_f + \frac{3}{4}v_f^2$$

$$\left(9.8 \text{ m/s}^2\right)h_i + \frac{3}{4}(0 \text{ m/s})^2 = \left(9.8 \text{ m/s}^2\right)(0 \text{ m}) + \frac{3}{4}(7 \text{ m/s})^2$$

$$\left(9.8 \text{ m/s}^2\right)h_i + 0 \text{ m}^2/\text{s}^2 = 0 \text{ m}^2/\text{s}^2 + 36.75 \text{ m}^2/\text{s}^2$$

$$\left(9.8 \text{ m/s}^2\right)h_i = 36.75 \text{ m}^2/\text{s}^2$$

$$h_i = 3.75 \text{ m}$$

Then use trigonometry to solve for the angle the plank makes with respect to the horizontal:

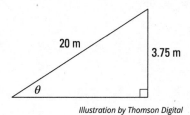

Illustration by Thomson Digital

$$\sin\theta = \frac{3.75 \text{ m}}{20 \text{ m}}$$

$$\sin\theta = 0.1875$$

$$\theta = \sin^{-1}(0.1875)$$

$$\theta = 10.8° \approx 11°$$

385.   0.19

Start by drawing a free-body diagram and using Newton's second law to find the relationships between the forces in the vertical (perpendicular to the ramp) and horizontal (parallel to the ramp) directions. (The solution given here uses "down the ramp" as the positive horizontal direction.)

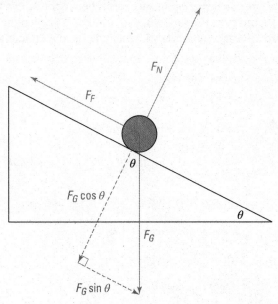

Illustration by Thomson Digital

$$F_{y,\text{net}} = \sum F_y$$
$$ma_y = F_N - F_G \cos\theta$$
$$ma_y = F_N - ma_G \cos\theta$$
$$ma_y = F_N - mg\cos\theta$$
$$m(0 \text{ m/s}^2) = F_N - m(9.8 \text{ m/s}^2)\cos 30°$$
$$m(0 \text{ m/s}^2) = F_N - m(9.8 \text{ m/s}^2)(0.866)$$
$$0 \text{ N} = F_N - (8.49 \text{ m/s}^2)m$$
$$(8.49 \text{ m/s}^2)m = F_N$$

$$F_{x,\text{net}} = \sum F_x$$
$$ma_x = F_G \sin\theta - F_F$$
$$ma = ma_G \sin\theta - \mu F_N$$
$$ma = mg\sin\theta - \mu F_N$$
$$ma = m(9.8 \text{ m/s}^2)\sin 30° - \mu\left((8.49 \text{ m/s}^2)m\right)$$
$$ma = m(9.8 \text{ m/s}^2)(0.5) - (8.49 \text{ m/s}^2)\mu m$$
$$ma = (4.9 \text{ m/s}^2)m - (8.49 \text{ m/s}^2)\mu m$$
$$a = (4.9 \text{ m/s}^2) - (8.49 \text{ m/s}^2)\mu$$

Then use the two formulas for torque, $\tau = rF \sin \theta$ and $\tau = I\alpha$, to relate the tangential (90 degrees to the disk's radius) force of friction to the acceleration. Use the formula for the moment of inertia for a solid disk $I = \frac{1}{2}mr^2$, and use the conversion between linear/tangential acceleration and angular acceleration $\frac{a}{r} = a$ to simplify the equation:

$$rF \sin \theta = \tau = I\alpha$$
$$rF_F \sin \theta = I\alpha$$
$$r(\mu F_N)\sin \theta = \left(\frac{1}{2}mr^2\right)\left(\frac{a}{r}\right)$$
$$r\mu F_N \sin \theta = \frac{1}{2}mra$$
$$2r\mu F_N \sin \theta = mra$$
$$2\mu F_N \sin \theta = ma$$
$$2\mu\left((8.49 \text{ m/s}^2)m\right)\sin 90° = ma$$
$$(16.98 \text{ m/s}^2)\mu m(1) = ma$$
$$(16.98 \text{ m/s}^2)\mu m = ma$$
$$(16.98 \text{ m/s}^2)\mu = a$$

Set the two expressions for a equal to one another to solve for μ:

$$(16.98 \text{ m/s}^2)\mu = a = (4.9 \text{ m/s}^2) - (8.49 \text{ m/s}^2)\mu$$
$$(16.98 \text{ m/s}^2)\mu = (4.9 \text{ m/s}^2) - (8.49 \text{ m/s}^2)\mu$$
$$(25.47 \text{ m/s}^2)\mu = (4.9 \text{ m/s}^2)$$
$$\mu = 0.19$$

386. **7.9 s**

Energy is conserved when a disk rolls down a plane, regardless of the presence of friction, as long the disk doesn't slip. First, use trigonometry to determine the lengths of the two ramps:

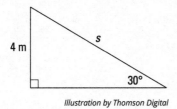

Illustration by Thomson Digital

$$\sin 30° = \frac{4 \text{ m}}{s}$$
$$0.5 = \frac{4 \text{ m}}{s}$$
$$0.5s = 4 \text{ m}$$
$$s = 8 \text{ m}$$

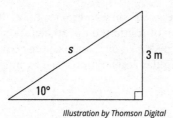

$$\sin 10° = \frac{3\,m}{s}$$

$$0.174 = \frac{3\,m}{s}$$

$$0.174\,s = 3\,m$$

$$s = 17.2\,m$$

For the first (left) ramp:

Write the formula for the conservation of mechanical energy, substitute the formula for a solid disk's moment of inertia — $I = \frac{1}{2}mr^2$ — and eventually solve for the final velocity:

$$U_i + K_i = U_f + K_f$$

$$mgh_i + \frac{1}{2}mv_i^2 + \frac{1}{2}I\omega_i^2 = mgh_f + \frac{1}{2}mv_f^2 + \frac{1}{2}I\omega_f^2$$

$$mgh_i + \frac{1}{2}mv_i^2 + \frac{1}{2}\left(\frac{1}{2}mr^2\right)\left(\frac{v_i^2}{r^2}\right) = mgh_f + \frac{1}{2}mv_f^2 + \frac{1}{2}\left(\frac{1}{2}mr^2\right)\left(\frac{v_f^2}{r^2}\right)$$

$$gh_i + \frac{1}{2}v_i^2 + \frac{1}{2}\left(\frac{1}{2}r^2\right)\left(\frac{v_i^2}{r^2}\right) = gh_f + \frac{1}{2}v_f^2 + \frac{1}{2}\left(\frac{1}{2}r^2\right)\left(\frac{v_f^2}{r^2}\right)$$

$$gh_i + \frac{1}{2}v_i^2 + \frac{1}{4}v_i^2 = gh_f + \frac{1}{2}v_f^2 + \frac{1}{4}v_f^2$$

$$gh_i + \frac{3}{4}v_i^2 = gh_f + \frac{3}{4}v_f^2$$

$$\left(9.8\,m/s^2\right)(4\,m) + \frac{3}{4}(0\,m/s)^2 = \left(9.8\,m/s^2\right)(0\,m) + \frac{3}{4}v_f^2$$

$$39.2\,m^2/s^2 + 0\,m^2/s^2 = 0\,m^2/s^2 + \frac{3}{4}v_f^2$$

$$39.2\,m^2/s^2 = \frac{3}{4}v_f^2$$

$$52.27\,m^2/s^2 = v_f^2$$

$$\sqrt{52.27\,m^2/s^2} = v_f$$

$$7.23\,m/s = v_f$$

Use the velocity displacement formula to solve for the disk's acceleration down the ramp:

$$v_f^2 = v_i^2 + 2as$$

$$(7.23\,m/s)^2 = (0\,m/s)^2 + 2a(8\,m)$$

$$52.27\,m^2/s^2 = 0m^2/s^2 + (16\,m)a$$

$$52.27m^2/s^2 = (16\,m)a$$

$$3.27\,m/s^2 = a$$

Then use that result in the velocity time formula to solve for the disk's time of transit down the ramp:

$$v_f = v_i + at$$
$$7.23 \text{ m/s} = 0 \text{ m/s} + \left(3.27 \text{ m/s}^2\right)t$$
$$7.23 \text{ m/s} = \left(3.27 \text{ m/s}^2\right)t$$
$$2.21 \text{ s} = t$$

For the flat section of the ramp, the net force and net torque on the disk is zero in the flat region, so it will traverse the flat region with a constant velocity and constant angular velocity. Use the displacement formula to find how long it takes the disk to traverse the flat region:

$$s = v_i t + \frac{1}{2}at^2$$
$$18 \text{ m} = (7.23 \text{ m/s})t + \frac{1}{2}\left(0 \text{ m/s}^2\right)t^2$$
$$18 \text{ m} = (7.23 \text{ m/s})t + 0 \text{ m}$$
$$18 \text{ m} = (7.23 \text{ m/s})t$$
$$2.49 \text{ s} = t$$

For the second (right) ramp, use conservation of energy one more time to solve for the final velocity of the disk when it reaches the top of the 3-meter-high second ramp:

$$U_i + K_i = U_f + K_f$$
$$mgh_i + \frac{1}{2}mv_i^2 + \frac{1}{2}I\omega_i^2 = mgh_f + \frac{1}{2}mv_f^2 + \frac{1}{2}I\omega_f^2$$
$$mgh_i + \frac{1}{2}mv_i^2 + \frac{1}{2}\left(\frac{1}{2}mr^2\right)\left(\frac{v_i^2}{r^2}\right) = mgh_f + \frac{1}{2}mv_f^2 + \frac{1}{2}\left(\frac{1}{2}mr^2\right)\left(\frac{v_f^2}{r^2}\right)$$
$$gh_i + \frac{1}{2}v_i^2 + \frac{1}{2}\left(\frac{1}{2}r^2\right)\left(\frac{v_i^2}{r^2}\right) = gh_f + \frac{1}{2}v_f^2 + \frac{1}{2}\left(\frac{1}{2}r^2\right)\left(\frac{v_f^2}{r^2}\right)$$
$$gh_i + \frac{1}{2}v_i^2 + \frac{1}{4}v_i^2 = gh_f + \frac{1}{2}v_f^2 + \frac{1}{4}v_f^2$$
$$gh_i + \frac{3}{4}v_i^2 = gh_f + \frac{3}{4}v_f^2$$
$$\left(9.8 \text{ m/s}^2\right)(0 \text{ m}) + \frac{3}{4}(7.23 \text{ m/s})^2 = \left(9.8 \text{ m/s}^2\right)(3 \text{ m}) + \frac{3}{4}v_f^2$$
$$0 \text{ m}^2/\text{s}^2 + 39.2 \text{ m}^2/\text{s}^2 = 29.4 \text{ m}^2/\text{s}^2 + \frac{3}{4}v_f^2$$
$$39.2 \text{ m}^2/\text{s}^2 = 29.4 \text{ m}^2/\text{s}^2 + \frac{3}{4}v_f^2$$
$$9.8 \text{ m}^2/\text{s}^2 = \frac{3}{4}v_f^2$$
$$13.07 \text{ m}^2/\text{s}^2 = v_f^2$$
$$\sqrt{13.07 \text{ m}^2/\text{s}^2} = v_f$$
$$3.62 \text{ m/s} = v_f$$

Then use the velocity displacement formula to solve for the disk's acceleration up the ramp:

$$v_f^2 = v_i^2 + 2as$$
$$(3.62 \text{ m/s})^2 = (7.23 \text{ m/s})^2 + 2a(17.2 \text{ m})$$
$$13.1 \text{ m}^2/\text{s}^2 = 52.27 \text{ m}^2/\text{s}^2 + (34.4 \text{ m})a$$
$$-39.17 \text{ m}^2/\text{s}^2 = (34.4 \text{ m})a$$
$$-1.14 \text{ m/s}^2 = a$$

Use that result in the velocity time formula to solve for the disk's time of transit down the ramp:

$$v_f = v_i + at$$
$$3.62 \text{ m/s} = 7.23 \text{ m/s} + \left(-1.14 \text{ m/s}^2\right)t$$
$$-3.61 \text{ m/s} = \left(-1.14 \text{ m/s}^2\right)t$$
$$3.17 \text{ s} = t$$

Add the times for the three sections to find the final transit time:

$$(2.21 \text{ s}) + (2.49 \text{ s}) + (3.17 \text{ s}) = 7.87 \approx 7.9 \text{ s}$$

387. $\frac{z}{4}$

Angular momentum is always conserved in the absence of external torques, so in the formula $L = I\omega$ (where L is angular momentum, I is inertia, and ω is angular velocity), L is a constant. Divide both sides of the formula by I, and you find that the angular velocity is inversely proportional to the moment of inertia:

$$\frac{L}{I} = \omega$$

A body's moment of inertia is always proportional to mr^2, where m is the body's mass and r is the distance from center of rotation. Therefore,

$$\frac{L}{mr^2} = \omega$$

So the angular velocity is inversely proportional to the *square* of the distance/radius. If the you multiplies the skater's r by 2, the skater's angular velocity — z — is *divided* by $2^2 = 4$.

388. $3.36 \times 10^{-3} \text{ kg} \cdot \text{m}^2/\text{s}$

First, convert the given into "correct" units:

$$(800 \text{ g})\left(\frac{1 \text{ kg}}{1,000 \text{ g}}\right) = 0.8 \text{ kg}$$
$$\left(12 \frac{\text{cm}}{\text{s}}\right)\left(\frac{1 \text{ m}}{100 \text{ cm}}\right) = 0.12 \text{ m/s}$$
$$(7 \text{ cm})\left(\frac{1 \text{ m}}{100 \text{ cm}}\right) = 0.07 \text{ m}$$

The formula for angular momentum is $L = I\omega$ (L is angular momentum, I is moment of inertia, and ω is angular velocity). The moment of inertia of a solid disk is $I = \frac{1}{2}mr^2$, where m is the plate's mass and r is its radius. In addition you can use the linear angular conversion formula for velocity ($v = r\omega$, where v is the linear velocity, ω is the angular velocity, and r is the radius of the plate) to make use of the tangential/linear velocity given in the question. Substitute, simplify, and solve the angular momentum equation for L:

$$L = I\omega$$
$$= \left(\frac{1}{2}mr^2\right)\left(\frac{v}{r}\right)$$
$$= \frac{1}{2}mrv$$
$$= \frac{1}{2}(0.8\text{ kg})(0.07\text{ m})(0.12\text{ m/s})$$
$$= 3.36 \times 10^{-3}\text{ kg} \cdot \text{m}^2/\text{s}$$

389. $\quad 2.7 \times 10^{40}\text{ kg} \cdot \text{m}^2/\text{s}$

First, convert the distance between Earth and the sun into meters:

$$\left(1.496 \times 10^8\text{ km}\right)\left(\frac{1 \times 10^3\text{ m}}{1\text{ km}}\right) = 1.496 \times 10^{11}\text{ m}$$

Convert the orbital period into seconds:

$$(365\text{ days})\left(\frac{24\text{ h}}{1\text{ day}}\right)\left(\frac{60\text{ min}}{1\text{ h}}\right)\left(\frac{60\text{ s}}{1\text{ min}}\right) = 3.154 \times 10^7\text{ s}$$

Then use the formula for angular momentum, $L = I\omega$, where L is the revolving object's angular momentum, I is its moment of inertia, and ω is its angular velocity. The moment of inertia of a point mass (despite its enormous size, Earth can be considered a point mass relative to even the more enormous distance between it and the sun) is $I = mr^2$, where m is Earth's mass and r is the distance from the sun. Therefore, Earth's moment of inertia is

$$I = mr^2$$
$$= \left(5.98 \times 10^{24}\text{ kg}\right)\left(1.496 \times 10^{11}\text{ m}\right)^2$$
$$= 1.338 \times 10^{47}\text{ kg} \cdot \text{m}^2$$

Use the definition of angular velocity $\omega = \frac{\theta}{t}$, where θ is angular displacement and t is the amount of time it takes to rotate through that displacement. In one revolution of a circular orbit around the sun, Earth has an angular displacement of 2π radians, so its angular velocity is

$$\omega = \frac{\theta}{t}$$
$$= \frac{2\pi\text{ rad}}{3.154 \times 10^7\text{ s}}$$
$$= 1.992 \times 10^{-7}\text{ rad/s}$$

Finally, substitute your results for the inertia and angular velocity back into formula for angular momentum to solve:

$$L = I\omega$$
$$= \left(1.338 \times 10^{47} \text{ kg} \cdot \text{m}^2\right)\left(1.992 \times 10^{-7} \text{ rad/s}\right)$$
$$= 2.665 \times 10^{40} \text{ kg} \cdot \text{m}^2/\text{s}$$
$$\approx 2.7 \times 10^{40} \text{ kg} \cdot \text{m}^2/\text{s}$$

390. 5.2 m/s

First, convert all the values that are not in "correct" units:

$$(15 \text{ rev})\left(\frac{2\pi \text{ rad}}{\text{rev}}\right) = 30\pi \approx 94.25 \text{ rad}$$

$$(20 \text{ cm})\left(\frac{1 \text{ m}}{100 \text{ cm}}\right) = 0.2 \text{ m}$$

Angular momentum is always conserved in the absence of external torques:

$$L_i = L_f$$
$$I_i\omega_i = I_f\omega_f$$

L is angular momentum, I is moment of inertia, and ω is angular velocity. Because you can approximate the man's body as a solid cylinder ($I = \frac{1}{2}mr^2$) with a mass of 80 kilograms and 0.2-meter radius, the initial moment of inertia is

$$I_i = \frac{1}{2}mr_i^2$$
$$= \frac{1}{2}(80 \text{ kg})(0.2 \text{ m})^2$$
$$= 1.6 \text{ kg} \cdot \text{m}^2$$

Use the angular velocity displacement formula, $\omega_f^2 = \omega_i^2 + 2\alpha\theta$, where ω_i and ω_f are the initial and final angular velocities, α is the angular acceleration, and θ is the angular displacement, to solve for the angular velocity after 15 revolutions:

$$\omega_f^2 = \omega_i^2 + 2\alpha\theta$$
$$\omega_f = \sqrt{\omega_i^2 + 2\alpha\theta}$$
$$= \sqrt{(0 \text{ rad/s})^2 + 2\left(3 \text{ rad/s}^2\right)(94.25 \text{ rad})}$$
$$= \sqrt{0 \text{ rad}^2/\text{s}^2 + 565.5 \text{ rad}^2/\text{s}^2}$$
$$= \sqrt{565.5 \text{ rad}^2/\text{s}^2}$$
$$= 23.78 \text{ rad/s}$$

This measure is the *initial* angular velocity with regard to the conservation of angular momentum that you're about to use to compare the "before" picture (man spinning with arms and legs in) to the "after" picture (man spinning with arms and legs out). The *final* angular velocity, as indicated in the question, is half this "initial" amount, or $\frac{1}{2}(23.78 \text{ rad/s}) = 11.89 \text{ rad/s}$.

The moment of inertia of a rod rotated about its end (as outstretched arms and legs are) is $I = \frac{1}{3}ml^2$, where m is the mass of the rod and l is its length. Therefore, the total moment of "final" inertia is

$$I_f = I_{torso} + I_{arms} + I_{legs}$$
$$= \frac{1}{2} m_{torso} r_{torso}^2 + \frac{1}{3} m_{arms} l_{arms}^2 + \frac{1}{3} m_{legs} l_{legs}^2$$

The question stipulates that the mass of the arms is 10 percent of the body's mass, or $(0.1)(80\,kg) = 8$ kilograms. Similarly, the mass of the legs is 20 percent, or 16 kilograms. The remaining 70 percent, or 56 kilograms, remains in the "torso" category. You're also told that the man's arms are 80 percent as long as his legs. Therefore,

$$l_{arms} = 0.8 l_{legs}$$
$$\frac{l_{arms}}{0.8} = l_{legs}$$
$$1.25 l_{arms} = l_{legs}$$

Substitute all this information into the inertia formula to formulate an expression for the final moment of inertia of the system in terms of the length of the man's arms — which is the region for which you want to determine the final tangential velocity:

$$I_f = \frac{1}{2} m_{torso} r_{torso}^2 + \frac{1}{3} m_{arms} l_{arms}^2 + \frac{1}{3} m_{legs} l_{legs}^2$$
$$= \frac{1}{2}(56\,kg)(0.2\,m)^2 + \frac{1}{3}(8\,kg)l_{arms}^2 + \frac{1}{3}(16\,kg)(1.25 l_{arms})^2$$
$$= 1.12\,kg \cdot m^2 + (2.67\,kg)l_{arms}^2 + (8.33\,kg)l_{arms}^2$$
$$= 1.12\,kg \cdot m^2 + (11\,kg)l_{arms}^2$$

Now substitute all the results for the initial and final angular velocities and moments of inertia into the conservation of angular momentum equation:

$$I_i \omega_i = I_f \omega_f$$
$$\left(1.6 kg \cdot m^2\right)(23.78\,rad/s) = \left(1.12\,kg \cdot m^2 + (11\,kg)l_{arms}^2\right)(11.89\,rad/s)$$
$$38.05\,kg \cdot m^2/s = 13.32\,kg \cdot m^2/s + (130.8\,kg/s)l_{arms}^2$$
$$24.73\,kg \cdot m^2/s = (130.8\,kg/s)l_{arms}^2$$
$$0.189\,m^2 = l_{arms}^2$$
$$\sqrt{0.189\,m^2} = l_{arms}$$
$$0.435\,m = l_{arms}$$

Finally, use the linear angular conversion formula for velocity, $v = r\omega$, to rewrite ω_i in terms of the tangential/linear velocity at the fingertips. (In this case, r and l are the same because the fingertips are l meters away from the center of rotation, and that is the "definition" of r.)

$$v = r\omega$$
$$v_{arms} = r_{arms}\omega_f$$
$$= l_{arms}\omega_f$$
$$= (0.435\,m)(11.89\,rad/s)$$
$$= 5.2\,m/s$$

391. 3.3×10^3 N/m

Hooke's law relates force to the spring constant and the displacement from equilibrium: $F = -kx$.

Solving this for the spring constant k gives $k = -\dfrac{F}{x}$.

You know that the force is in the direction opposite the spring displacement. If you choose the direction of the force to be positive, the direction of the spring displacement must be negative. This means that

$$x = -3.0 \text{ cm} \times \frac{1 \text{ m}}{100 \text{ cm}}$$
$$= -0.030 \text{ m}$$

and $F = 100$ newtons. Solving Hooke's law for the spring constant k and inserting these values gives

$$F = -kx$$
$$k = -\frac{F}{x}$$
$$= -\frac{100 \text{ N}}{-0.030 \text{ m}}$$
$$= 3.3 \times 10^3 \text{ N/m}$$

392. 48 newtons in the negative x direction

Hooke's law relates force to the spring constant and the displacement from equilibrium: $F = -kx$.

Inserting the given values of $k = 4,000$ newtons per meter and

$$x = 1.2 \text{ cm} \times \frac{1 \text{ m}}{100 \text{ cm}}$$
$$= 0.012 \text{ m}$$

gives

$$F = -kx$$
$$= -4,000 \text{ N/m} \times 0.012 \text{ m}$$
$$= -48 \text{ N}$$

The minus sign means that the force acts in the negative x direction.

393. 0.53 m

After the spring combination is stretched, it no longer moves, so the forces pulling to the left and right at each point in the springs have the same magnitude. This means that a force of 500 newtons is pulling to both the left and the right at each point in the two springs. Apply this logic to the end of spring two. A force of 500 newtons is pulling to the right on the spring, so it must respond with a force of 500 newtons pulling to the left. Calling left the negative direction, this means that spring two applies a force $F_2 = -500$ newtons. Solving Hooke's law for the stretch from equilibrium of spring two gives

$$F_2 = -k_2 x_2$$
$$x_2 = -\frac{F_2}{k_2}$$
$$= -\frac{-500 \text{ N}}{1{,}500 \text{ N/m}}$$
$$= 0.33 \text{ m}$$

Apply the same logic to the right end of spring one. Spring two pulls this point to the right with a force of magnitude of 500 newtons, and spring one responds with a force of $F_1 = -500$ newtons (in other words, to the left). Hooke's law for this point gives

$$F_1 = -k_1 x_1$$
$$x_1 = -\frac{F_1}{k_1}$$
$$= -\frac{-500 \text{ N}}{2{,}500 \text{ N/m}}$$
$$= 0.200 \text{ m}$$

The total stretch x of the two-spring combination is

$$x = x_1 + x_2$$
$$= 0.20 \text{ m} + 0.33 \text{ m}$$
$$= 0.53 \text{ m}$$

---

394. 5.5 kg

Apply Hooke's law: $F = -kx$. The force applied to the spring is $F_{applied} = -mg$, where the minus sign means that this force acts downward. By Newton's third law, the spring must exert a force of equal magnitude but in the upward direction, so

$$F = -F_{applied}$$
$$= -(-mg)$$
$$= mg$$

Insert this into Hooke's law and solve for the mass m:

$$F = -kx$$
$$mg = -kx$$
$$m = -\frac{kx}{g}$$

Insert the known value of $k = 1{,}200$ newtons per meter and

$$x = -4.5 \text{ cm} \times \frac{1 \text{ m}}{100 \text{ cm}}$$
$$= -0.045 \text{ m}$$

where the minus sign indicates that the displacement of the end of the spring is in the downward direction. The result is

$$m = -\frac{kx}{g}$$
$$= -\frac{1{,}200 \text{ N/m} \times (-0.045 \text{ m})}{9.8 \text{ m/s}^2}$$
$$= 5.5 \text{ kg}$$

395. 4.5×10^2 N/m

Apply Hooke's law: $F = -kx$. The force applied to the spring is $F_{applied} = -mg$, where the minus sign means that this force acts downward. By Newton's third law, the spring must exert a force of equal magnitude but in the upward direction, so

$$F = -F_{applied}$$
$$= -(-mg)$$
$$= mg$$

The displacement of the spring is downward (negative), which gives

$$x = -4.8 \text{ cm} \times \frac{1 \text{ m}}{100 \text{ cm}}$$
$$= -0.048 \text{ m}$$

The mass is $m = 2.2$ kilograms. Insert these values into Hooke's law to find the spring constant k:

$$F = -kx$$
$$mg = -kx$$
$$k = -\frac{mg}{x}$$
$$= -\frac{2.2 \text{ kg} \times 9.8 \text{ m/s}^2}{-0.048 \text{ m}}$$
$$= 4.5 \times 10^2 \text{ N/m}$$

396.   0.71 s

The maximum of the sine function is +1, so the maximum position occurs when $\sin\left(2.2 \text{ s}^{-1} \times t\right) = 1$. This occurs when

$$2.2 \text{ s}^{-1} \times t = \frac{\pi}{2}$$
$$t = \frac{\pi}{2 \times 2.2 \text{ s}^{-1}}$$
$$= 0.71 \text{ s}$$

397.   13

The period of a sine wave is $T = \frac{2\pi}{\omega}$. In this case, $\omega = 1.8 \times 10^{13} \text{ s}^{-1}$. The separation between the atoms goes through a maximum once per period, so you just need to know how many periods fit in the time $t_0 = 4.6 \times 10^{-12}$ s. This is given by

$$N = \frac{t_0}{T}$$
$$= \frac{\omega t_0}{2\pi}$$
$$= \frac{1.8 \times 10^{13} \text{ s}^{-1} \times 4.6 \times 10^{-12} \text{ s}}{2\pi}$$
$$= 13$$

398. 1.1×10^6 rad

In the reference circle, the oscillator has moved through the angle given by the argument of the sine function, which is ωt. In this case, you know $\omega = 3.5 \times 10^5 \, \text{s}^{-1}$ and $t = 3.2 \, \text{s}$. Therefore, the angle is

$$\begin{aligned}
\alpha &= \omega t \\
&= 3.5 \times 10^5 \, \text{s}^{-1} \times 3.2 \, \text{s} \\
&= 1.1 \times 10^6 \, \text{rad}
\end{aligned}$$

399. 2.6 cm

The maximum of the cosine function is +1, and its minimum is −1. Therefore, the maximum position is $y_{max} = 1.3 \, \text{cm}$, and the minimum position is $y_{min} = -1.3 \, \text{cm}$.

The difference is

$$\begin{aligned}
\Delta y &= y_{max} - y_{min} \\
&= 1.3 \, \text{cm} - (-1.3 \, \text{cm}) \\
&= 2.6 \, \text{cm}
\end{aligned}$$

400. 0.023 s

The tip is at 3.9 centimeters when $3.9 \, \text{cm} = (4.9 \, \text{cm}) \sin \left(98 \, \text{s}^{-1} \times t \right)$. Solving this for the time gives

$$3.9 \, \text{cm} = (4.9 \, \text{cm}) \sin \left(98 \, \text{s}^{-1} \times t \right)$$

$$\frac{3.9 \, \text{cm}}{4.9 \, \text{cm}} = \sin \left(98 \, \text{s}^{-1} \times t \right)$$

$$\arcsin \left(\frac{3.9 \, \text{cm}}{4.9 \, \text{cm}} \right) = 98 \, \text{s}^{-1} \times t$$

$$\begin{aligned}
t &= \frac{1}{98 \, \text{s}^{-1}} \times \arcsin \left(\frac{3.9 \, \text{cm}}{4.9 \, \text{cm}} \right) \\
&= \begin{cases} \dfrac{1}{98 \text{s}^{-1}} \times (0.92 + 2\pi n) \\[2mm] \dfrac{1}{98 \text{s}^{-1}} \times (\pi - 0.92 + 2\pi n) \end{cases}
\end{aligned}$$

where $n = 0, 1, 2, \ldots$. The upper solution corresponds to the spring tip moving up through the point 3.9 centimeters. The lower solution corresponds to the tip moving down through the point 3.9 centimeters. Thus, use $n = 0$ to find the first two times the tip moves through the point 3.9 centimeters. The result is

$$t = \begin{cases} 0.0094 \, \text{s} \\ 0.023 \, \text{s} \end{cases}$$

So the second time the tip moves through the point 3.9 centimeters is at 0.023 seconds.

401. 46 s

The *frequency of rotation* is the number of turns the merry-go-round makes per unit of time. This is

$$f = \frac{13}{10 \text{ min}} \times \frac{1 \text{ min}}{60 \text{ s}}$$

$$= 0.0217 \text{ Hz}$$

The *period* is the inverse of the frequency, so

$$T = \frac{1}{f}$$

$$= \frac{1}{0.0217 \text{ Hz}}$$

$$= 46 \text{ s}$$

402. 5.5 rad/s

Each face of the cube reflects the laser beam, so the reflected laser beam has a frequency of $f_{laser} = 4f_{cube}$ where f_{cube} is the cube's frequency of rotation. Thus, the cube's angular frequency is

$$\omega_{cube} = 2\pi f_{cube}$$

$$= 2\pi \times \frac{f_{laser}}{4}$$

$$= \frac{\pi f_{laser}}{2}$$

Using $f_{laser} = 3.5 \text{ Hz}$ gives

$$\omega_{cube} = \frac{\pi f_{laser}}{2}$$

$$= \frac{\pi \times 3.5 \text{ Hz}}{2}$$

$$= 5.5 \text{ rad/s}$$

403. 34

In each period, the branch moves twice through its equilibrium position: once going up and once going down. Thus, in 1 minute it moves through its equilibrium position 34 times.

$$N = 2 \times \frac{1 \text{ min}}{T}$$

$$= 2 \times \frac{1 \text{ min}}{3.5 \text{ s}} \times \frac{60 \text{ s}}{1 \text{ min}}$$

$$= 34$$

404. 2.6 m/s

Insert the given time into the equation for velocity to find

$$v = 2.8 \text{ m/s} \times \cos\left(3.5 \text{ s}^{-1} \times t\right)$$

$$= 2.8 \text{ m/s} \times \cos\left(3.5 \text{ s}^{-1} \times 45 \text{ s}\right)$$

$$= 2.6 \text{ m/s}$$

405. $7.3 \times 10^2 \text{ cm/s}^2$

The oscillator's angular speed is $\omega = 14 \text{ s}^{-1}$. The oscillator's acceleration is

$$a = -3.7 \text{ cm} \times \omega^2 \times \sin\left(14 \text{ s}^{-1} \times t\right)$$
$$= -3.7 \text{ cm} \times \left(14 \text{ s}^{-1}\right)^2 \times \sin\left(14 \text{ s}^{-1} \times t\right)$$
$$= -7.3 \times 10^2 \text{ cm/s}^2 \times \sin\left(14 \text{ s}^{-1} \times t\right)$$

The minimum value of the cosine function is -1, so the maximum acceleration is $7.3 \times 10^2 \text{ cm/s}^2$.

406. $4.5 \times 10^{-3} \text{ m}$

The displacement of an oscillator is $y = A \sin(\omega t)$. The maximum of the sine function is A, so A is the oscillator's maximum displacement.

The acceleration of an oscillator is given by $a = -A\omega^2 \sin(\omega t)$. The maximum of the sine function is 1, so the maximum acceleration of the oscillator is $a_{max} = A\omega^2$. You know that $\omega = 54$ radians per second and that $a_{max} = 13$ meters per second squared. Solving the equation for a_{max} for A gives

$$a_{max} = A\omega^2$$
$$A = \frac{a_{max}}{\omega^2}$$
$$= \frac{13 \text{ m/s}^2}{(54 \text{ rad/s})^2}$$
$$= 4.5 \times 10^{-3} \text{ m}$$

So the maximum displacement of the oscillator is 4.5×10^{-3} meters.

407. $1.0 \times 10^3 \text{ m/s}^2$

The acceleration is $a = -A\omega^2 \sin(\omega t)$. From the equation for displacement of an oscillator, you have

$$A = 4.5 \text{ m}$$
$$\omega = 23 \text{ s}^{-1}$$

Inserting these values and $t = 4.8$ into the equation for acceleration gives

$$a = -A\omega^2 \sin(\omega t)$$
$$= -4.5 \text{ m} \times \left(23 \text{ s}^{-1}\right)^2 \sin\left(23 \text{ s}^{-1} \times 4.8 \text{ s}\right)$$
$$= 1.0 \times 10^3 \text{ m/s}^2$$

408. 0.33 s

You know that $m = 3.6$ kilograms and $k = 1,300$ newtons per meter. Inserting these values into the equation for the period of a mass oscillating on a spring gives

$$T = 2\pi \sqrt{\frac{m}{k}}$$
$$= 2\pi \sqrt{\frac{3.6 \text{ kg}}{1,300 \text{ N/m}}}$$
$$= 0.33 \text{ s}$$

409.   0.81 kg

The period of oscillation is $T = \frac{1}{f}$. Solve the equation for the period of a mass on a spring for the mass m:

$$T = 2\pi \sqrt{\frac{m}{k}}$$

$$\frac{T}{2\pi} = \sqrt{\frac{m}{k}}$$

$$\left(\frac{T}{2\pi}\right)^2 = \frac{m}{k}$$

$$m = k\left(\frac{T}{2\pi}\right)^2$$

$$= k\left(\frac{1}{2\pi f}\right)^2$$

Inserting the given values of $f = 4.5$ hertz and $k = 650$ newtons per meter gives

$$m = k\left(\frac{1}{2\pi f}\right)^2$$

$$= 650 \text{ N/m} \times \left(\frac{1}{2\pi \times 4.5 \text{ Hz}}\right)^2$$

$$= 0.81 \text{ kg}$$

410. 4.0×10^2 N/m

The spring completes a half oscillation in 0.50 seconds, so the period is

$$T = \frac{0.50 \text{ s}}{1/2}$$

$$= 1.0 \text{ s}$$

Solving the equation for the period for the spring constant gives

$$T = 2\pi \sqrt{\frac{m}{k}}$$

$$\frac{T}{2\pi} = \sqrt{\frac{m}{k}}$$

$$\left(\frac{T}{2\pi}\right)^2 = \frac{m}{k}$$

$$k = m\left(\frac{2\pi}{T}\right)^2$$

Inserting the period from above and the given mass $m = 10$ kilograms gives a spring constant of

$$k = m\left(\frac{2\pi}{T}\right)^2$$

$$= 10 \text{ kg} \times \left(\frac{2\pi}{1.0 \text{ s}}\right)^2$$

$$= 4.0 \times 10^2 \text{ N/m}$$

411.   **30**

The frequency of a mass oscillating on a spring is $f = \frac{1}{2\pi}\sqrt{\frac{k}{m}}$. You know that $f_1 = 5.5 \times f_2$. The frequency of each spring is then

$$f_1 = \frac{1}{2\pi}\sqrt{\frac{k_1}{m}}$$
$$= 5.5 \times f_2$$
$$f_2 = \frac{1}{2\pi}\sqrt{\frac{k_2}{m}}$$

Taking the ratio of these two equations and solving for k_1/k_2 gives

$$\frac{5.5 \times f_2}{f_2} = \frac{\frac{1}{2\pi}\sqrt{\frac{k_1}{m}}}{\frac{1}{2\pi}\sqrt{\frac{k_2}{m}}}$$
$$5.5 = \sqrt{\frac{k_1}{k_2}}$$
$$\frac{k_1}{k_2} = (5.5)^2$$
$$= 30$$

412.   **13 J**

The total energy of a spring system is its kinetic energy plus its potential energy. Its potential energy is $PE = \frac{1}{2}kx^2$. At equilibrium, $x = 0$ meters, so the potential energy is zero.

The kinetic energy is $KE = \frac{1}{2}mv^2$. Inserting $m = 2.1$ kilograms and $v = 3.5$ meters per second gives

$$KE = \frac{1}{2}mv^2$$
$$= \frac{1}{2} \times 2.1\,\text{kg} \times (3.5\,\text{m/s})^2$$
$$= 13\,\text{J}$$

413.   **2.0×10^2 J**

The potential energy of a spring is $PE = \frac{1}{2}kx^2$. For $k = 1{,}300$ newtons per meter and $x = 0.55$ meters, the potential energy is

$$PE = \frac{1}{2} \times 1{,}300\,\text{N/m} \times (0.55\,\text{m})^2$$
$$= 2.0 \times 10^2\,\text{J}$$

414.   6.8 J

The potential energy of a spring-mass system is $PE = \frac{1}{2}kx^2$. At maximum compression, you have $PE_{max} = \frac{1}{2}kx^2_{max}$, where PE_{max} 12 joules. Solving for x_{max} gives

$$PE_{max} = \frac{1}{2}kx^2_{max}$$

$$\frac{2PE_{max}}{k} = x^2_{max}$$

$$x_{max} = \pm\sqrt{\frac{2PE_{max}}{k}}$$

At three quarters of the maximum position, you have $PE_{3/4} = \frac{1}{2}k\left(\frac{3}{4}x_{max}\right)^2$. Plugging in the result for x_{max} from above gives

$$PE_{3/4} = \frac{1}{2}k\left(\frac{3}{4}x_{max}\right)^2$$

$$= \frac{1}{2}k\left[\frac{3}{4}\left(\pm\sqrt{\frac{2PE_{max}}{k}}\right)\right]^2$$

$$= \frac{9}{32}k \times \frac{2PE_{max}}{k}$$

$$= \frac{9}{16} \times PE_{max}$$

$$= \frac{9}{16} \times 12\,J$$

$$= 6.8\ J$$

415. 2.7×10^5 J

The frequency of an oscillating spring-mass system is $f = \frac{1}{2\pi}\sqrt{\frac{k}{m}}$. You can solve for the spring constant of the system. This gives

$$f = \frac{1}{2\pi}\sqrt{\frac{k}{m}}$$

$$2\pi f = \sqrt{\frac{k}{m}}$$

$$(2\pi f)^2 = \frac{k}{m}$$

$$m(2\pi f)^2 = k$$

Insert this result into the equation for the potential energy of the system to find

$$PE = \frac{1}{2}kx^2$$

$$= \frac{1}{2}m(2\pi f)^2 x^2$$

$$= \frac{1}{2} \times 4\pi^2 \times 2.3\,kg \times (35\,Hz)^2 \times (2.2\,m)^2$$

$$= 2.7 \times 10^5\ J$$

416.　0.36

At equilibrium, the total energy of a spring is just its kinetic energy, which is

$$E = KE$$
$$= \frac{1}{2}mv_{max}^2$$

At the maximum extension, the total energy is just the potential energy, so you have

$$E = PE$$
$$= \frac{1}{2}kx_{max}^2$$

Because the total energy remains constant, you have

$$\frac{1}{2}kx_{max}^2 = \frac{1}{2}mv_{max}^2$$
$$kx_{max}^2 = mv_{max}^2$$

Applying this equation to spring one and spring two and taking the ratio gives

$$\frac{k_1 x_{max,1}^2}{k_2 x_{max,2}^2} = \frac{mv_1^2}{mv_2^2}$$
$$= \frac{v_{max,1}^2}{v_{max,2}^2}$$

Solving this for the ratio k_1/k_2 gives

$$\frac{k_1 x_{max,1}^2}{k_2 x_{max,2}^2} = \frac{v_{max,1}^2}{v_{max,2}^2}$$
$$\frac{k_1}{k_2} = \frac{v_{max,1}^2}{v_{max,2}^2} \frac{x_{max,2}^2}{x_{max,1}^2}$$
$$= \frac{(13\,\text{m/s})^2}{(8.5\,\text{m/s})^2} \times \frac{(1.3\,\text{m})^2}{(3.3\,\text{m})^2}$$
$$= 0.36$$

417.　1.2 Hz

The frequency of a pendulum is $f = \frac{1}{2\pi}\sqrt{\frac{g}{L}}$. Changing the mass doesn't affect the frequency. Increasing the length of the pendulum arm by a factor of 3 gives a new frequency of

$$f' = \frac{1}{2\pi}\sqrt{\frac{g}{3L}}$$
$$= \frac{1}{\sqrt{3}}\frac{1}{2\pi}\sqrt{\frac{g}{L}}$$
$$= \frac{1}{\sqrt{3}}f$$

Plugging in $f = 2.1$ hertz gives

$$f' = \frac{1}{\sqrt{3}}f$$
$$= \frac{1}{\sqrt{3}} \times 2.1\,\text{Hz}$$
$$= 1.2\,\text{Hz}$$

418.   1.9 s

The equation for the period of a pendulum is $T = 2\pi\sqrt{\frac{L}{g}}$. Let $g' = 1.6$ meters per second squared be the acceleration due to gravity and $T' = 4.8$ seconds be the period on the moon. Apply the equation for the period on the moon and on Earth and take the ratio to find the period on Earth. The result is

$$\left. \begin{array}{l} T = 2\pi\sqrt{\frac{L}{g}} \\[2mm] T' = 2\pi\sqrt{\frac{L}{g'}} \end{array} \right\} \quad \frac{T}{T'} = \frac{\sqrt{L/g}}{\sqrt{L/g'}}$$

$$= \sqrt{\frac{g'}{g}}$$

$$T = T'\sqrt{\frac{g'}{g}}$$

$$= 4.8\,\text{s} \times \sqrt{\frac{1.6\,\text{m/s}^2}{9.8\,\text{m/s}^2}}$$

$$= 1.9\,\text{s}$$

419. 1.4×10^2 rad/s²

The angular displacement of a pendulum is given by $\theta = A\sin(\omega t)$. Because the maximum value of the sine function is 1, the maximum angular displacement is A, which you are told is 0.13 radians. The angular acceleration of the pendulum is $\alpha = -A\omega^2\sin(\omega t)$.

The minimum value of the sine function is –1, so the maximum angular acceleration is $\alpha_{\text{max}} = A\omega^2$.

Inserting $A = 0.13$ radians and $\omega = 33$ radians per second gives

$$\alpha_{\text{max}} = A\omega^2$$
$$= 0.13\,\text{rad} \times (33\,\text{rad/s})^2$$
$$= 1.4 \times 10^2\,\text{rad/s}^2$$

420.   9

The period of a pendulum is given by $T = 2\pi\sqrt{\dfrac{L}{g}}$. The desired period is $T' = \dfrac{T}{3}$, which means that $T = 3T'$. Inserting this in the equation for the period gives

$$T = 2\pi\sqrt{\frac{L}{g}}$$

$$3T' = 2\pi\sqrt{\frac{L}{g}}$$

$$T' = \frac{2\pi}{3}\sqrt{\frac{L}{g}}$$

$$= 2\pi\sqrt{\frac{L}{9g}}$$

Thus, the required acceleration due to gravity is 9 times the acceleration due to gravity near the surface of Earth.

421.   22°C

To convert from Fahrenheit to Celsius, first subtract 32 degrees from the Fahrenheit temperature. Then multiply by $\dfrac{5}{9}$:

$$72° - 32° = 40°$$

$$\frac{5}{9}\left(40°\right) = 22°$$

422.   4.4°C

To convert from Fahrenheit to Celsius, first subtract 32 degrees from the Fahrenheit temperature. Then multiply by $\dfrac{5}{9}$:

$$40° - 32° = 8°$$

$$\frac{5}{9}\left(8°\right) = 4.4°$$

423.   0°C

The Celsius temperature scale is based on the freezing and boiling points of pure water at sea level. Water freezes at 0 degrees Celsius and boils at 100 degrees Celsius.

424.   273.15 K

The Kelvin temperature scale is based on *absolute zero*, the temperature at which all molecular motion stops, which is much colder than the freezing point of water. Water freezes at 273.15 kelvins.

425.   1,811 K

To convert from degrees Fahrenheit to kelvins, first convert to Celsius, and then add 273.15:

$$\frac{5}{9}\left(2{,}800°F - 32°\right) = 1{,}538°C$$

$$1{,}538°C + 273.15 \approx 1{,}811\,K$$

426. Temperature measures speed, and 0 K indicates no motion.

Zero kelvins is also known as *absolute zero* because the temperature of a material is, in effect, a description of how fast the smallest particles of the material are moving. Zero kelvins is the temperature referring to no motion at all, and you can't move more slowly than stopped!

427. 10.03 m

This one is a plug-and-chug problem, using the formula given in the problem itself: $\Delta L = \alpha L_o \Delta T$. In other words, the change in length is equal to the coefficient of expansion multiplied by the original length times the change in temperature.

According to the question, the original length is 10 meters, the coefficient of linear expansion is 0.0000222, and the change in temperature is $423\,K - 273\,K = 150\,K$. Plug these values into the formula and chug away:

$$\Delta L = \alpha L_o \Delta T$$
$$= (0.0000222)(10.0\,m)(150.0\,K)$$
$$= 0.033\,m$$

Now add this change in length to the initial length of 10 meters to get the final length:

$$10.0\,m + 0.033\,m \approx 10.03\,m$$

428. 208°C

You need to modify the formula for thermal expansion to solve for temperature:

$$\Delta L = \alpha L_o \Delta T$$
$$\Delta T = \frac{\Delta L}{\alpha L_o}$$

Then plug in the new information:

$$\Delta T = \frac{0.005\,m}{\left(0.000012\,^\circ C^{-1}\right)(2\,m)}$$
$$\Delta T = 208\,^\circ C$$

With a 5-millimeter gap, the homeowners have room for a 208 degree Celsius temperature range. That should be plenty!

429. 12.5 m

To find the original length, you need to solve the formula for calculating linear thermal expansion for L_o:

$$\Delta L = \alpha L_o \Delta T$$
$$L_o = \frac{\Delta L}{\Delta T \cdot \alpha}$$

Then plug in the new information:

$$L_o = \frac{0.0172\,m}{\left(115\,^\circ C\right)(0.000012)}$$
$$L_o = 12.5\,m$$

The original length of the rod was 12.5 meters.

430. 0.04 L

The formula for thermal volume expansion is $\Delta V = \beta V_0 \Delta T$. Because you're solving for the change in volume, you don't need to make any changes to the equation; just plug and chug:

$$\Delta V = \left(0.000526°C^{-1}\right)(2.0\,\text{L})\left(40°C\right)$$
$$\Delta V = 0.04\,\text{L}$$

The soda bottle needs about 0.04 liters of extra volume to handle the expected expansion.

431. 9.3°C

The formula for thermal volume expansion is $\Delta V = \beta V_0 \Delta T$. The question asks for a final temperature, so you need to know the change in temperature to add to the initial temperature of 22 degrees Celsius. Solve the formula for ΔT and plug in your known values:

The change in volume is $11.92\,\text{oz} - 12\,\text{oz} = -0.08\,\text{oz}$.

$$\Delta T = \frac{\Delta V}{\beta \bullet V_0}$$
$$\Delta T = \frac{(-0.08\,\text{oz})}{\left(5.26 \times 10^{-4}°C^{-1}\right)(12.0\,\text{oz})}$$
$$\Delta T = -12.67°C$$

Now you can add the initial temperature to the change in temperature to find the final temperature of the cooler:

$$T_F = T_0 + \Delta T$$
$$T_F = 22.0°C + \left(-12.67°C\right)$$
$$T_F = 9.3°C$$

432. 0.067 gal

This one is challenging. First you need to calculate the volume expansion of the can, and then calculate the expansion of the paint. Finally, compare the difference between them to the initial empty volume.

Start with the change in volume of the can, using the formula $\Delta V = \beta V_0 \Delta T$:

$$\Delta V = \left(3.6 \times 10^{-5}°C^{-1}\right)(1.0\,\text{gal})\left(33°C - 23°C\right)$$
$$\Delta V = 3.6 \times 10^{-4}\,\text{gal}$$

Now find the change in volume of the paint the same way:

$$\Delta V = \left(4.05 \times 10^{-4}°C^{-1}\right)(0.93\,\text{gal})\left(33°C - 23°C\right)$$
$$\Delta V = 3.8 \times 10^{-3}\,\text{gal}$$

The difference in volume between them is the change in empty space at the top of the can. (Note that the volume of the empty space decreases because the increase in paint volume is larger than the increase in can volume.)

$$3.8 \times 10^{-3} \text{ gal} - 3.6 \times 10^{-4} \text{ gal} = 3.4 \times 10^{-3} \text{ gal}$$

Because the can starts with 1 gal − 0.930 gal = 0.070 gal of initial empty space, you can subtract the change in empty space from that to get the final unused volume in the can: 0.070 gal − 0.0034 gal = 0.067 gal.

433. 72,240 J

To find the needed amount of thermal energy, use the formula $Q = cm\Delta T$, where Q is the thermal energy, c is the specific heat capacity of aluminum, m is the mass in kilograms, and ΔT is the change in temperature:

$$Q = \left(\frac{903 \text{ J}}{\text{kg} \cdot {}^{\circ}\text{C}} \right) (2 \text{ kg}) (40^{\circ}\text{C})$$
$$Q = 72,240 \text{ J}$$

To increase the temperature of 2 kilograms of aluminum, you need 72,240 joules of heat energy.

434. $\dfrac{853 \text{ J}}{\text{kg} \cdot {}^{\circ}\text{C}}$

You can find the specific heat capacity, c, by solving the thermal energy formula $Q = mc\Delta T$ for c: $c = \dfrac{Q}{m\Delta T}$.

Use this version of the formula with the values given in the problem:

$$c = \frac{12,800 \text{ J}}{1 \text{ kg} \cdot 15^{\circ}\text{C}}$$
$$c = \frac{853 \text{ J}}{\text{kg} \cdot {}^{\circ}\text{C}}$$

The specific heat capacity of the material is $\dfrac{853 \text{ J}}{\text{kg} \cdot {}^{\circ}\text{C}}$.

435. 292 kg

To calculate the mass, solve the thermal energy formula for mass, $m = \dfrac{Q}{c\Delta T}$, and plug in the values from the question:

$$m = \frac{8.79 \times 10^{7} \text{ J}}{\left(\dfrac{4,180 \text{ J}}{\text{kg} \cdot {}^{\circ}\text{C}} \right) (72.0^{\circ}\text{C})}$$
$$m = 292 \text{ kg}$$

The water has a mass of 292 kilograms.

436. 58°C

Remember that the energy added to the mixture by the hotter coffee is equal to the energy absorbed by the cooler coffee, and the final temperature for both volumes is the same. Start with the formula for calculating thermal energy change, $\Delta Q = mc(T - T_0)$; set the energy gain from one volume of coffee equal to the loss of the other; and simplify using the values from the question:

$$m_1 c_1 (T - T_{1,0}) = -(m_1 c_1 (T - T_{2,0}))$$

$$m_1 c_1 (T - T_{1,0}) = (m_1 c_1 (-T + T_{2,0}))$$

$$(0.5 \text{ kg}) \left(\frac{4,180 \text{ J}}{\text{kg} \bullet °C} \right) (T - 88°C) = (1.2 \text{ kg}) \left(\frac{4,180 \text{ J}}{\text{kg} \bullet °C} \right) (-T + 46°C)$$

$$\left(\frac{2,090 \text{ J}}{°C} \right) (T - 88°C) = \left(\frac{5,016 \text{ J}}{°C} \right) (-T + 46°C)$$

$$(2,090 \text{ J})(T - 88°C) = (5,016 \text{ J})(-T + 46°C)$$

$$(T - 88°C) = (2.4)(-T + 46°C)$$

$$T - 88°C = -2.4T + 110.4°C$$

$$3.4T = 198.4°C$$

$$T = 58°C$$

The mixture reaches equilibrium at 58 degrees Celsius.

437. 4.00×10^6 J

To calculate the thermal energy needed to turn the water into steam, multiply the latent heat of vaporization of water, 2.26×10^6 J/kg, by the given mass of the water, 1.77 kilograms:

$$\Delta Q = \left(\frac{2.26 \times 10^6 \text{ J}}{\text{kg}} \right) (1.77 \text{ kg})$$

$$\Delta Q = 4.00 \times 10^6 \text{ J}$$

It takes 4.00×10^6 J to vaporize the water.

438. $\dfrac{2.09 \times 10^5 \text{ J}}{\text{kg}}$

To calculate the latent heat of fusion, you need to solve the formula of heat energy of fusion for L, the latent heat: $L = \frac{\Delta Q}{m}$. Then plug in the given values for mass and heat energy and solve:

$$L = \frac{1.55 \times 10^6 \text{ J}}{7.40 \text{ kg}}$$

$$L = \frac{2.09 \times 10^5 \text{ J}}{\text{kg}}$$

The latent heat of fusion of copper is $\frac{2.09 \times 10^5 \text{ J}}{\text{kg}}$.

439.   22.7 kg

To calculate the mass, you need to solve the formula for heat energy of vaporization for mass, m: $m = \frac{\Delta Q}{L}$. Then use the given values for L and heat energy to solve:

$$m = \frac{5.126 \times 10^7 \text{ J}}{\left(\frac{2.258 \times 10^6 \text{ J}}{\text{kg}} \right)}$$

$$m = 22.70 \text{ kg}$$

The stated 5.126×10^7 J of heat energy vaporizes 22.7 kilograms of boiling-temperature water.

440.   0°C

This is a bit of a trick question. It hardly seems possible to add a million joules of heat energy to the ice without changing the temperature at all, but you can verify this by using the formula for latent heat of fusion, $Q = mL$:

$$Q = (3 \text{ kg})(3.35 \times 10^5 \text{ J/kg})$$

$$Q = 1,005,000 \text{ J}$$

It actually takes just a bit *more* than 1 million joules of heat energy to melt the block of ice, let alone heat up the water afterward.

441.   12 kg

To solve this problem, first find the heat energy the water loses to drop in temperature by 100 degrees Celsius, because it goes from boiling temperature to freezing temperature, and then calculate how much ice melts with that much heat energy.

To calculate the heat energy the water gives up, first find the mass of the water (a mass of 2.5 gallons at 3.8 kilograms per gallon):

$$m = (2.5 \text{ gal})\left(\frac{3.8 \text{ kg}}{\text{gal}} \right) = 9.5 \text{ kg}$$

Now you have m for your formula for the change in heat energy, $\Delta Q = mc(T - T_0)$. The question gives you the specific heat of water, c, and the temperature change, so you have:

$$\Delta Q = \left(\frac{4,180 \text{ J}}{\text{kg} \cdot °\text{C}} \right)(9.5 \text{ kg})(100°\text{C})$$

$$\Delta Q = 4.0 \times 10^6 \text{ J}$$

To calculate the mass of the ice that melts, you need to solve the formula for heat energy of fusion for mass, m: $m = \frac{\Delta Q}{L}$. Then use the given value for L, the latent heat of fusion of water, and the heat energy ΔQ you just calculated:

$$m = \frac{4.0 \times 10^6 \text{ J}}{\left(\frac{3.35 \times 10^5 \text{ J}}{\text{kg}} \right)}$$

$$m = 12 \text{ kg}$$

You can melt 12 kilograms of the ice with the boiling water.

442. heat, fluid

Convection is a means of transferring heat through a fluid.

443. buoyancy

Buoyancy is the force that causes less-dense liquids, objects, or materials to rise above more-dense fluids.

444. Heating makes the molecules of the fluid move more quickly.

When energy (heat) is added to the molecules of a fluid, the molecules become more energetic, moving around more quickly and with more force. More-energetic molecules take up more space, so fewer can fit into the same space, and the fluid becomes less dense.

445. A standard oven uses natural convection.

Both standard and convection ovens rely on convection to move heated air around the food inside. A standard oven relies on *natural* convection to move the air next to the heating element away so the cooler air can be warmed. A convection oven uses fans to create *forced* convection to move the air.

446. just energy; the molecules of the material generally stay in the same area

When molecules of a substance are heated, they move around more. When more active molecules (like the ones on the bottom of a pan) bump into less active molecules farther away from the heat source, they share some of their extra energy. In this way, heat energy can be transferred even through solid material.

447. The filling conducts heat better.

Right out of the oven, the crust and the filling are the same temperature, but the filling is a much more efficient heat conductor, so it transfers extra heat to your tongue more quickly than the crust does.

Later on, the filling is actually hotter than the crust because it has a greater thermal heat capacity and therefore cools more slowly.

448. 1.0×10^6 J

When everything else is the same, the amount of heat energy conducted through a given material is directly proportional to the time period in which it transfers.

Because heat energy is proportional to time, set up an equation relating the initial heat and time to the proposed time and unknown heat, x:

$$\frac{600,000 \text{ J}}{42 \text{ s}} = \frac{x \text{ J}}{70 \text{ s}}$$

$$\frac{(600,000)(70)}{42} = x$$

$$1,000,000 = x$$

$$x = 1.0 \times 10^6$$

The pan conducts 1.0×10^6 of heat energy in 70 seconds.

449. 1.7×10^4 J

Generally speaking, thermal conductivity is directly proportional to the cross-section area and inversely proportional to length. The first bar has a length of 16 centimeters and an area of 2 centimeters squared. The second is only 8 centimeters long (half as long), with an area of only 1 centimeter squared (half the cross-section area).

Cutting the length in half doubles the conductivity, but cutting the area in half also reduces the conductivity by half. So the same amount of energy is transferred in the same time period.

450. 1.9 J

The formula for calculating heat transfer by conduction through a material is $Q = \frac{kA\Delta Tt}{l}$. All the values you need are given in the question: k, the thermal conductivity constant; A, the cross-section area; ΔT, the change in temperature; and t, the time in seconds:

$$Q = \frac{kA\Delta Tt}{l}$$

$$Q = \frac{\frac{0.17 \text{ J}}{\text{s} \bullet \text{m} \bullet °\text{C}} \left(0.0002 \text{ m}^2\right)\left(100°\text{C}\right)(90 \text{ s})}{0.16 \text{ m}}$$

$$Q = \frac{\frac{0.17 \text{ J}}{1}(0.0002)(100)(90)}{0.16}$$

$$Q = 1.9 \text{ J}$$

The dowel transfers 1.9 joules of heat energy in 90 seconds.

451. 210,000 J

Use the formula for heat transfer by conduction, $Q = \frac{kA\Delta Tt}{l}$, where $k = 0.80\frac{\text{J}}{\text{S} \bullet \text{m} \bullet °\text{C}}$, $A = 1.0 \text{ m}^2$, $\Delta T = 22 - 0 = 22°\text{C}$, $t = 60$ s, and $l = 0.005$ m.

$$Q = \frac{\left(\frac{0.80 \text{ J}}{\text{s} \bullet \text{m} \bullet °\text{C}}\right)\left(1.0 \text{ m}^2\right)\left(22°\text{C}\right)(60 \text{ s})}{0.005 \text{ m}}$$

$$Q = 210,000 \text{ J}$$

452. about 30 hours

This is a two-part problem. First you need to find out how much heat energy is conducted along the bar each second, and then how much heat energy is required to melt the ice so you can calculate the number of seconds required.

Start with the formula for heat transfer by conduction, $Q = \frac{kA\Delta Tt}{l}$, where $k = 14.0\frac{\text{J}}{\text{s} \bullet \text{m} \bullet °\text{C}}$, $A = 2.0 \times 10^{-4} \text{ m}^2$, $\Delta T = 1,100°\text{C} - 0°\text{C} = 1$, $t = 1$ s, and $l = 0.5$ m.

$$Q = \frac{\left(\frac{14.0 \text{ J}}{\text{s} \bullet \text{m} \bullet °\text{C}}\right)\left(2 \times 10^{-4} \text{ m}^2\right)\left(1,100°\text{C}\right)(1 \text{ s})}{0.5 \text{ m}}$$

$$Q = 6.16 \text{ J/s}$$

Now use $Q = mL$, where Q is heat energy in joules, m is mass in kilograms, and L is the latent heat of fusion of water.

$$Q = (2\,\text{kg})\left(3.35 \times 10^5 \text{ J/kg}\right)$$
$$Q = 670{,}000 \text{ J}$$

Now divide the number of joules of heat energy needed, 670,000 joules, by the heat energy transferred per second, 6.16 joules, to get the time. The ice melts in approximately 109,000 seconds, or about $109.000 \text{ s} \bullet \dfrac{1\,\text{h}}{3{,}600\,\text{s}} = 30\,\text{h}$, or 30 hours.

453. all of it

The space between the sun and Earth is a near-perfect vacuum, so there is no mass to conduct heat and no fluid for convection. All the energy Earth receives from the sun is transferred via radiation.

454. 9.4×10^{30} W

The heat energy an object radiates is proportional to T^4, the surface temperature to the fourth power. Set up a proportion with the values from the question:

$$\frac{(x)}{(3{,}400\,\text{K})^4} = \frac{3.2 \times 10^{30}\,\text{W}}{(2{,}600\,\text{K})^4}$$
$$x = \frac{\left(3.2 \times 10^{30}\,\text{W}\right)\left((3{,}400)^4\right)}{(2{,}600)^4}$$
$$x = 9.4 \times 10^{30}\,\text{W}$$

The hotter star radiates about 9.4×10^{30} watts, which is about three times the energy of the cooler star.

455. 140 W

This question looks harder than it is. Just be careful to substitute the correct values into the formula.

Convert temperature from degrees Celsius to kelvins:

$$22°\text{C} = (22 + 273.15)\,\text{K} = 295.15\,\text{K}$$
$$33°\text{C} = (33 + 273.15)\,\text{K} = 306.15\,\text{K}$$
$$\frac{Q}{t} = e\sigma A\left(T_{hot}^4 - T_{cold}^4\right)$$
$$\frac{Q}{t} = (0.97)\left(\frac{5.67 \times 10^{-8}\,\text{J}}{\text{s} \bullet \text{m}^2 \bullet \text{K}^4}\right)\left(2.15\,\text{m}^2\right)\left(306.15^4\,\text{K}^4 - 295.15^4\,\text{K}^4\right)$$
$$\frac{Q}{t} = 141.44 \text{ watts} \approx 140 \text{ watts}$$

456. converting AMU to grams

Avogadro's number, 6.02×10^{23}, is the number of atoms of an element that has a mass in grams equal to the molecular mass of the element in AMU.

457.   18 g

One mole of any atom has a mass in grams equal to the average atomic mass in AMU. Because hydrogen has an average atomic mass of 1 AMU (as noted in the question), 1 mole of hydrogen atoms has a mass of 1 gram. Also as noted in the question, oxygen has an average atomic mass of 16 AMU, so 1 mole of oxygen atoms has a mass of 16 grams.

Water (H_2O) molecules each contain two hydrogen atoms and one oxygen atom, so 1 mole of H_2O has a mass of $2(1\,g)+1(16\,g)=18\,g$.

458.   The product of pressure and volume is conserved.

Boyle's law states that if all other factors remain constant, the product of pressure and volume is conserved.

459. 1.3×10^5 Pa

Solve Boyle's law for P_1:

$$P_1 \bullet V_1 = P_2 \bullet V_2$$
$$P_1 = \frac{P_2 \bullet V_2}{V_1}$$

Substitute the values from the question and simplify.

$$P_2 = 3.0 \times 10^5 \,\text{Pa}$$
$$V_2 = 1.25\,\text{L}$$
$$V_1 = 3.0\,\text{L}$$
$$P_1 = \frac{\left(3.0 \times 10^5 \,\text{Pa}\right)(1.25\,\text{L})}{3.0\,\text{L}}$$
$$P_1 = 1.3 \times 10^5 \,\text{Pa}$$

The initial pressure was $P_1 = 1.3 \times 10^5$ Pa.

460.   0.5 atm

The temperature of the sample is constant, so Boyle's law applies. Solve Boyle's law for

$$P_1 \bullet V_1 = P_2 \bullet V_2$$
$$P_2 = \frac{P_1 \bullet V_1}{V_2}$$

Substitute the values from the question and simplify.

$$P_1 = 4.2\,\text{atm}$$
$$V_1 = 0.3\,\text{m}^3$$
$$V_2 = 2.6\,\text{m}^3$$
$$P_2 = \frac{(4.2\,\text{atm})\left(0.3\,\text{m}^3\right)}{2.6\,\text{m}^3}$$
$$P_2 = 0.5\,\text{atm}$$

The pressure of the sample in the larger container would be 0.5 atmospheres.

461. $1.512 \times 10^{-3} \, \text{m}^3$

This is an application of Boyle's law because the temperature of the sample is constant. Solve the formula for V_2:

$$P_1 \bullet V_1 = P_2 \bullet V_2$$
$$V_2 = \frac{P_1 \bullet V_1}{P_2}$$

The initial pressure is given in atmospheres, and the final pressure in pascals, so convert atmospheres to pascals using the scale given in the question: $1 \, \text{atm} = 1.013 \times 10^5 \, \text{Pa}$.

$$P_1 = 1.013 \times 10^5 \, \text{Pa}$$
$$V_1 = 22.40 \, \text{L}$$
$$P_2 = 1.500 \times 10^6 \, \text{Pa}$$
$$V_2 = \frac{\left(1.013 \times 10^5 \, \text{Pa}\right)\left(22.40 \, \text{L}\right)}{1.500 \times 10^6 \, \text{Pa}}$$
$$V_2 = 1.512 \, \text{L}$$

Finally, convert to m³: 1.512 liters $= 1.512 \times 10^{-3} \, \text{m}^3$.

462. $0.04 \, \text{m}^3$

This is a perfect application of Charles's law. Solve for V_f and use the remaining values from the question:

$$V_i = 0.02 \, \text{m}^3$$
$$T_i = 225 \, \text{K}$$
$$T_f = 395 \, \text{K}$$
$$V_f = \frac{V_i \bullet T_f}{T_i}$$
$$V_f = \frac{\left(0.02 \, \text{m}^3\right)\left(395 \, \text{K}\right)}{225 \, \text{K}}$$
$$V_f = 0.04 \, \text{m}^3$$

The final volume will be 0.04 m³.

463. $23°\text{C}$

This is a great application of Charles's law, as pressure is constant. Note that you need to convert from degrees Celsius to kelvins to calculate, and then go back to degrees Celsius to get your final answer. Begin by solving Charles's law for final temperature, T_f:

$$T_f = \frac{V_f \bullet T_i}{V_i}$$
$$T_f = \frac{(8.23 \, \text{l}) \bullet (-56 + 273.15 \, \text{K})}{6.04 \, \text{L}}$$
$$T_f = 295.88 \, \text{K}$$
$$T_f = (295.88 - 273.15)°\text{C}$$
$$T_f = 23°\text{C}$$

The final temperature would be $23°\text{C}$.

464. 0.0003 m³

Don't let this one throw you; it's just another application of Charles's law. Solve the formula for V_f, and be sure to convert milliliters to cubic meters and degrees Celsius to kelvins:

$$V_f = \frac{V_i \bullet T_f}{T_i}$$

$$V_f = \frac{\left(0.0003 \text{ m}^3\right)(-64.00 + 273.15 \text{ K})}{-32.00 + 273.15 \text{ K}}$$

$$V_f = 0.0003 \text{ m}^3$$

Despite what appears to be a significant drop in temperature, the volume of the sample doesn't decrease by that much.

465. 580 K

The question essentially asks what happens to the temperature of an ideal gas if the volume is reduced by half.

According to the ideal gas law, the pressure of an ideal gas is inversely proportional to volume: Half the volume equals twice the pressure. Additionally, the temperature of an ideal gas is directly proportional to pressure: Twice the pressure equals twice the temperature. So, half the volume equals twice the temperature, resulting in a final temperature of $2(290 \text{ K}) = 580 \text{ K}$.

466. 92.4 mol

According to the ideal gas law, the temperature of an ideal gas is directly proportional to pressure: Twice the pressure equals twice the temperature. Additionally (assuming constant volume), pressure is directly proportional to the number of moles of gas molecules: Twice the number of moles of molecules in the same volume equals twice the pressure.

Because volume is constant here, the increase in temperature is proportional to the increase in moles. Set up a proportion:

$$\left(\frac{x}{449 \text{ K}}\right) = \left(\frac{56.0 \text{ mol}}{272 \text{ K}}\right)$$

$$x = 449 \bullet \left(\frac{56 \text{ mol}}{272}\right)$$

$$x = 92.4 \text{ mol}$$

If there are 56.0 moles of ideal gas molecules at 272 kelvins, there are 92.4 moles at 449 kelvins, if volume is constant.

467. Increase the volume to 4.6 liters.

To maintain a constant temperature, the pressure of the gas must remain constant. To increase the amount of the gas without increasing pressure, the volume must increase by the same factor as the amount.

The amount increases by a factor of 2, from 5.4 moles to 10.8 moles, so the volume must also be increased by a factor of 2, from 2.3 liters to 4.6 liters.

468. 2.1 atm

Recall that the amount of gas is directly proportional to pressure, and volume is inversely proportional to pressure, so you know the pressure will increase twice.

The amount of gas increases by a factor of 1.5:

$$\Delta A = \frac{17.4 \, \text{mol}}{11.6 \, \text{mol}}$$

$$\Delta A = 1.5$$

The volume of the container decreases by a factor of 1.4:

$$\Delta vol = \frac{3.3 \, \text{liters}}{4.6 \, \text{liters}}$$

$$\Delta vol = 0.72$$

Now multiply the initial pressure by the factor of increase in amount, and then divide by the decrease in volume (because pressure is *inversely* proportional to volume):

$$\frac{(1.0 \, \text{atm})(1.5)}{0.72} = 2.1 \, \text{atm}$$

Increasing the amount of gas by a factor of 1.5 and multiplying the volume by a factor of 0.72 results in an increase in pressure to 2.1 atm.

469. 443 m/s

You should recognize the formula for kinetic energy, though you probably remember it as $KE = \left(\frac{1}{2}\right)mv^2$. Conveniently, the problem includes it solved for v, velocity. Substitute the values for m = mass and KE = kinetic energy from the problem into the kinetic energy formula (solved for v):

$$v = \sqrt{\frac{2KE}{m}}$$

$$v = \sqrt{\frac{2(4.55 \times 10^{-21} \, \text{J})}{4.65 \times 10^{-26} \, \text{kg}}}$$

$$v = \sqrt{\frac{1.96 \times 10^5 \, \text{J}}{\text{kg}}}$$

$$v = \sqrt{\frac{1.96 \times 10^5 \, \text{kg} \bullet \text{m}^2}{\text{kg} \bullet \text{s}^2}} \left(\text{Recall that} 1 \, \text{J} = \frac{1 \, \text{kg} \bullet \text{m}^2}{\text{s}^2} \right)$$

$$v = 443 \, \text{m/s}$$

The average speed of the molecules is 443 meters per second.

470. 4.2×10^4 J

As suggested by the hint, the thermal energy of the sample is a pretty surprisingly straightforward calculation with $KE = \left(\frac{3}{2}\right)nRT$. Use $n = 8.5 \, \text{mol}$, $T = (125 + 273.15) \, \text{K}$, and $R = 8.31 \, \text{J/mol–K}$:

$$KE = \frac{3}{2}nRT$$
$$KE = \left(\frac{3}{2}\right)(8.5 \text{ mol})(8.31 \text{ J/mol} \bullet \text{K})((125+273.15) \text{ K})$$
$$KE = 4.2 \times 10^4 \text{ J}$$

The total thermal (kinetic) energy of the sample is about 4.2×10^4 joules.

---

471. **The change in internal energy (ΔU) is equal to heat gained or lost (Q) minus work done (W).**

The first law of thermodynamics relates heat, work, and internal energy. According to the law, the change in internal energy is equal to the heat energy the system absorbs or emits minus the work done by the system on its surroundings (or the work done on the system by its surroundings).

---

472. **–3,700 J**

Before you substitute values into the first law formula, $\Delta U = Q - W$, consider the signs of the values. The motor is doing work on its surroundings, so you know that W is positive. The motor is also giving off heat, which reduces its energy, so Q is negative. Now you know the correct signs, so substitute the values from the question and solve for ΔU:

$$\Delta U = Q - W$$
$$\Delta U = (-1,200 \text{ J}) - (2,500 \text{ J})$$
$$\Delta U = -3,700 \text{ J}$$

---

473. **The system does 700 J of work.**

Consider the signs of the values first. The system emits 2,150 joules of heat, so Q is negative. The internal energy of the system decreases by 2,850 joules, so ΔU is also negative. Substitute the values into the formula and solve:

$$\Delta U = Q - W$$
$$-2,850 \text{ J} = -2,150 \text{ J} - W$$
$$W = +700 \text{ J}$$

Because W is positive, you know that the system did work on its surroundings. So the system process emits 2,150 joules of heat as it does 700 joules of work.

---

474. **System x: –407 J; system y: –572 J**

Begin by working out the signs:

System x emits heat, so Q_x is negative; it has work done on it, so W_y is negative.

System y emits heat, so Q_x is negative; it has work done on it, so W_y is negative.

Substitute the values into the formula with the proper signs and solve for ΔU for each system:

$$\Delta U_x = Q_x - W_x$$
$$\Delta U_x = -1{,}235 \text{ J} - (-828) \text{ J}$$
$$\Delta U_x = -407 \text{ J}$$

$$\Delta U_y = Q_y - W_y$$
$$\Delta U_y = -2{,}120 \text{ J} - (-1{,}548) \text{ J}$$
$$\Delta U_y = -572 \text{ J}$$

System x suffers a loss in internal energy of 407 joules, and system y suffers a loss of 572 joules.

475. Volume increases.

Recall from the ideal gas law that an increase in temperature indicates more activity in the molecules or atoms composing the gas. Greater activity either results in greater pressure on the container holding the gas or results in the gas requiring greater volume to maintain the same pressure. Because an isobaric process indicates no change in pressure, increasing the temperature of a given amount of gas results in an increase in volume.

476. 58.7 J

Because pressure is constant, you can use the work formula $W = P\Delta V$. Here, pressure, P, is 1.35×10^6 pascals, and the change in volume, ΔV, is 4.35×10^{-5}.

$$W = P\Delta V$$
$$W = \left(1.35 \times 10^6 \text{ Pa}\right)\left(4.35 \times 10^{-5} \text{ m}^3\right)$$
$$W = 58.7 \text{ J}$$

The gas does 58.7 joules of work on the atmosphere as it expands.

477. The internal energy of the sample is increased by 245 J.

To calculate the change in internal energy using $\Delta U = Q - W$, you first need to know W because Q is given in the hint. Use the work formula $W = P\Delta V$:

$$W = P\Delta V$$
$$W = \left(2.02 \times 10^5 \text{ Pa}\right)\left(-2.30 \times 10^{-3} \text{ m}^3\right)$$
$$W = -465 \text{ J}$$

Now substitute this value for W into the internal energy formula:

$$\Delta U = Q - W$$
$$\Delta U = -220 \text{ J} - (-465 \text{ J})$$
$$\Delta U = 245 \text{ J}$$

478. The system is isothermal.

The question specifies that the temperature is held constant, meaning that the gas maintains the *same* (iso-) *temperature* (thermal) value. This system is an isothermal system.

479. 5.94×10^4 J

In an isothermal process, the internal energy, ΔU, of a sample remains unchanged because internal energy is a function of temperature: no temperature change = no internal energy change. Use $\Delta U = Q - W$, solved for W (work), with $\Delta U = 0$ and $Q = 5.94 \times 10^4$ joules:

$$\Delta U = Q - W$$
$$W = Q - \Delta U$$
$$W = \left(5.94 \times 10^4 \text{ J}\right) - 0$$
$$W = 5.94 \times 10^4 \text{ J}$$

The work done is 5.94×10^4 joules, the same as the heat energy added to the system.

480. 278 K

The work done by an expanding gas that is held at a constant temperature is: $W = nRT \cdot \ln\left(\dfrac{V_f}{V_i}\right)$.

Solve for T to get: $T = \dfrac{W}{nR \cdot \ln\left(\dfrac{V_f}{V_i}\right)}$, where:

T = temperature

W = work done: 1.12×10^4 J

n = number of moles: 7.00 mol

R = the gas constant: $\dfrac{8.31 \text{ J}}{(\text{mol} \cdot \text{K})}$

V_f = final volume: 2.00 m³

V_i = initial volume: 1.00 m³

$$T = \frac{W}{nR \cdot \ln\left(\dfrac{V_f}{V_i}\right)}$$

$$T = \frac{1.12 \times 10^4 \text{ J}}{(7.00 \text{ mol})\left(8.31 \dfrac{\text{J}}{\text{mol} \cdot \text{K}}\right) \cdot \ln\left(\dfrac{2.00 \text{ m}^3}{1.00 \text{ m}^3}\right)}$$

$$T = 278 \text{ K}$$

481. 8,600 J

This process is isothermal because the temperature is constant. To calculate heat energy, Q, use the fact that $\Delta U = 0$ in an isothermal system; because $\Delta U = Q - W$, then $Q = W$. So calculate W using the familiar formula $W = nRT \cdot \ln\left(\dfrac{V_f}{V_i}\right)$. The answer is equal to Q:

$$W = nRT \cdot \ln\left(\frac{V_f}{V_i}\right)$$

$$W = (3.5 \text{ mol})\left(8.31 \frac{\text{J}}{\text{mol} \cdot \text{K}}\right)(55 + 273.15 \text{ K}) \cdot \ln\left(\frac{0.0017 \text{ m}^3}{0.0042 \text{ m}^3}\right)$$

$$W = -8,600 \text{ J}$$

If $W = -8,600$, then $Q = -8,600$ also. The negative sign indicates the system emits energy.

482. 0 J

The question specifies that this process is isochoric, meaning that the volume is constant. $W = P\Delta V$, so if $\Delta V = 0$, then $W = 0$ also.

A gas does no work during an isochoric process.

483. No, because the volume would increase with the temperature.

The universal gas law states that volume is proportional to temperature if amount and pressure are constant. The question states that you are working with a single sample held at constant pressure, so you know that the volume must increase with the temperature. An isochoric process requires constant volume, so this process can't be isochoric.

484. –237 J

Compressed air is stored in rigid containers capable of sustaining the pressure of the air inside, so you know that the volume of the container is constant; this process is isochoric. The question specifies that the gas emits heat energy, so Q is negative. Use $\Delta U = Q - W$, where $W = 0$ and $Q = -237$ J:

$$\Delta U = -237 \text{ J} - 0 \text{ J}$$
$$\Delta U = -237 \text{ J}$$

485. 510 J

The tank is rigid, so the process is isochoric; thus, $\Delta U = Q$. Use $Q = cm\Delta T$ as the question suggests to find Q:

$$Q = cm\Delta T$$
$$Q = \left(2{,}020\frac{\text{J}}{\text{kg}\bullet\text{K}}\right)(0.00326 \text{ kg})(78 \text{ K})$$
$$Q = 510 \text{ J}$$

Because $\Delta U = Q$, all that remains is to verify the sign of the 510-joule change in energy. The question specifies that the gas is heated, indicating that energy was added to it. The internal change in energy is an increase of 510 joules.

486. The system is adiabatic.

In an adiabatic process, no heat flows to or from the system. Because Q in the $\Delta U = Q - W$ formula represents heat energy, $Q = 0$ in an adiabatic process; therefore, $\Delta U = -W$.

487. 0.46 atm

Use the formula $P_i V_i^{5/3} = P_f V_f^{5/3}$, solved for P_f. Initial pressure is $P_i = 1.0$ atmosphere. Initial volume is $V_i = 2.2$ liters. Final volume is $V_f = 3.5$ liters.

$$P_iV_i^{5/3} = P_fV_f^{5/3}$$

$$P_f = \left(\frac{P_iV_i^{5/3}}{V_f^{5/3}} \right)$$

$$P_f = \left(\frac{1.0 \text{ atm} \cdot 2.2 \text{ L}^{5/3}}{3.5 \text{ L}^{5/3}} \right)$$

$$P_f = 0.46 \text{ atm}$$

The final pressure is 0.46 atmospheres.

488.   3.2 L

Simply use the formula $P_iV_i^{5/3} = P_fV_f^{5/3}$, solved for V_i. However, the math is a bit tricky to solve because of the fractional exponent, so take it a step at a time:

$$P_iV_i^{5/3} = P_fV_f^{5/3}$$

$$V_i^{5/3} = \frac{P_fV_f^{5/3}}{P_i}$$

The trick now is to remember that you multiply when you raise a power to a power. You know that $\frac{5}{3} \times \frac{3}{5} = 1$, so just raise both sides of the equation to the power of $\frac{3}{5}$ to get V_i:

$$\left(V_i^{5/3} \right)^{\frac{3}{5}} = \left(\frac{P_fV_f^{5/3}}{P_i} \right)^{\frac{3}{5}}$$

$$V_i = \left(\frac{P_fV_f^{5/3}}{P_i} \right)^{\frac{3}{5}}$$

Now you can substitute the values from the question and solve. Final pressure is $P_f = 0.65$ atmosphere. Final volume is $V_f = 7.8$ liters. Initial pressure is $P_i = 2.8$ atmospheres.

$$V_i = \left(\frac{P_fV_f^{5/3}}{P_i} \right)^{\frac{3}{5}}$$

$$V_i = \left(\frac{0.65 \text{ atm} \cdot 7.8 \text{ liters}^{5/3}}{2.8 \text{ atm}} \right)^{\frac{3}{5}}$$

$$V_i = 3.2 \text{ liters}$$

The initial volume was 3.2 liters.

489. 6.10×10^{-4} L

This question requires a bit of unit conversion to get the final answer. Begin by solving the adiabatic pressure/volume proportion for final volume V_f:

$$P_iV_i^{5/3} = P_fV_f^{5/3}$$

$$V_f = \left(\frac{P_iV_i^{5/3}}{P_f} \right)^{\frac{3}{5}}$$

Now substitute the values from the problem. You can calculate with the given units, as long as they're consistent throughout the calculation:

$$V_f = \left(\frac{P_i V_i^{5/3}}{P_f} \right)^{\frac{3}{5}}$$

$$V_f = \left[\frac{\left(3.04 \times 10^5 \text{ Pa}\right)\left(\left(1.18 \times 10^{-6} \text{ m}^3\right)^{5/3}\right)}{3\left(3.04 \times 10^5 \text{ Pa}\right)} \right]^{\frac{3}{5}}$$

$$V_f = 6.10 \times 10^{-7} \text{ m}^3$$

Now just convert the final volume into liters:

$$\frac{6.10 \times 10^{-7} \text{ m}^3}{1} \times \frac{1000 \text{ L}}{1 \text{ m}^3} = 6.10 \times 10^{-4} \text{ L}$$

490. $\dfrac{\text{work}}{\text{heat input}}$

The efficiency of any engine is the ratio of work done to energy used to do it. Because a heat engine uses heat input as an energy source, a heat engine's efficiency is work over heat input.

491. 7.0×10^3 J

This problem is a great example of the simple application of the heat engine efficiency formula. Just solve for W and pay attention to correct scientific notation:

$$\text{efficiency} = \frac{W}{Q_h}$$
$$\text{efficiency} \bullet Q_h = W$$
$$0.25 \bullet 2.8 \times 10^4 = W$$
$$W = 7{,}000 \text{ J} = 7.0 \times 10^3 \text{ J}$$

492. 1.37×10^8 J

Use the heat engine efficiency formula, solved for Q_h, the heat energy input:

$$\text{efficiency} = \frac{W}{Q_h}$$
$$Q_h = \frac{W}{\text{efficiency}}$$
$$Q_h = \frac{8.22 \times 10^7}{0.60}$$
$$Q_h = 1.37 \times 10^8 \text{ J}$$

So, theoretically, a gallon of diesel contains about 1.37×10^8 joules of available energy.

493.   584 K

To determine the temperature increase, first calculate how much energy is added to the heat sink per kilogram of fuel. The heat added to the heat sink, Q_c, equals the energy not converted to work: $Q_c = Q_h - W$. In this case, that is $100\% - 54\% = 46\%$ of the input heat energy, or:

$$Q_c = 0.46 \bullet 2.53 \times 10^7 \text{ J}$$
$$Q_c = 1.16 \times 10^7 \text{ J}$$

Now calculate the change in temperature using $Q = mc\Delta T$:

$$\Delta T = \frac{Q}{mc}$$
$$\Delta T = \frac{1.16 \times 10^7 \text{ J}}{(5.2 \text{ kg}) \left(3820 \frac{\text{J}}{\text{kg} \bullet \text{K}} \right)}$$
$$\Delta T = 584 \text{ K}$$

So the temperature of the heat sink increases by 584 kelvins per kilogram of burned fuel.

494.   It is reversible.

A Carnot engine is a theoretical engine that is completely reversible. In other words, all the energy drained from the input is recovered as output to be used again, without loss.

495.   60%

The maximum theoretical efficiency of a heat engine is the efficiency it would have if no heat loss occurred at all, also known as a Carnot engine. A Carnot engine's efficiency is calculated by dividing the heat sink temperature T_c by the heat source temperature T_h and subtracting the quotient from 1: efficiency $= 1 - (T_c/T_h)$. Use this formula, substituting the values from the question:

$$T_c = 100 \text{ K}$$
$$T_h = 250 \text{ K}$$

$$efficiency \text{ (\%)} = 1 - \frac{T_c}{T_h}$$
$$efficiency \text{ (\%)} = 1 - \frac{100 \text{ K}}{250 \text{ K}}$$
$$efficiency \text{ (\%)} = 1 - 0.4$$
$$efficiency = 60\%$$

496. **153 K**

Use the equation for maximum possible efficiency, solved for $T_h : T_h = T_c / (1 - (\text{efficiency}\%))$. Substitute 0.43 for efficiency and 87 kelvins for T_c:

$$T_h = \frac{T_c}{(1 - \text{efficiency}\,(\%))}$$
$$T_h = \frac{87\,\text{K}}{(1 - (0.43))}$$
$$T_h = \frac{87\,\text{K}}{.57}$$
$$T_h = 153\,\text{K}$$

497. **–146°C**

Don't fall into the trap of trying to use all the information you're given. You know that $T_c = T_h \times (1 - (\text{efficiency}\%))$, and you have the efficiency and the heat source temperature, T_h. The heat energy is extraneous here.

$$T_c = 363.15\,\text{K}(1 - 0.65)$$
$$T_c = 127\,\text{K} = -146°\text{C}$$

498. **Heat pumps pull extra heat energy from a heat source.**

Heat pumps seem impossibly efficient because they make use of heat energy from other sources, such as the outside air. Even if the source air is colder than the target air, the heat energy still exists and can be harnessed.

499. **147 K**

All the information you need to solve this one is in the question, but you do need to begin by solving the equation for T_c, the temperature of the heat source:

$$W = Q_h \left(1 - \left(\frac{T_c}{T_h} \right) \right)$$
$$\frac{W}{Q_h} = 1 - \left(\frac{T_c}{T_h} \right)$$
$$\frac{T_c}{T_h} = 1 - \frac{W}{Q_h}$$
$$T_c = (T_h) \left(1 - \frac{W}{Q_h} \right)$$

Now just substitute and solve:

$$T_c = (T_h) \left(1 - \frac{W}{Q_h} \right)$$
$$T_c = (210\,\text{K}) \left(1 - \frac{750\,\text{J}}{2,500\,\text{J}} \right)$$
$$T_c = 210\,\text{K}\,(0.7)$$
$$T_c = 147\,\text{K}$$

500. 882 J

You're given the coefficient of performance, COP, and heat transferred, Q_h, in the question. Solve the formula for W, substitute, and solve:

$$COP = \frac{Q_h}{W}$$

$$W = \frac{Q_h}{COP}$$

$$W = \frac{12,350 \text{ J}}{14}$$

$$W = 882 \text{ J}$$

501. 32,600 J

To find Q_h, the heat energy transferred, you need the COP. To find the COP, you can use $COP = \frac{1}{1 - \frac{T_c}{T_h}}$; just remember to convert degrees Celsius to kelvins first:

$$COP = \frac{1}{1 - \frac{T_c}{T_h}}$$

$$COP = \frac{1}{1 - \frac{(-35 + 273.15)\text{ K}}{(33 + 273.15)\text{ K}}}$$

$$COP = \frac{1}{1 - 0.78}$$

$$COP = 4.5$$

Now use the COP with $COP = \frac{Q_h}{W}$ (solved for Q_h):

$$Q_h = COP \bullet W$$

$$Q_h = 4.5 \bullet 7,250 \text{ J}$$

$$Q_h \approx 32,600 \text{ J}$$

Theoretically, you can pump about 32,600 joules.

Index

B

banked turns, 242–247
Bernoulli's Equation, 267–270
Boyle's Law, 400–401
buoyancy, defined, 397
buoyant force, 51, 263–264

C

Carnot engine, 410
Celsius temperature scale, 391
center of circle, acceleration toward, 236–238
centripetal acceleration
  acceleration toward center, 236–238
  calculating, 333–334
  formula for, 238
  tension and, 332
centripetal force, 237–239, 243
Charles's Law, 401–402
circles
  acceleration toward center, 236–238
  angular acceleration, 235
  angular velocity, 232–234
  circumference of, 251, 357
  radians, 233
circular motion
  acceleration toward center, 236–238
  applying law of universal gravitation to stars, 248–250
  banked turns, 242–247
  constant speed around circle, 229
  finding speed of satellites, 253–256
  overview, 39, 45
  speed required to avoid falling off vertical loop, 257–259
  speeding up and down around circle, 234–235
  time to travel around, 255–257
  traveling with angular velocity, 232–233
  turning with friction, 241–242
circumference of circles, 251, 357
coefficient of friction, 182–186

coefficient of performance (COP), 412
collisions
  conserving momentum during, 311
  elastic
    along line, 319–324
    two-dimensional, 324–331
  finding initial velocity for, 317–319
  inelastic, 311–317
  initial velocity of, 317–319
components
  breaking vectors into, 139–142
  defined, 17
  reassembling vector from, 141–144
  separating motion in, 216–224
compressed air, 407
compression, spring, 380–381
conduction, thermal, 397–399
conservation of energy principle, 368
conservation-of-mechanical-energy formula, 300
constant velocity, 192
continuity equation, 270
convection, 397
converting units, 114
COP (coefficient of performance), 500
cosine function, 383
cylinders, 360, 370

D

density, fluid, 260
depth, calculating pressure at, 261–262
digits, rounding, 116–117
displacement
  angular velocity displacement formula, 357
  defined, 11
  describing in two dimensions, 143–150
  equilibrium from, 380
  finding acceleration with time and, 132–133
  finding acceleration with velocity and, 134–135

F

Fahrenheit temperature scale, 391
feet (measurement), 114
femtoseconds, 115
first law formula, 404
floating, 263–264
flow
  speed of, 266–267
  types of, 265
  volume flow rate formula, 271
fluids
  applying pressure with force, 261
  Archimedes' Principle and, 263–264
  Bernoulli's Equation and, 267–269
  calculating pressure at depth, 261–262
  density, 260–261
  flow speed, 266–267
  flow types, 265
  overview, 51
  Pascal's Principle and, 262–263
  pipes and, 268–272
force
  acceleration of massive objects, 156–157
  applying at angle, 275–279
  applying in direction of movement, 273–275
  applying in opposite direction of motion, 279–282
  applying pressure with, 261
  balancing to find equilibrium, 173–176
  banked turns, 242–247
  buoyant, 51
  calculating impulse, 306
  centripetal, 238–240, 247
  defined, 23
  finding from impulse and momentum, 310–311
  finding needed force to speed up, 162–165
  free-body diagrams, 166
  impulse and, 65
  moving distance with net force, 160–162

net
  adding forces for, 157–160
  moving distance with, 160–162
  work and, 57
  normal, 242
  overcoming friction, 168–169
  pairing equal and opposite, 166–168
  restoring, for springs, 380–381
  shooting objects straight up, 206–209
  tension, 156
  torque and, 360
  torque formula, 363
  vectors and, 278
  work and, 57
formulas
  angular acceleration, 234
  angular momentum, 376
  angular velocity, 377
  angular velocity displacement, 357
  angular work, 363, 365
  Bernoulli's Equation, 267–269
  centripetal acceleration, 238, 333
  conservation of kinetic energy, 324
  conservation of momentum, 317
  conservation-of-mechanical-energy, 300
  continuity, 270
  converting between linear and angular velocities, 334
  converting vectors to component form, 145
  displacement, 124, 155
  distance, 121
  energy of spring system, 387
  first law formula, 404
  force torque, 363
  frequency of pendulum, 389
  friction, 186
  gravitational potential energy, 288–289
  heat energy of fusion, 395
  heat energy of vaporization, 396
  heat transfer by conduction, 398
  impulse-momentum theorem, 309

S

satellites
  finding speed in circular orbits, 253–256
  time traveling around orbit, 255–257
scalars, 139
scientific notation, 115
shooting objects
  at angle, 216–224
  straight up
    maximum height for, 206–213
    time to go up and down, 213–216
SI (International System of Units), 156
significant digits, 116
sine function, 382
sine waves
  of harmonic motion, 383
  period of, 382
sliding, 369–370
slopes
  finding gravity along, 176–178
  moving with gravity down, 178–182
  non-slippery
    covering distance, 197–206
    pushing/pulling on, 192–197
solid cylinders, 364
solid disks, 373
specific gravity, 260
speed
  acceleration, 126–129
  acceleration around circle, 235
  average, 122–123
  average versus average velocity,
    123–127
  Bernoulli's Equation and, 267–270
  calculating momentum, 306–308
  constant around circle, 229
  defined, 11, 229
  finding needed force for, 162–165
  flow, 266–267

  increasing power using, 303–306
  instantaneous, 122–123
  required to avoid falling off vertical loop,
    257–259
  of satellites in circular orbits, 253–256
spinning, 370
springs
  finding period of mass on, 385–387
  general discussion, 87
  overcoming equilibrium, 381–382
  restoring force, 380–381
stars
  acceleration with gravity near planet surface,
    250–253
  applying law of universal gravitation to,
    248–250
static friction, 185–188
stationary items, elastic collisions with, 319
stretching of springs, 380–381

T

tangential acceleration, 356, 373
tangential motion, 331–333
tangential velocity, 233, 371
temperature
  Boyle's Law and, 400–401
  calculating increase of, 410
  changing with energy flow,
    394–395
  convection, 397
  converting Fahrenheit to/from
    Celsius, 391
  defined, 93
  expansion from heat rise,
    392–394
  ideal gas law and, 402–403
  maintaining constant, 405–406
  measuring, 391
  radiation, 399
  thermal conduction, 397–399

Publisher's Acknowledgments

Executive Editor: Lindsay Lefevere

Project Editors: Jennifer Tebbe, Tim Gallan, and Donna Wright

Copy Editors: Krista Hansing, Todd Lothery, Ashley Petry

Contributor: LearningMate Solutions

Contributing Editor: Dan Wohns

Technical Editors: Matt Cannon, Laurie Fuhr

Art Coordinator: Alicia B. South

Managing Editor: Michelle Hacker

Illustrator: Thomson Digital

Production Editor: Mohammed Zafar Ali

Cover Image: Monitor © spaxiax/Adobe Stock

Take dummies with you everywhere you go!

Whether you are excited about e-books, want more from the web, must have your mobile apps, or are swept up in social media, dummies makes everything easier.

Find us online!

dummies
A Wiley Brand

Leverage the power

Dummies is the global leader in the reference category and one of the most trusted and highly regarded brands in the world. No longer just focused on books, customers now have access to the dummies content they need in the format they want. Together we'll craft a solution that engages your customers, stands out from the competition, and helps you meet your goals.

Advertising & Sponsorships

Connect with an engaged audience on a powerful multimedia site, and position your message alongside expert how-to content. Dummies.com is a one-stop shop for free, online information and know-how curated by a team of experts.

- Targeted ads
- Video
- Email Marketing

- Microsites
- Sweepstakes sponsorship

20 **MILLION** PAGE VIEWS **EVERY SINGLE MONTH**

15 MILLION **UNIQUE** **VISITORS PER MONTH**

43% OF ALL VISITORS ACCESS THE SITE **VIA THEIR MOBILE DEVICES**

700,000 NEWSLETTER SUBSCRIPTION **TO THE INBOXES OF**

300,000 UNIQUE INDIVIDUALS EVERY WEEK

# of dummies

Custom Publishing

Reach a global audience in any language by creating a solution that will differentiate you from competitors, amplify your message, and encourage customers to make a buying decision.

- Apps
- Books
- eBooks
- Video
- Audio
- Webinars

Brand Licensing & Content

Leverage the strength of the world's most popular reference brand to reach new audiences and channels of distribution.

For more information, visit **dummies.com/biz**

PERSONAL ENRICHMENT

9781119187790	9781119179030	9781119293354	9781119293347	9781119310068	9781119235606
USA $26.00	USA $21.99	USA $24.99	USA $22.99	USA $22.99	USA $24.99
CAN $31.99	CAN $25.99	CAN $29.99	CAN $27.99	CAN $27.99	CAN $29.99
UK £19.99	UK £16.99	UK £17.99	UK £16.99	UK £16.99	UK £17.99

9781119251163	9781119235491	9781119279952	9781119283133	9781119287117	9781119130246
USA $24.99	USA $26.99	USA $24.99	USA $24.99	USA $24.99	USA $22.99
CAN $29.99	CAN $31.99	CAN $29.99	CAN $29.99	CAN $29.99	CAN $27.99
UK £17.99	UK £19.99	UK £17.99	UK £17.99	UK £16.99	UK £16.99

PROFESSIONAL DEVELOPMENT

9781119311041	9781119255796	9781119293439	9781119281467	9781119280651	9781119251132	9781119310563
USA $24.99	USA $39.99	USA $26.99	USA $26.99	USA $29.99	USA $24.99	USA $34.00
CAN $29.99	CAN $47.99	CAN $31.99	CAN $31.99	CAN $35.99	CAN $29.99	CAN $41.99
UK £17.99	UK £27.99	UK £19.99	UK £19.99	UK £21.99	UK £17.99	UK £24.99

9781119181705	9781119263593	9781119257769	9781119293477	9781119265313	9781119239314	9781119293323
USA $29.99	USA $26.99	USA $29.99	USA $26.99	USA $24.99	USA $29.99	USA $29.99
CAN $35.99	CAN $31.99	CAN $35.99	CAN $31.99	CAN $29.99	CAN $35.99	CAN $35.99
UK £21.99	UK £19.99	UK £21.99	UK £19.99	UK £17.99	UK £21.99	UK £21.99

Learning Made Easy

ACADEMIC

9781119293576
USA $19.99
CAN $23.99
UK £15.99

9781119293637
USA $19.99
CAN $23.99
UK £15.99

9781119293491
USA $19.99
CAN $23.99
UK £15.99

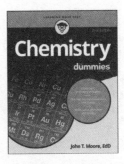

9781119293460
USA $19.99
CAN $23.99
UK £15.99

9781119293590
USA $19.99
CAN $23.99
UK £15.99

9781119215844
USA $26.99
CAN $31.99
UK £19.99

9781119293378
USA $22.99
CAN $27.99
UK £16.99

9781119293521
USA $19.99
CAN $23.99
UK £15.99

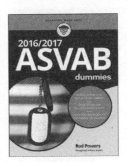

9781119239178
USA $18.99
CAN $22.99
UK £14.99

9781119263883
USA $26.99
CAN $31.99
UK £19.99

Available Everywhere Books Are Sold

Small books for big imaginations